Lecture Notes in Computer Scie...

Marina L. Gavrilova C.J. Kenneth Tan
Yingxu Wang Yiyu Yao Guoyin Wang (Eds.)

Transactions on Computational Science II

Springer

Editors-in-Chief

Marina L. Gavrilova
University of Calgary, Department of Computer Science
2500 University Drive N.W., Calgary, AB, T2N 1N4, Canada
E-mail: marina@cpsc.ucalgary.ca

C.J. Kenneth Tan
OptimaNumerics Ltd.
Cathedral House, 23-31 Waring Street, Belfast BT1 2DX, UK
E-mail: cjtan@optimanumerics.com

Guest Editors

Yingxu Wang
University of Calgary, Schulich School of Engineering
Department of Electrical and Computer Engineering
2500 University Drive N.W., Calgary, AB, T2N 1N4, Canada
E-mail: yingxu@ucalgary.ca

Yiyu Yao
University of Regina, Department of Computer Science
Regina, SK, S4S 0A2, Canada
E-mail: yyao@cs.uregina.ca

Guoyin Wang
Chongqing University of Posts and Telecommunications
Institute of Computer Science and Technology, Chongqing 400065, China
E-mail: wanggy@cqupt.edu.cn

Library of Congress Control Number: 2008935543

CR Subject Classification (1998): F, D, C.2-3, G, E.1-2
LNCS Sublibrary: SL 1 – Theoretical Computer Science and General Issues

ISSN 0302-9743 (Lecture Notes in Computer Science)
ISSN 1866-4733 (Transactions on Computational Science)
ISBN 978-3-540-87562-8 Springer Berlin Heidelberg New York

Springer is a part of Springer Science+Business Media

springer.com

Typesetting: Camera-ready by author, data conversion by Scientific Publishing Services, Chennai, India
Printed on acid-free paper SPIN: 12446056 06/3180 5 4 3 2 1 0

LNCS Transactions on Computational Science

Computational science, an emerging and increasingly vital field, is now widely recognized as an integral part of scientific and technical investigations, affecting researchers and practitioners in areas ranging from aerospace and automotive research to biochemistry, electronics, geosciences, mathematics, and physics. Computer systems research and the exploitation of applied research naturally complement each other. The increased complexity of many challenges in computational science demands the use of supercomputing, parallel processing, sophisticated algorithms, and advanced system software and architecture. It is therefore invaluable to have input by systems research experts in applied computational science research.

Transactions on Computational Science focuses on original high-quality research in the realm of computational science in parallel and distributed environments, also encompassing the underlying theoretical foundations and the applications of large-scale computation. The journal offers practitioners and researchers the possibility to share computational techniques and solutions in this area, to identify new issues, and to shape future directions for research, and it enables industrial users to apply leading-edge, large-scale, high-performance computational methods.

In addition to addressing various research and application issues, the journal aims to present material that is validated – crucial to the application and advancement of the research conducted in academic and industrial settings. In this spirit, the journal focuses on publications that present results and computational techniques that are verifiable.

Scope

The scope of the journal includes, but is not limited to, the following computational methods and applications:

- Aeronautics and Aerospace
- Astrophysics
- Bioinformatics
- Climate and Weather Modeling
- Communication and Data Networks
- Compilers and Operating Systems
- Computer Graphics
- Computational Biology
- Computational Chemistry
- Computational Finance and Econometrics
- Computational Fluid Dynamics
- Computational Geometry

- Computational Number Theory
- Computational Physics
- Data Storage and Information Retrieval
- Data Mining and Data Warehousing
- Grid Computing
- Hardware/Software Co-design
- High-Energy Physics
- High-Performance Computing
- Numerical and Scientific Computing
- Parallel and Distributed Computing
- Reconfigurable Hardware
- Scientific Visualization
- Supercomputing
- System-on-Chip Design and Engineering

Preface

The denotational and expressive needs in cognitive informatics, computational intelligence, software engineering, and knowledge engineering have led to the development of new forms of mathematics collectively known as denotational mathematics. *Denotational mathematics* is a category of mathematical structures that formalize rigorous expressions and long-chain inferences of system compositions and behaviors with abstract concepts, complex relations, and dynamic processes. Typical paradigms of denotational mathematics are concept algebra, system algebra, Real-Time Process Algebra (RTPA), Visual Semantic Algebra (VSA), fuzzy logic, and rough sets. A wide range of applications of denotational mathematics have been identified in many modern science and engineering disciplines that deal with complex and intricate mathematical entities and structures beyond numbers, Boolean variables, and traditional sets.

This issue of Springer's Transactions on Computational Science on *Denotational Mathematics for Computational Intelligence* presents a snapshot of current research on denotational mathematics and its engineering applications. The volume includes selected and extended papers from two international conferences, namely IEEE ICCI 2006 (on *Cognitive Informatics*) and RSKT 2006 (on *Rough Sets and Knowledge Technology*), as well as new contributions. The following four important areas in denotational mathematics and its applications are covered:

- *Foundations and applications of denotational mathematics,* focusing on: a) contemporary denotational mathematics for computational intelligence; b) denotational mathematical laws of software; c) a comparative study of STOPA and RTPA; and d) a denotational mathematical model of abstract games.
- *Rough and fuzzy set theories,* focusing on: a) transformation of vague sets to fuzzy sets; b) reduct construction algorithms; and c) attribute set dependence in reduct computation.
- *Granular computing,* focusing on: a) mereological theories of concepts; and b) rough logic and reasoning.
- *Knowledge and information modeling,* focusing on: a) semantic consistency of knowledge bases; b) contingency matrix theory; and c) analysis of information tables containing stochastic values.

The editors believe that the readers of the *Transactions on Computational Science* (TCS) series will benefit from the papers presented in this special issue on the latest advances in denotational mathematics and applications in cognitive informatics, natural intelligence, computational intelligence, and AI.

Acknowledgments

The guest editors of this *Special Issue on Denotational Mathematics for Computational Intelligence* in Springer's *Transactions on Computational Science* series, would like to thank all authors for submitting their latest interesting work. We are grateful to the program committee members of the IEEE ICCI 2006 and RSKT 2006 conferences, as well as the reviewers, for their great contributions to this special issue. We would like to thank the Editors-in-Chief of TCS, Dr. Marina L. Gavrilova and Dr. Chih Jeng Kenneth Tan, for their advice, vision, and support. We also thank the editorial staff at Springer for their professional help during the publication of this special issue.

June 2008

Yingxu Wang
Yiyu Yao
Guoyin Wang

LNCS Transactions on Computational Science – Editorial Board

Table of Contents

Regular Papers

Perspectives on Denotational Mathematics: New Means of Thought 1
Yingxu Wang, Yiyu Yao, and Guoyin Wang

On Contemporary Denotational Mathematics for Computational Intelligence 6
Yingxu Wang

Mereological Theories of Concepts in Granular Computing 30
Lech Polkowski

On Mathematical Laws of Software 46
Yingxu Wang

Rough Logic and Its Reasoning 84
Qing Liu and Lan Liu

On Reduct Construction Algorithms 100
Yiyu Yao, Yan Zhao, and Jue Wang

Attribute Set Dependence in Reduct Computation 118
Pawel Terlecki and Krzysztof Walczak

A General Model for Transforming Vague Sets into Fuzzy Sets 133
Yong Liu, Guoyin Wang, and Lin Feng

Quantifying Knowledge Base Inconsistency Via Fixpoint Semantics 145
Du Zhang

Contingency Matrix Theory I: Rank and Statistical Independence in a Contigency Table 161
Shusaku Tsumoto and Shoji Hirano

Applying Rough Sets to Information Tables Containing Possibilistic Values 180
Michinori Nakata and Hiroshi Sakai

Toward a Generic Mathematical Model of Abstract Game Theories 205
Yingxu Wang

A Comparative Study of STOPA and RTPA 224
Natalia Lopez, Manuel Núñez, and Fernando L. Pelayo

Author Index 247

Perspectives on Denotational Mathematics: New Means of Thought

Yingxu Wang[1], Yiyu Yao[2], and Guoyin Wang[3]

[1] Dept. of Electrical and Computer Engineering, University of Calgary, Canada
[2] Dept. of Computer Science, University of Regina, Canada
[3] Institute of Computer Science and Technology,
Chongqing University of Posts and Telecommunications, China

Abstract. The denotational and expressive needs in cognitive informatics, computational intelligence, software engineering, and knowledge engineering lead to the development of new forms of mathematics collectively known as denotational mathematics. Denotational mathematics is a category of mathematical structures that formalize rigorous expressions and long-chain inferences of system compositions and behaviors with abstract concepts, complex relations, and dynamic processes. Typical paradigms of denotational mathematics are such as concept algebra, system algebra, Real-Time Process Algebra (RTPA), Visual Semantic Algebra (VSA), fuzzy logic, and rough sets. A wide range of applications of denotational mathematics have been identified in many modern science and engineering disciplines that deal with complex and intricate mathematical entities and structures beyond numbers, Boolean variables, and traditional sets.

Keywords: Cognitive informatics, computational intelligence, denotational mathematics, concept algebra, system algebra, process algebra, RTPA, visual semantic algebra, rough set, granular computing, knowledge engineering, AI, natural intelligence.

1 Introduction

Recent transdisciplinary researches in cognitive informatics, natural intelligence, artificial intelligence, computing science, software science, computational intelligence, knowledge science, and system science have led to an interesting discovery that a new form of mathematics is needed, which is collectively known as *denotational mathematics*, in order to deal with the complex mathematical entities beyond numbers, Boolean variables, and traditional sets. In denotational mathematics, not only the mathematical entities are greatly complicated, but also the mathematical structures and methodologies are intricately expanded from simple relations and individual functions to embedded relations and series of dynamic functions.

The history of sciences and engineering shows that many branches of mathematics have been created in order to meet their abstract, rigorous, and expressive needs. These phenomena may be conceived as that new problems require new forms of mathematics [1], [2], [26], and the maturity of a scientific discipline is characterized

M.L. Gavrilova et al. (Eds.): Trans. on Comput. Sci. II, LNCS 5150, pp. 1–5, 2008.

by the maturity of its mathematical means [15]. Upon the identification of more and more new complex mathematical entities and structures in the aforementioned disciplines, novel denotational mathematical forms are yet to be sought.

The lasting vigor of automata theory [8], Turing machines [7], and formal inference methodologies [2], [6] reveals that suitable mathematical means such as tuples, processes, and symbolic logics are the essences of computing science and computational intelligence. However, although these profound mathematical structures underlie the modeling of natural and machine intelligence, the level of their mathematical entities as characterized by integers, real numbers, Boolean variables, and simple sets is too low to be able to process concepts, knowledge, and series of behavioral processes. The requirements for reduction of complex knowledge onto the low-level data objects in conventional computing technologies and their associated analytic mathematical means have greatly constrained the inference and computing ability toward the development of intelligent knowledge processors known as cognitive computers [14]. This has triggered the current transdisciplinary investigation into novel mathematical structures for computational intelligence in the category of denotational mathematics.

2 What Is Denotational Mathematics?

Applied mathematics can be classified into two categories known as *analytic* and *denotational* mathematics [15]. The former are mathematical structures that deal with functions of variables as well as their operations and behaviors. The latter are mathematical structures that formalize rigorous expressions and inferences of system architectures and behaviors with abstract concepts, complex relations, and dynamic processes. Denotational mathematics is a collection of higher order functions on complex mathematical entities. Given a certain mathematical structure, when both its functions and inputs are varying in a series, it belongs to the category of denotational mathematics; otherwise, it falls into the category of conventional analytic mathematics.

Denotational mathematics is a category of expressive mathematical structures that deals with high-level mathematical entities beyond numbers and sets, such as abstract objects, complex relations, perceptual information, abstract concepts, knowledge, intelligent behaviors, behavioral processes, and systems.

The term denotational mathematics is first introduced by Yingxu Wang in an emerging discipline of *cognitive informatics* [10]. It is then formally described in his latest book, *Software Engineering Foundations: A Software Science Perspective* [15]. Denotational mathematics may be viewed as a new way of formal inference on both complex architectures and intelligent behaviors to meet modern challenges in understanding, describing, and modeling natural and machine intelligence in general, and software and knowledge engineering in particular. As a counterpart of the conventional analytic mathematics, denotational mathematics concerns new forms of mathematical structures for dealing with complex mathematical entities emerged in cognitive informatics, computational intelligence, software engineering, and knowledge engineering.

The convergence of mathematics and computing science is evident ever since the introduction of computers. On one hand, mathematics provides computer science with a formal foundation, a rigorous approach of exploration, an abstraction power of induction, and a systematic generalization means for enabling deduction in applications. On the other hand, computer science raises many challenges to classical mathematics, brings new ways of mathematical inferences, and offers help for tackling intricate problems.

A great extent of effort has been put on extending the capacity of set theory and mathematical logic in dealing with problems in cognitive informatics and computational intelligence. The former are represented by the proposals of *fuzzy sets* [26] and *rough sets* [4], [9], [24]. The latter are enabled by the development of *temporal logic* [5] and *fuzzy logic* [26]. New mathematical structures are created for dealing with new problems such as *embedded relations*, *incremental relations*, and the *big-R notation* [15], [18]. More systematically, a set of novel denotational mathematical forms [16], [18], [22] are developed known as *concept algebra* [19], *system algebra* [20], *Real-Time Process Algebra* (RTPA) [11], [21], and *Visual Semantic Algebra* (VSA) [23].

3 Why Denotational Mathematics Is Needed?

Christopher Strachey (1965), the founder of the Programming Research Group (PRG) in the Computing Laboratory at Oxford University, wrote: "It has long been my personal view that the separation of practical and theoretical work is artificial and injurious. Much of the practical work done in computing, both in software and in hardware design, is unsound and clumsy because the people who do it have not any clear understanding of the fundamental design principles of their work. Most of the abstract mathematical and theoretical work is sterile because it has no point of contact with real computing." The succeeding director of PRG and the Oxford Computing Laboratory, C.A.R. Hoare, asserted that software is a mathematical entity that may be treated by process algebra, particularly his CSP [3]. Then, Yingxu Wang, a visiting professor working with C.A.R. Hoare in 1995, proved recently that there exists a generic mathematical model of abstract program systems, based on it any concrete program instance and application software system can be derived or treated as an instance of the generic program model [15].

The emergence of denotational mathematics is driven by the practical needs in cognitive informatics, computational intelligence, computing science, software science, and knowledge engineering, because all these modern disciplines study complex human and machine intelligence and their rigorous treatments. Among the new forms of denotational mathematics, *concept algebra* is designed to deal with the new abstract mathematical structure of concepts and their representation and manipulation in knowledge engineering [19]. *System algebra* is created for a rigorous treatment of abstract systems and their algebraic relations and operations [20]. RTPA is developed to deal with series of behavioral processes and architectures of software and intelligent systems [21]. VSA is introduced for the formal modeling and manipulation of abstract visual objects and patterns for robots, machine visions, and computational intelligence [23].

Denotational mathematics has gain a wide range of real-world applications in cognitive informatics and computational intelligence [10], [12], [13], [14], [15], [17], from the cognitive processes of the brain to the generic model of software systems, from rigorous system manipulation to knowledge network modeling, and from autonomous machine learning to cognitive computers [14].

Acknowledgement. The authors would like to acknowledge the Natural Science and Engineering Council of Canada (NSERC) for its partial support to this work. We would like to thank Co-Editor-in-Chiefs of TCS, Dr. Marina L. Gavrilova and Dr. Chih Jeng Kenneth Tan, for their comments and support for the *Special Issue on Denotational Mathematics for Computational Intelligence* in Springer *Transaction of Computational Science* (TCS), Vol. II.

References

1. Bender, E.A.: Mathematical Methods in Artificial Intelligence. IEEE CS Press, Los Alamitos (1996)
2. Boole, G.: The Laws of Thought, vol. 1854. Prometheus Books, NY (2003)
3. Hoare, C.A.R.: Communicating Sequential Processes. Prentice-Hall International, London (1985)
4. Pawlak, Z.: Rough Logic. Bulletin of the Polish Academy of Science, Technical Science 5-6, 253–258 (1987)
5. Pnueli, A.: The Temporal Logic of Programs. In: Proc. 18th IEEE Symposium on Foundations of Computer Science, pp. 46–57. IEEE, Los Alamitos (1977)
6. Russell, B.: The Principles of Mathematics, vol. 1903. W.W. Norton & Co., NY (1996)
7. Turing, A.M.: Computing Machinery and Intelligence. Mind 59, 433–460 (1950)
8. von Neumann, J.: The Principles of Large-Scale Computing Machines. Annals of History of Computers 3(3), 263–273 (reprinted, 1946)
9. Wang, G., Peters, J.F., Skowron, A., Yao, Y.Y. (eds.): RSKT 2006. LNCS (LNAI), vol. 4062. Springer, Heidelberg (2006)
10. Wang, Y.: On Cognitive Informatics, Keynote. In: Proc. 1st IEEE International Conference on Cognitive Informatics (ICCI 2002), Calgary, Canada, pp. 34–42. IEEE CS Press, Los Alamitos (2002)
11. Wang, Y.: The Real-Time Process Algebra (RTPA). Annals of Software Engineering: An International Journal, USA 14, 235–274 (2002)
12. Wang, Y.: On Cognitive Informatics. Brain and Mind: A Transdisciplinary Journal of Neuroscience and Neurophilisophy, USA 4(3), 151–167 (2003)
13. Wang, Y., Wang, Y.: On Cognitive Informatics Models of the Brain. IEEE Transactions on Systems, Man, and Cybernetics (C) 36(2), 16–20 (2006)
14. Wang, Y.: Keynote: Cognitive Informatics - Towards the Future Generation Computers that Think and Feel. In: Proc. 5th IEEE International Conference on Cognitive Informatics (ICCI 2006), Beijing, China, pp. 3–7. IEEE CS Press, Los Alamitos (2006)
15. Wang, Y.: Software Engineering Foundations: A Software Science Perspective, CRC Series in Software Engineering, vol. II. Auerbach Publications, NY, USA (2007)
16. Wang, Y.: Keynote: On Theoretical Foundations of Software Engineering and Denotational Mathematics. In: Proc. 5th Asian Workshop on Foundations of Software, Xiamen, China, pp. 99–102 (2007)

17. Wang, Y.: The Theoretical Framework of Cognitive Informatics. International Journal of Cognitive Informatics and Natural Intelligence 1(1), 1–27 (2007)
18. Wang, Y.: On Contemporary Denotational Mathematics for Computational Intelligence. In: Gavrilova, M.L., et al. (eds.) Transactions on Computational Science, II. LNCS, vol. 5150, pp. 6–29. Springer, Heidelberg (2008)
19. Wang, Y.: On Concept Algebra: A Denotational Mathematical Structure for Knowledge and Software Modeling. International Journal of Cognitive Informatics and Natural Intelligence 2(2), 1–19 (2008)
20. Wang, Y.: On System Algebra: A Denotational Mathematical Structure for Abstract System modeling. International Journal of Cognitive Informatics and Natural Intelligence 2(2), 20–42 (2008)
21. Wang, Y.: RTPA: A Denotational Mathematics for Manipulating Intelligent and Computational Behaviors. International Journal of Cognitive Informatics and Natural Intelligence 2(2), 44–62 (2008)
22. Wang, Y.: Keynote: On Denotational Mathematics Foundations of Abstract Intelligence. In: Proc. 7th IEEE International Conference on Cognitive Informatics (ICCI 2008), Stanford University, CA, USA, pp. 3–8. IEEE CS Press, Los Alamitos (2008)
23. Wang, Y.: On Visual Semantic Algebra (VSA) and the Cognitive Process of Pattern Recognition. In: Proc. 7th International Conference on Cognitive Informatics (ICCI 2008), Stanford University, CA. IEEE CS Press, Los Alamitos (2008)
24. Yao, Y.Y.: A Comparative Study of Formal Concept Analysis and Rough Set Theory in Data Analysis. In: Tsumoto, S., Słowiński, R., Komorowski, J., Grzymała-Busse, J.W. (eds.) RSCTC 2004. LNCS (LNAI), vol. 3066, pp. 59–68. Springer, Heidelberg (2004)
25. Yao, Y.Y., Shi, Z., Wang, Y., Kinsner, W.(eds.): Cognitive Informatics: In: Proc. 5th IEEE International Conference (ICCI 2006), Beijing, China, vol. I and II. IEEE CS Press, Los Alamitos (2006)
26. Zadeh, L.A.: Fuzzy Sets and Systems. In: Fox, J. (ed.) Systems Theory, pp. 29–37. Polytechnic Press, Brooklyn (1965)

On Contemporary Denotational Mathematics for Computational Intelligence

Yingxu Wang

Theoretical and Empirical Software Engineering Research Centre (TESERC)
International Center for Cognitive Informatics (ICfCI)
Dept. of Electrical and Computer Engineering
Schulich School of Engineering, University of Calgary
2500 University Drive, NW, Calgary, Alberta, Canada T2N 1N4
Tel.: (403) 220 6141, Fax: (403) 282 6855
yingxu@ucalgary.ca

Abstract. Denotational mathematics is a category of expressive mathematical structures that deals with high-level mathematical entities beyond numbers and sets, such as abstract objects, complex relations, behavioral information, concepts, knowledge, processes, intelligence, and systems. New forms of mathematics are sought, collectively known as denotational mathematics, in order to deal with complex mathematical entities emerged in cognitive informatics, computational intelligence, software engineering, and knowledge engineering. The domain and architecture of denotational mathematics are presented in this paper. Three paradigms of denotational mathematics, known as concept algebra, system algebra, and Real-Time Process Algebra (RTPA), are introduced. Applications of denotational mathematics in cognitive informatics and computational intelligence are elaborated. A set of case studies is presented on the modeling of iterative and recursive systems architectures and behaviors by RTPA, the modeling of autonomic machine learning by concept algebra, and the modeling of granular computing by system algebra.

Keywords: Denotational mathematics, concept algebra, system algebra, process algebra, RTPA, cognitive informatics, computational intelligence, software engineering, knowledge engineering, embedded relations, incremental relations, big-R notation.

1 Introduction

The history of sciences and engineering shows that many branches of mathematics have been created in order to meet their *abstract, rigorous,* and *expressive* needs. These phenomena may be conceived as that new problems require new forms of mathematics [5], [26]. It also indicates that the maturity of a new discipline is characterized by the maturity of its theories denoted in rigorous and efficient mathematical means [42], [43]. Therefore, the entire computing theory, as Lewis and Papadimitriou perceived, is about mathematical models of computers and algorithms [19]. Hence, the entire theory of cognitive informatics, computational intelligence, and software

M.L. Gavrilova et al. (Eds.): Trans. on Comput. Sci. II, LNCS 5150, pp. 6–29, 2008.

science is about new mathematical structures for natural and machine intelligence and efficient mathematical means.

Applied mathematics can be classified into two categories known as *analytic* and *denotational* mathematics [37], [42], [43], [46]. The former are mathematical structures that deal with functions of variables and their operations and behaviors; while the latter are mathematical structures that formalize rigorous expressions and inferences of system architectures and behaviors with abstract concepts and dynamic processes. It is observed that all existing mathematics, continuous or discrete, are mainly analytic, seeking unknown variables from known factors according to certain functions. Modern sciences have been mainly using analytic methodologies and mathematics in theory development and problem solving. However, in cognitive informatics and computational intelligence, the need is to formally describe and manipulate software and instructional behaviors in terms of operational logic, timing, and memory manipulation. Therefore, denotational mathematics are sought [37], [42], [43], [46], [48-52], which are able to describe software and intelligent architectures and behaviors rigorously, precisely, and expressively.

Definition 1. *Denotational mathematics* is a category of expressive mathematical structures that deals with high-level mathematical entities beyond numbers and sets, such as abstract objects, complex relations, behavioral information, concepts, knowledge, processes, intelligence, and systems.

The utility of denotational mathematics serves as the means and rules to rigorously and explicitly express design notions and conceptual models of abstract architectures and interactive behaviors of complex systems at the highest level of abstraction, in order to deal with the problems of cognitive informatics and computational intelligence characterized by large scales, complex architectures, and long chains of computing behaviors. Therefore, denotational mathematics is a system level mathematics, in which detailed individual computing behaviors may still be manipulated by conventional analytical mathematics. Typical forms of denotational mathematics [43], [46] are concept algebra [49], system algebra [50], and Real-Time Process Algebra (RTPA) [37], [41], [43], [51], [52].

It is observed in formal linguistics that human and system behaviors can be classified into three categories: to *be*, to *have*, and to *do* [6], [30], [41], [43]. All mathematical means and forms, in general, are an abstract description of these three categories of human and system behaviors and common rules of them. Taking this view, as shown in Table 1, mathematical logic may be perceived as the abstract means for describing "*to be*," set theory for describing "*to have*," and functions for describing "*to do*" in classic mathematics.

Table 1 summarized the usages of classic and denotational mathematics, which presents a fundamental view toward the modeling and expression of natural and machine intelligence in general, and software system in particular. Table 1 also indicates that only the logic- and set-based approaches are inadequate to deal with the entire problems in complex software and intelligent systems.

Table 1. Basic Expressive Power and Mathematical Means in System Modeling

Basic expressive power in system modeling	Mathematical means		Usage
	Classic mathematics	Denotational mathematics	
To *be*	Logic	Concept algebra	Identify *objects* and *attributes*
To *have*	Set theory	System algebra	Describe *relations* and *possession*
To *do*	Functions	RTPA	Describe *status* and *behaviors*

This paper presents the contemporary denotational mathematical structures for cognitive informatics and computational intelligence beyond classic mathematical entities, such as information, concepts, knowledge, processes, behaviors, intelligence, systems, distributed objects, and complex relations. The emergence and domain of denotational mathematics are described in Section 2. The paradigms of denotational mathematics, such as concept algebra, system algebra, and RTPA, are introduced in Section 3. Applications of denotational mathematics are demonstrated in Section 4, which covers the modeling of iterative and recursive systems architectures and behaviors by RTPA, the modeling of autonomic machine learning by concept algebra, and the modeling of granular computing by system algebra.

2 The Emergence and Development of Denotational Mathematics

The emergence of denotational mathematics is driven by the practical needs in cognitive informatics, computational intelligence, computing science, software science, and knowledge engineering, because all these modern disciplines study complex human and machine behaviors and their rigorous treatments. This section analyzes the fundamental elements in modeling computing systems and explains why these complex mathematical entities cannot be modeled by simple numbers and sets. This leads to the requirements for denotational mathematics that extends both entities and their manipulations in conventional mathematics. Then, the domain and architecture of denotational mathematics are summarized.

2.1 Fundamental Elements in Modeling Cognitive and Intelligent Systems

It is recognized that the behavioral space of any system or human action is three-dimensional, which encompasses the dimensions of *action*, *time*, and *space* [43]. Correspondingly, there are three fundamental categories of computational behaviors in a software system: a) Computational operations for variable manipulations, b) Timing operations for event manipulation, and c) Space operations for memory manipulation. Therefore, the behavior of a software or intelligent system can, in general, be viewed as a set of behavioral processes with computational operations on time and memory.

Definition 2. A *behavior* of a software or intelligent system, *B*, is a tuple of its computing operations *OPs* and observable outcomes and effects that affect or change the states of a system in the environment modeled by all variables and input/output events, as well as related memory structures *M* over time *T*, i.e.:

$$\begin{aligned} B &\triangleq (OP,\ T,\ M) \\ &= OP \times T \times M \end{aligned} \tag{1}$$

Behaviors of generic computing systems can be classified as static and dynamic ones as shown in Table 2. In Table 2, a *static behavior* of computing is a process that can be determined at design or compile time; while a *dynamic behavior* of computing is a process specified by given timing requirements that may only be determined at run-time.

Table 2. Characteristics of Computing System Behaviors

No.	Behaviors	Static	Dynamic	Behavioral category
1	System architectures	✓	✓	To be / to have
2	Data objects	✓	✓	To be / to have
3	Dynamic memory allocation		✓	To do
4	Timing		✓	To do
5	Input/output manipulations		✓	To do
6	Event handling		✓	To do
7	Mathematical operations	✓	✓	To do

It is noteworthy in Table 2 that most system behaviors are dynamic or both dynamic and static. Set theories and mathematical logic are found capable to deal with the 'to be' and 'to have' type static behaviors. However, the dynamic 'to do' behaviors in computing have to be manipulated by process algebras, e.g., RTPA. Even for the first two categories of behavioral problems in software and intelligent systems, concept algebra and system algebra are capable to deal with the problems more efficiently than logic and set theories, because they work at a higher level of mathematical entities known as abstract concepts and systems rather than numbers and sets.

2.2 New Problems Require New Forms of Mathematics

The history of sciences and engineering shows that new problems require new forms of mathematics. Software science and computational intelligence are emerging transdisciplinary enquiries that encompass a wide range of contemporary mathematical entities, which cannot be adequately described by conventional analytic mathematics. Therefore, new forms of mathematics are sought, collectively known as denotational mathematics.

The discussions in Section 2.1 indicate that classic mathematical forms such as sets, logic, and functions are inadequate to deal with the complex and dynamic behavioral problems of software and intelligent systems. The weaknesses of classic mathematics are in both of their expressive power and manipulation efficiency in the three categories of system descriptivity. The profound problems can be analogized to the evolutions of computing architectures, where, although Turing machines [19] are the most fundamental models for any computing need, von Neumann machines [36] and cognitive machines [38], [39], [44] are required to solve the problems of complex data processing and knowledge processing more efficiently and expressively.

A great extent of effort has been put on extending the capacity of sets and mathematical logic in dealing with the above problems in cognitive informatics and computational intelligence. The former are represented by the proposals of fuzzy sets [60, 62] and rough sets [27]. The letter are represented by the development of temporal logic [29] and fuzzy logic [60]. New mathematical structures are created such as *embedded relations*, *incremental relations*, and the *big-R notation* [37], [48]. More systematically, a set of new denotational mathematical forms [42], [43], [46] are developed known as concept algebra [49], system algebra [50], and RTPA [37], [40], [41], [43], [51], [52]. These new mathematical structures are introduced below, while the three paradigms of denotational mathematics will be elaborated in Section 3.

2.2.1 The Big-R Notation

The *big-R notation* is introduced to deal with the fundamental requirement in computing and software engineering [48], which is proposed first in RTPA [37]. In order to develop a general mathematical model for unifying the syntaxes and semantics of iterations and recursions, their inductive nature may be analyzed as follows.

Definition 3. An *iteration* of a process P can be defined as a series of $n+1$ repetitions, R_i, $1 \leq i \leq$ n+1, of P by mathematical induction, i.e.:

$$\begin{aligned} R_0 &= \otimes, \\ R_1 &= P \rightarrow R_0, \\ &\ldots \\ R_{n+1} &= P \rightarrow R_n, \ n \geq 0 \end{aligned} \tag{2}$$

Where ⊗ denotes a skip, or doing nothing but exit.

Based on Definition 3, the big-R notation can be introduced below.

Definition 4. The *big-R notation,* R, is a generic mathematical calculus in computing that is used to denote: (a) a finite set of *repetitive* behaviors, or (b) a finite set of recurring architectural constructs, in the following forms, respectively:

$$\text{(a)} \quad \mathop{\mathrm{R}}_{\mathrm{exp}\mathbf{BL}=\mathbf{T}}^{\mathbf{F}} P \tag{3}$$

$$\text{(b)} \quad \mathop{\mathrm{R}}_{i\mathbf{N}=1}^{n} P(i\mathbf{N}) \tag{4}$$

where **BL** and **N** are the type suffixes of Boolean and natural numbers, respectively; **T** and **F** are the Boolean constants true and false, respectively.

The big-R notation is a new denotational mathematical structure that enables efficient representation and manipulation of iterative and recursive behaviors in system modeling and description. Further description of the type system and a summary of all type suffixes of RTPA will be presented in Section 3.3.

2.2.2 The Embedded Relations

Definition 5. An *embedded relation* r_{ij} is a sequence of left-associated cumulative relations among a series of computing behaviors p_i and p_j, $1 \leq i < n$, $1 < j \leq m = n+1$, i.e.:

$$(...(((p_1)\ r_{12}\ p_2)\ r_{23}\ p_3)\ ...\ r_{n-1,n}\ p_n) = \mathop{\mathcal{R}}_{i=1}^{n-1}(p_i\ r_{ij}\ p_j), j = i+1 \tag{5}$$

where $r_{ij} \in \Re$, which is a set of relational process operators of RTPA that will be formally defined in Lemma 6.

The embedded relational operation is a new denotational mathematical structure, which provides a generic mathematic model for any program and software system in computing and intelligent system modeling.

2.2.3 The Incremental Relations

Definition 6. An *incremental union* of two sets of relations R_1 and R_2, denoted by $\boxplus$, are a union of R_1 and R_2 plus a newly generated incremental set of relations ΔR_{12}, i.e.:

$$R_1 \boxplus R_2 \triangleq R_1 \cup R_2 \cup \Delta R_{12} \tag{6}$$

where $\Delta R_{12} \not\subseteq R_1 \wedge \Delta R_{12} \not\subseteq R_2$ and $\Delta R_{12} = 2(\#C_1 \bullet \#C_2) \subseteq R_1 \boxplus R_2$.

The incremental relational operation is a new denotational mathematical structure, which provides a generic mathematical model for revealing the fusion principle and system gains during system unions and compositions.

2.3 The Domain and Architecture of Denotational Mathematics

The emergence of new mathematical entities in computing, cognitive informatics, and computational intelligence, as well as the requirements for new mathematical calculi, are the driving forces for seeking new mathematical structures and forms known collectively as denotational mathematics. The domain and architecture of denotational mathematics are illustrated in Fig. 1.

Denotational mathematics is usually in the form of abstract algebra that is a form of mathematics in which a system of abstract notations is adopted to denote relations of abstract mathematical entities and their algebraic operations based on given axioms and laws. Denotational mathematics may be used to denote complex behaviors of humans and intelligent systems, as well as long sequences of inference processes. A wide range of applications of denotational mathematics has been identified, such as cognitive informatics, computational intelligence, software engineering, knowledge engineering, information engineering, autonomic computing, autonomous machine learning, and neural informatics.

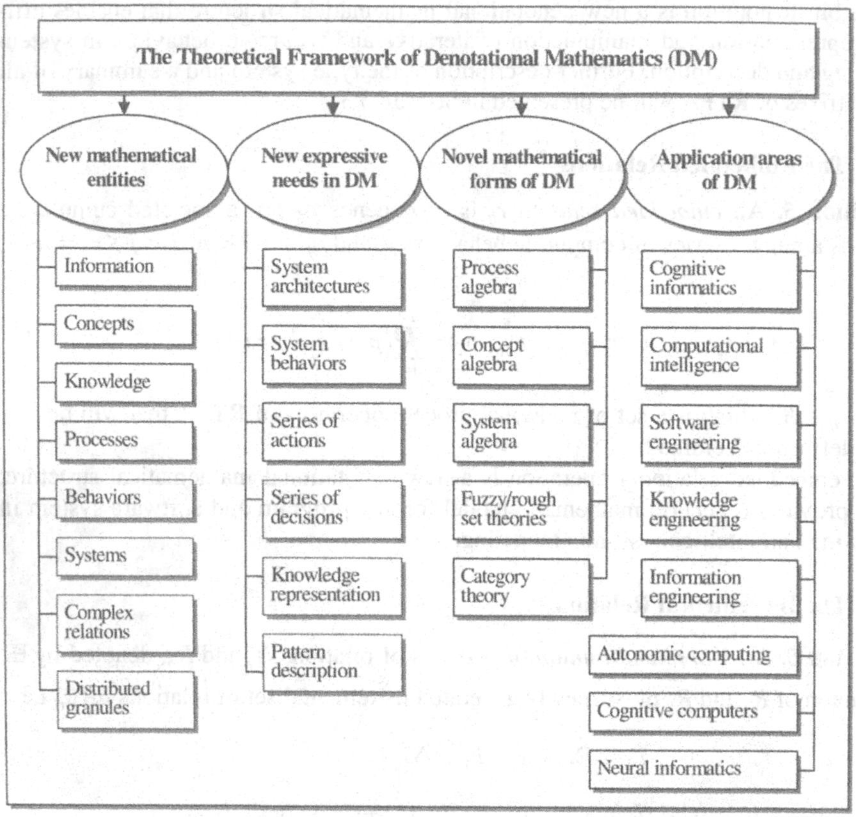

Fig. 1. Architecture of Denotational Mathematics

The following sections will introduce three paradigms of contemporary denotational mathematics. Their applications in cognitive informatics and computational intelligence will be demonstrated, which show how denotational mathematics may greatly improve the expressive power and efficiency in complex system modeling and manipulations.

3 Paradigms of Denotational Mathematics

Algebra is a branch of mathematics in which a system of symbolic abstractions and algebraic operations are adopted to denote variables, relations and their manipulation rules. Extensions of conventional algebra onto more complicated mathematical entities beyond numbers lead to the contemporary denotational mathematics. Three new denotational mathematical forms are created in exploring the mathematical foundations of cognitive informatics and computational intelligence [42], [43], [46]. Within the new forms of descriptive mathematics, *concept algebra* is designed to deal with the new abstract mathematical structure of concepts and their representation and manipulation. RTPA is developed to deal with series of behavioral processes and architectures of software and intelligent systems. *System algebra* is created for the rigorous treatment of abstract systems and their algebraic operations.

3.1 Concept Algebra

In cognitive informatics, logic, linguistics, psychology, software engineering, knowledge engineering, and computational intelligence, concepts are identified as the basic unit of both knowledge and reasoning [2], [8], [11], [12], [17], [22], [25], [41], [45], [49]. The rigorous modeling and formal treatment of concepts are at the center of theories for knowledge presentation and manipulation [7], [25], [34], [55], [59]. A *concept* in linguistics is a noun or noun-phrase that serves as the subject or object of a *to-be* statement [17], [37], [43]. Concepts in denotational mathematics [37], [49] are an abstract structure that carries certain meaning in almost all cognitive processes such as thinking, learning, and reasoning.

Definition 7. A *concept* is a cognitive unit to identify and/or model a real-world concrete entity and a perceived-world abstract object.

This section describes the formal treatment of abstract concepts and a new mathematical structure known as concept algebra in cognitive informatics and computational intelligence. Before an abstract concept is defined, the semantic environment or context [11], [12], [17], [23], [59] in a given language, is introduced.

Definition 8. Let $\mathcal{O}$ denote a finite nonempty set of *objects*, and $\mathcal{A}$ be a finite nonempty set of *attributes*, then a *semantic environment* or *context* Θ_C is denoted as a triple, i.e.:

$$\begin{aligned}\Theta_C &\triangleq (\mathcal{O},\mathcal{A},\mathcal{R}) \\ &= \mathcal{R}: \mathcal{O}\rightarrow\mathcal{O} \mid \mathcal{O}\rightarrow\mathcal{A} \mid \mathcal{A}\rightarrow\mathcal{O} \mid \mathcal{A}\rightarrow\mathcal{A}\end{aligned} \tag{7}$$

where $\mathcal{R}$ is a set of relations between $\mathcal{O}$ and $\mathcal{A}$, and | demotes alternative relations.

Definition 9. An *abstract concept* c on Θ_C is a 5-tuple, i.e.:

$$c \triangleq (O, A, R^c, R^i, R^o) \tag{8}$$

where

- O is a finite nonempty set of objects of the concept, $O = \{o_1, o_2, ..., o_m\} \subseteq$ Þ$\mathcal{O}$, where Þ$\mathcal{O}$ denotes a power set of $\mathcal{O}$.
- A is a finite nonempty set of attributes, $A = \{a_1, a_2, ..., a_n\} \subseteq$ Þ$\mathcal{A}$.
- $R^c = O \times A$ is a set of internal relations.
- $R^i \subseteq A' \times A$, $A' \sqsubseteq C' \wedge A \sqsubseteq c$, is a set of input relations, where C′ is a set of external concepts, $C' \subseteq \Theta_C$. For convenience, $R^i = A' \times A$ may be simply denoted as $R^i = C' \times c$.
- $R^o \subseteq c \times C'$ is a set of output relations.

Concept algebra is an abstract mathematical structure for the formal treatment of concepts and their algebraic relations, operations, and associative rules for composing complex concepts.

Definition 10. A *concept algebra CA* on a given semantic environment Θ_C is a triple, i.e.:

$$CA \triangleq (C, OP, \Theta_C) = (\{O, A, R^c, R^i, R^o\}, \{\bullet_r, \bullet_c\}, \Theta_C) \tag{9}$$

where $OP = \{\bullet_r, \bullet_c\}$ are the sets of *relational* and *compositional* operations on abstract concepts.

Lemma 1. The *relational operations* $\bullet_r$ in concept algebra encompass 8 comparative operators for manipulating the algebraic relations between concepts, i.e.:

$$\bullet_r \triangleq \{\leftrightarrow, \not\leftrightarrow, \prec, \succ, =, \cong, \sim, \triangleq\} \tag{10}$$

where the relational operators stand for *related, independent, subconcept, superconcept, equivalent, consistent, comparison,* and *definition*, respectively.

Lemma 2. The *compositional operations* $\bullet_c$ in concept algebra encompass 9 associative operators for manipulating the algebraic compositions among concepts, i.e.:

$$\bullet_c \triangleq \{\Rightarrow, \overset{-}{\Rightarrow}, \overset{+}{\Rightarrow}, \overset{\sim}{\Rightarrow}, \uplus, \pitchfork, \Lleftarrow, \vdash, \mapsto\} \tag{11}$$

where the compositional operators stand for *inheritance, tailoring, extension, substitute, composition, decomposition, aggregation, specification,* and *instantiation*, respectively.

Concept algebra provides a denotational mathematical means for algebraic manipulations of abstract concepts. Concept algebra can be used to model, specify, and manipulate generic "*to be*" type problems, particularly system architectures, knowledge bases, and detail-level system designs, in cognitive informatics, computational intelligence, computing science, software engineering, and knowledge engineering. Detailed relational and compositional operations of concept algebra may be referred to [42], [49].

3.2 System Algebra

Systems are the most complicated entities and phenomena in abstract, physical, information, and social worlds across all science and engineering disciplines. The system concept can be traced back to the 17th Century when R. Descartes (1596-1650) noticed the interrelationships among scientific disciplines as a system. Then, the general system notion was proposed by Ludwig von Bertalanffy in the 1920s [35], followed by the theories of system science [3], [4], [9], [13], [18], [31]. Further, there are proposals of complex systems theories [18], [61], fuzzy theories [60], [61], and chaos theories [10], [32]. Yingxu Wang found that, because of their extremely wide and frequent usability, systems may be treated rigorously as a new mathematical structure beyond conventional mathematical entities known as the *abstract systems* [50]. Based on this view, the concept of abstract systems and their mathematical models are introduced below.

Definition 11. An *abstract system* is a collection of coherent and interactive entities that has stable functions and a clear boundary with the external environment.

An abstract system forms the generic model of various real-world systems and represents the most common characteristics and properties of them. For instance, the granularity of granular computing can be explained by the following lemma in the abstract system theory.

Lemma 3. The *generality principle of system abstraction* states that a system can be represented as a whole in a given level k of reasoning, $1 \leq k \leq n$, without knowing the details at levels below k.

Definition 12. Let $\mathcal{C}$ be a finite nonempty set of *components*, and $\mathcal{B}$ a finite nonempty set of *behaviors*, then the *universal system environment* $\mathfrak{U}$ is denoted as a triple, i.e.:

$$\begin{aligned} \mathfrak{U} &\triangleq (\mathcal{C}, \mathcal{B}, \mathcal{R}) \\ &= \mathcal{R}: \mathcal{C} \to \mathcal{C} \mid \mathcal{C} \to \mathcal{B} \mid \mathcal{B} \to \mathcal{C} \mid \mathcal{B} \to \mathcal{B} \end{aligned} \tag{12}$$

where $\mathcal{R}$ is a set of relations between $\mathcal{C}$ and $\mathcal{B}$, and | demotes alternative relations.

Abstract systems can be classified into two categories known as the *closed* and *open* systems. Most practical and useful systems in nature are open systems in which there are interactions between the system and its environment. That is, they need to interact with external world known as the *environment* Θ, $\Theta \sqsubseteq \mathfrak{U}$, in order to exchange energy, matter, and/or information. Such systems are called open systems. Typical interactions between an open system and the environment are inputs and outputs.

Definition 13. An *open system* S on $\mathfrak{U}$ is a 7-tuple, i.e.:

$$S \triangleq (C, R^c, R^i, R^o, B, \Omega, \Theta) \tag{13}$$

where

- C is a finite nonempty set of *components* of the system, $C = \{c_1, c_2, ..., c_n\} \subseteq Þ\mathcal{C} \sqsubseteq \mathfrak{U}$.
- R is a finite nonempty set of *relations* between pairs of the components in the system, $R = \{r_1, r_2, ..., r_m\} \subseteq C \times C$.
- $R^c = C \times C$ is a set of *internal relations.*
- $R^i \subseteq C_\Theta \times C$ is a set of external *input relations*, $C_\Theta \subseteq Þ\mathcal{C} \sqsubseteq \mathfrak{U}$.
- $R^o \subseteq C \times C_\Theta$ is a set of external *output relations.*
- B is a set of *behaviors* (or functions), $B = \{b_1, b_2, ..., b_p\} \subseteq Þ\mathcal{B} \sqsubseteq \mathfrak{U}$.
- Ω is a set of *constraints* on the memberships of components, the conditions of relations, and the scopes of behaviors, $\Omega = \{\omega_1, \omega_2, ..., \omega_q\}$.
- Θ is the *environment* of S with a nonempty set of components C_Θ outside C, i.e., $\Theta = C_\Theta \subseteq Þ\mathcal{C} \sqsubseteq \mathfrak{U}$.

System algebra is an abstract mathematical structure for the formal treatment of abstract and general systems as well as their algebraic relations, operations, and associative rules for composing and manipulating complex systems [43], [50].

Definition 14. A *system algebra SA* on a given universal system environment $\mathfrak{U}$ is a triple, i.e.:

$$SA \triangleq (S, OP, \Theta) = (\{C, R^c, R^i, R^o, B, \Omega\}, \{\bullet_r, \bullet_c\}, \Theta) \tag{14}$$

where $OP = \{\bullet_r, \bullet_c\}$ are the sets of *relational* and *compositional* operations, respectively, on abstract systems as defined below.

Definition 15. The *relational operations* $\bullet_r$ in system algebra encompass 6 comparative operators for manipulating the algebraic relations between abstract systems, i.e.:

$$\bullet_r \triangleq \{\not\leftrightarrow, \leftrightarrow, \prod, =, \sqsubseteq, \sqsupseteq\} \tag{15}$$

where the relational operators stand for *independent, related, overlapped, equivalent, subsystem,* and *supersystem*, respectively.

Definition 16. The *compositional operations* $\bullet_c$ in system algebra encompass 9 associative operators for manipulating the algebraic compositions among abstract systems, i.e.:

$$\bullet_c \triangleq \{\Rightarrow, \overset{-}{\Longrightarrow}, \overset{+}{\Longrightarrow}, \overset{\sim}{\Longrightarrow}, \boxminus, \uplus, \pitchfork, \Lleftarrow, \vdash\} \tag{16}$$

where the compositional operators stand for system *inheritance, tailoring, extension, substitute, difference, composition, decomposition, aggregation,* and *specification*, respectively.

System algebra provides a denotational mathematical means for algebraic manipulations of all forms of abstract systems. System algebra can be used to model, specify, and manipulate generic "*to be*" and "*to have*" type problems, particularly system architectures and high-level system designs, in cognitive informatics, computational intelligence, computing science, software engineering, and system engineering. It will be demonstrated in Section 4.3 that the abstract system and system algebra are an ideal model for rigorously describing both the structures and behaviors of granules in granular computing. Detailed relational and compositional operations on abstract systems may be referred to [50].

3.3 Real-Time Process Algebra (RTPA)

A key metaphor in system modeling, specification, and description is that a software and intelligent system can be perceived and described as the *composition* of a set of interacting *processes*. Hoare [15], [16], Milner [24], and others developed various algebraic approaches to represent communicating and concurrent systems, known as process algebra. A *process algebra* is a set of formal notations and rules for describing algebraic relations of software engineering processes. RTPA [37], [40], [43], [51], [52] is a real-time process algebra that can be used to formally and precisely describe and specify architectures and behaviors of human and software systems.

Definition 17. A *process P* is an embedded relational composition of a list of *n* meta-statements p_i and p_j, $1 \leq i < n$, $1 < j \leq m = n+1$, according to certain composing relations r_{ij}, i.e.:

$$P = \mathop{\mathcal{R}}_{i=1}^{n-1}(p_i \; r_{ij} \; p_j), j = i+1 \qquad (17)$$
$$= (...(((p_1) \; r_{12} \; p_2) \; r_{23} \; p_3) \; ... \; r_{n-1,n} \; p_n)$$

where the big-R notation [48] is adopted that describes the nature of processes as the building blocks of programs.

Definition 17. indicates that the mathematical model of a process is a cumulative relational structure among basic computing operations, where the simplest process is a single computational statement.

Definition 18. *RTPA* is a denotational mathematical structure for algebraically denoting and manipulating system behavioural processes and their attributes by a triple, i.e.:

$$RTPA \triangleq (\mathfrak{T}, \mathfrak{P}, \mathfrak{N}) \qquad (18)$$

where $\mathfrak{T}$ is a set of 17 primitive types for modeling system architectures and data objects, $\mathfrak{P}$ a set of 17 meta-processes for modeling fundamental system behaviors, and $\mathfrak{R}$ a set of 17 relational process operations for constructing complex system behaviors.

Lemma 4. The *primary types of computational objects* state that the *RTPA type system* $\mathfrak{T}$ encompasses 17 primitive types elicited from fundamental computing needs, i.e.:

$$\mathfrak{T} \triangleq \{\mathbf{N}, \mathbf{Z}, \mathbf{R}, \mathbf{S}, \mathbf{BL}, \mathbf{B}, \mathbf{H}, \mathbf{P}, \mathbf{TI}, \mathbf{D}, \mathbf{DT}, \mathbf{RT}, \mathbf{ST}, @e\mathbf{S}, @t\mathbf{TM}, @int\odot, ⓢs\mathbf{BL}\} \qquad (19)$$

where the primary types of RTPA denote *natural number, integer, real number, string, Boolean variable, byte, hyper-decimal, pointer, time, date, date-time, run-time type, system type, event, timing-event, interrupt-event,* and *system status,* respectively.

Definition 19. A *meta-process* in RTPA is a primitive computational operation that cannot be broken down to further individual actions or behaviors.

A meta-process is an elementary process that serves as a basic building block for modeling software behaviors. *Complex processes* can be composed from meta- processes using *process relational operations.* In RTPA, a set of 17 meta-processes has been elicited from essential and primary computational operations commonly identified in existing formal methods and modern programming languages [1], [14], [21], [56], [57].

Lemma 5. The *RTPA meta-process system* $\mathfrak{P}$ encompasses 17 fundamental computational operations elicited from the most basic computing needs, i.e.:

$$\mathfrak{P} \triangleq \{:=, \blacklozenge, \Rightarrow, \Leftarrow, \not\Leftarrow, \succ, \prec, |\!\succ, |\!\prec, @, \triangleq, \uparrow, \downarrow, !, \otimes, \boxtimes, \S\} \quad (20)$$

where the meta-processes of RTPA stand for *assignment, evaluation, addressing, memory allocation, memory release, read. write, input, output, timing, duration, increase, decrease, exception detection, skip, stop*, and *system*, respectively.

Definition 20. A *process relation* in RTPA is an algebraic operation and a compositional rule between two or more meta-processes in order to construct a complex process. A set of 17 fundamental process relations has been elicited from fundamental algebraic and relational operations in computing in order to build and compose complex processes in the context of real-time software systems.

Lemma 6. The software composing rules state that the *RTPA process relation system* $\mathfrak{R}$ encompasses 17 fundamental algebraic and relational operations elicited from basic computing needs, i.e.:

$$\mathfrak{R} \triangleq \{\rightarrow, \curvearrowright, |, |\ldots|\ldots, R^{*}, R^{+}, R^{i}, \circlearrowleft, \rightarrowtail, ||, \oint\!\!\oint, |||, \gg, \lightning, \hookrightarrow_{t}, \hookrightarrow_{e}, \hookrightarrow_{i}\} \quad (21)$$

where the relational operators of RTPA stand for *sequence, jump, branch, while-loop, repeat-loop, for-loop, recursion, function call, parallel, concurrence, interleave, pipeline, interrupt, time-driven dispatch, event-driven dispatch,* and *interrupt-driven dispatch*, respectively.

The generic program model can be established by a formal treatment of statements, processes, and complex processes from the bottom-up in the program hierarchy.

Definition 21. A *program* $\wp$ is a composition of a finite nonempty set of *m* processes according to the time-, event-, and interrupt-based process dispatching rules, i.e.:

$$\wp = \mathop{R}_{k=1}^{m}(@\, e_k \hookrightarrow P_k) \quad (22)$$

Definitions 17 and 21 indicate that a program is an *embedded relational algebraic* entity as follows.

Theorem 1. The *Embedded Relational Model (ERM)* of programs states that a software system or a program $\wp$ is a set of complex embedded relational processes, in which all previous processes of a given process form the context of the current process, i.e.:

$$\begin{aligned} \wp &= \mathop{R}_{k=1}^{m}(@\, e_k \hookrightarrow P_k) \\ &= \mathop{R}_{k=1}^{m}[@\, e_k \hookrightarrow \mathop{R}_{i=1}^{n-1}(p_i(k)\; r_{ij}(k)\; p_j(k))],\; j = i+1,\; p_i \in \mathfrak{P}, r_{ij} \in \mathfrak{R} \end{aligned} \quad (23)$$

Proof. Theorem 1 can be directly proven on the basis of Definitions 17 and 21. Substituting P_k in Definition 21 with Eq. 17, a generic program $\wp$ obtains the form as a series of embedded relational processes as presented in Theorem 1.

The ERM model provides a unified mathematical treatment of programs, which reveals that a program is a finite nonempty set of embedded binary relations between a current statement and all previous ones that form the semantic context or environment of computing.

RTPA provides a coherent notation system and a formal engineering methodology for modeling both software and intelligent systems. RTPA can be used to describe both *logical* and *physical* models of systems, where logic views of the architecture of a software system and its operational platform can be described using the same set of

Table 3. Taxonomy of Denotational Mathematics

Operations		Concept Algebra	System Algebra	Real-Time Process Algebra			
				Meta-processes		Relational operations	
Relational operations	Related/independent	$\leftrightarrow / \nleftrightarrow$	$\sqsupseteq / \sqsubseteq$	Assignment	$:=$	Sequence	$\rightarrow$
	Super/sub relation	$\succ / \prec$	$\leftrightarrow / \nleftrightarrow$	Evaluation	$\blacklozenge$	Jump	$\curvearrowright$
	Equivalent	$=$	$=$	Addressing	$\Rightarrow$	Branch	$\mid$
	Consistent	$\cong$		Memory allocation	$\Leftarrow$	Switch	$\mid \ldots \mid \ldots$
	Overlapped		Π	Memory release	$\not\Leftarrow$	While-loop	R^*
	Comparison	$\sim$		Read	$\gtrdot$	Repeat-loop	R^+
	Definition	$\triangleq$		Write	$\lessdot$	For-loop	R^i
Compositional operations	Inheritance	$\Rightarrow$	$\Rightarrow$	Input	$\mid\gtrdot$	Recursion	$\circlearrowleft$
	Tailoring	$\overset{-}{\Rightarrow}$	$\overset{-}{\Rightarrow}$	Output	$\mid\lessdot$	Procedure call	$\rightarrowtail$
	Extension	$\overset{+}{\Rightarrow}$	$\overset{+}{\Rightarrow}$	Timing	$\underline{@}$	Parallel	$\parallel$
	Substitute	$\overset{\sim}{\Rightarrow}$	$\overset{\sim}{\Rightarrow}$	Duration	$\triangleq$	Concurrence	$\oiint$
	Composition	$\uplus$	$\uplus$	Increase	$\uparrow$	Interleave	$\mid\mid\mid$
	Decomposition	$\pitchfork$	$\pitchfork$	Decrease	$\downarrow$	Pipeline	»
	Aggregation/	$\Lleftarrow$	$\Lleftarrow$	Exception detection	!	Interrupt	↯
	Specification	$\vdash$	$\vdash$	Skip	$\oslash$	Time-driven dispatch	↳
	Instantiation	$\mapsto$	$\mapsto$	Stop	$\boxtimes$	Event-driven dispatch	↳
	Difference		$\boxminus$	System	§	Interrupt-driven dispatch	↳

notations. When the system architecture is formally modelled, the static and dynamic behaviors that perform on the system architectural model, can be specified by a three-level refinement scheme at the system, class, and object levels in a top-down approach. Detailed syntaxes and formal semantics of RTPA meta-processes and process relations may be referred to [37], [41], [43], [51], [52].

A summary of the algebraic operations and their notations in concept algebra, system algebra, and RTPA is provided in Table 3.

4 Applications of Denotational Mathematics

A wide range of applications of denotational mathematics have been identified, which encompass concept algebra for knowledge manipulations, system algebra for system architectural manipulations, and RTPA for system behavioral manipulations. This section presents some typical applications of denotational mathematics for the modeling of iterative and recursive systems architectures and behaviors by RTPA, the modeling of autonomous machine learning by concept algebra, and the modeling of granular computing by system algebra.

4.1 The Big-R Notation of RTPA for Modeling Iterative and Recursive System Architectures and Behaviors

The most generic and fundamental operations in system and human behavioral modeling are iterations and recursions. Because a variety of iterative constructs are provided in different programming languages, the notation for repetitive, cyclic, recursive behaviors and architectures in computing need to be unified.

The mechanism of the big-R notation can be analogized with the mathematical notations Σ or Π. To a great extent, Σ and Π can be treated as special cases of the big-R for repetitively doing additions and multiplications, respectively.

Example 1. The *big-*Σ notation $\sum_{i=1}^{n} x_i$ is a widely used calculus for denoting repetitive additions. Assuming that the operation of addition is represented by *sum*(x), the mechanism of big-Σ can be expressed more generally by the big-R notation, i.e.:

$$\sum_{i=1}^{n} x_i = \mathop{\mathrm{R}}_{\mathrm{i}=1}^{n} sum(x_i) \tag{24}$$

According to Definition 4, the big-R notation can be used to denote not only repetitive operational behaviors in computing, but also recurring constructs of architectures and data objects as shown below.

Example 2. The architecture of a two-dimensional array with $n \times m$ integer elements, A_{nm}, can be denoted by the big-R notation as follows:

$$A_{nm} = \mathop{\mathrm{R}}_{\mathrm{i}=0}^{n-1} \mathop{\mathrm{R}}_{\mathrm{j}=0}^{m-1} A[i, j]\mathbf{N} \tag{25}$$

Because the big-R notation provides a powerful and expressive means for denoting iterative and recursive behaviors, and architectures of systems or human beings, it is a universal mathematical means for system modeling in terms of repetitive behaviors and architectures, respectively.

Definition 22. *An infinitive iteration* can be denoted by the big-R notation as:

$$\mathbf{R}\, P \triangleq \gamma \bullet P \curvearrowright \gamma \tag{26}$$

where γ is a label that denotes the fix (rewinding) point of a loop, and $\curvearrowright$ denotes a jump in RTPA.

The infinitive iteration may be used to formally describe an everlasting behavior of systems.

Example 3. A simple everlasting clock, *CLOCK,* which does nothing but tick as C.A.R. Hoare proposed [16], i.e.:

$$CLOCK \triangleq tick \rightarrow tick \rightarrow tick \rightarrow \tag{27}$$

can be efficiently denoted by the big-R notation as simply as follows:

$$CLOCK \triangleq \mathbf{R}\, tick \tag{28}$$

A more generic and useful iterative construct is the conditional iteration.

Definition 23. *A conditional iteration* can be denoted by the big-R notation as:

$$\begin{aligned} \overset{\mathbf{F}}{\underset{exp\mathbf{BL}=\mathbf{T}}{\mathbf{R}}} P \triangleq \gamma \bullet (\ & \blacklozenge\ exp\mathbf{BL} = \mathbf{T} \\ & \quad \rightarrow P \\ & \quad \curvearrowright \gamma \\ & \mid \blacklozenge \sim \\ & \quad \rightarrow \varnothing \\ &) \end{aligned} \tag{29}$$

where $\varnothing$ denotes a skip.

The conditional iteration is frequently used to formally describe repetitive behaviors on given conditions. Eq. 29 expresses that the iterative execution of *P* will go on as long as the evaluation $\blacklozenge$ of the conditional expression is true ($exp\mathbf{BL} = \mathbf{T}$), until $exp\mathbf{BL} = \mathbf{F}$ abbreviated by '~'.

Definition 24. *Recursion* $\overset{\circlearrowleft}{\mathbf{R}} P^i$ is a multi-layered, embedded process relation in which a process *P* at layer *i* of embedment, P^i, calls itself at an inner layer *i-1*, P^{i-1}, $0 \le i \le n$. The termination of P^i depends on the termination of P^{i-1} during its execution, i.e.:

$$
\overset{\circlearrowleft}{R} P^i \triangleq \underset{i\mathbf{N}=n\mathbf{N}}{\overset{0}{R}} \begin{pmatrix} \blacklozenge\, i\mathbf{N} > 0 \\ \quad \to P^{i\mathbf{N}} := P^{i\mathbf{N}-1} \\ \mid \blacklozenge \sim \\ \quad \to P^{0} \end{pmatrix} \tag{30}
$$

where n is the *depth of recursion* or embedment that is determined by an explicitly specified conditional expression $exp\mathbf{BL} = \mathbf{T}$ inside the body of P.

Example 4. Using the big-R notation, the algorithm of the factorial function can be recursively defined as shown below:

$$
\begin{aligned}
(n\mathbf{N})! &\triangleq \overset{\circlearrowleft}{R}\, (n\mathbf{N})\,! \\
&= \underset{i\mathbf{N}=n\mathbf{N}}{\overset{0}{R}} \begin{pmatrix} \blacklozenge\, i\mathbf{N} > 0 \\ \quad \to (i\mathbf{N})! := i\mathbf{N} \bullet (i\mathbf{N}-1)! \\ \mid \blacklozenge \sim \\ \quad \to (i\mathbf{N})! := 1 \end{pmatrix}
\end{aligned} \tag{31}
$$

The big-R notation of RTPA captures a fundamental and widely applied mathematical concept in computing and human behavior description, which demonstrates that a convenient mathematical calculus and notation may dramatically reduce the difficulty and complexity in expressing a frequently used and highly recurring concept and notion in computing.

4.2 Autonomous Machine Learning Using Concept Algebra

Cognitive informatics [38], [39], [44] defines learning as a cognitive process at the higher cognitive function layer (Layer 7) according to the Layered Reference Model of the Brain (LRMB) [53]. The learning process is interacting with multiple fundamental cognitive processes such as *object identification, abstraction, search, concept establishment, comprehension, memorization,* and *retrievably testing*. Learning is closely related to other higher cognitive processes of inferences such as *deduction, induction, abduction, analogy, explanation, analysis, synthesis, creation, modeling,* and *problem solving*.

Definition 25. *Learning* is a higher cognitive process of the brain at the higher cognitive layer of LRMB that gains knowledge of something or acquires skills in some actions by updating the cognitive models in long-term memory.

According to the *Object-Attribute-Relation* (OAR) model [45], results of learning can be embodied by the updating of the existing *OAR* in the brain. In other words, learning is a dynamic composition of the currently created sub-*OAR* and the existing *OAR* in long-term memory (LTM) as expressed below.

Theorem 2. The *representation of learning result* states that the internal memory in the form of the OAR structure can be updated by a conjunction between the existing OAR and the newly created sub-*OAR* (sOAR), i.e.:

$$\begin{aligned} OAR'\mathbf{ST} &\triangleq OAR\mathbf{ST} \uplus sOAR\mathbf{ST} \\ &= OAR\,\mathbf{ST} \uplus (O_s, A_s, R_s) \end{aligned} \quad (32)$$

where the composition operation $\uplus$ in concept algebra is defined below.

Definition 26. A *composition* of concept *c* from *n* subconcepts $c_1, c_2, \ldots, c_n$, denoted by $\uplus$, is an integration of them that creates the new super concept *c* via concept conjunction, and establishes new associations between them, i.e.:

$$\begin{aligned} & c(O, A, R^c, R^i, R^o) \uplus \mathop{R}_{i=1}^{n} c_i \triangleq \\ & \quad c(O, A, R^c, R^i, R^o \mid O = \bigcup_{i=1}^{n} O_{c_i}, A = \bigcup_{i=1}^{n} A_{c_i}, \\ & \quad R^c = \bigcup_{i=1}^{n} (R^c{}_{c_i} \cup \{(c, c_i), (c_i, c)\}, R^i = \bigcup_{i=1}^{n} R^i{}_{c_i}, R^o = \bigcup_{i=1}^{n} R^o{}_{c_i}) \\ & \parallel \mathop{R}_{i=1}^{n} c_i(O_i, A_i, R^c{}_i, R^{i'}{}_i, R^{o'}{}_i \mid R^{i'}{}_i = R^i{}_i \cup \{(c, c_i)\}, R^{o'}{}_i = R^o{}_i \cup \{(c_i, c)\}) \end{aligned} \quad (33)$$

As specified in Eq. 33, the composition operation results in the generation of new internal relations $\Delta R^c = \bigcup_{i=1}^{n} \{(c, c_i), (c_i, c)\}$ that is not belongs to any of its subconcepts. It is also noteworthy that, during learning by concept composition, the existing knowledge in forms of the individual *n* concepts is changed and updating concurrently via the newly created input/output relations with the newly generated concept.

Corollary 1. The *learning process* is a cognitive composition of a piece of newly acquired information and the existing knowledge in LTM in the form of the OAR-based knowledge networks.

The cognitive process of learning can be formally modeled using concept algebra and RTPA as given in Fig. 2. The center of the cognitive process of learning is that knowledge about the learn objects and intermediate results are represented internally in the brain as a sub-OAR model. According to the LRMB model [53] and the OAR model [45] of internal knowledge representation in the brain, the temporal result of learning in short-term memory (STM) is a new sub-OAR model, which will be used to update the entire OAR model of knowledge in LTM as permanent learning result.

```
The Learning Process

Learning (I:: OS; O:: OAR'ST) ≙
{I. Identify object
↣ ObjectIdentification (I:: OS; O:: AS)
   // AST – a set of attributes of OS

II. Concept establishment
↣ ConceptEstablishment (I:: OS, AS; O:: c(OS, AS, RS)ST)

III. Comprehension
↣ Comprehension (I:: c(OS, AS, RS)ST; O:: sOAR'ST)

IV. Memorization
↣ Memorization (I:: sOARST; O:: OAR'ST)

V. Rehearsal
→ (◆ Rehearsal BL = T
         ( ( ↣ ConceptEstablishment  (I:: sOARST;
                                      O:: c(OS, AS, RS)ST)
           || ↣ Comprehension (I:: sOARST; O:: sOARST)
           )
           ↣ Memorization (I:: sOARST; O:: OAR'ST)
         )
      |◆ ~
         → ⊗
   )
}
```

Fig. 2. Formal description of the learning process in concept algebra and RTPA

According to the formal model of the learning process, autonomic machine learning can be carried out by the following steps: 1) *Identify object*: This step identifies the learning object O; 2) *Concept establishment*: This step establishes a concept model for the learning object O, $c(A, R, O)$, by searching related attributes A, relations R, and instances O; 3) *Comprehension:* This step comprehends the concept and represents the concept with a sub-OAR model in STM; 4) *Memorization:* This step associates the learnt sub-OAR of the learning object with the entire OAR knowledge, and retains it in LTM; 5) *Rehearsal test:* This step checks if the learning result needs to be rehearsed. If yes, it continues to parallel execution of Steps (6) and (7); otherwise, it exits; 6) *Re-establishment of concept:* This step recalls the concept establishment process to rehearse the learning result; 7) *Re-comprehension:* This step recalls the comprehension process to rehearse the learning result.

The formalization of the cognitive process of learning does not only reveal the mechanisms of human learning, but also explain how machine may gain the capability of autonomous learning. Based on the rigorous syntaxes and semantics of RTPA, the formal learning process can be implemented by computers in order to form the core of machine intelligence [47].

4.3 Granular Computing Using System Algebra

The term *granule* is originated from Latin *granum*, i.e., grain, to denote a small compact particle in physics and in the natural world. The *taxonomy of granules* in computing can be classified into the data granule, information granule, concept granule, computing granule, cognitive granule, and system granule [20], [28], [33], [54], [58], [62].

Definition 27. A *computing granule*, shortly a *granule*, is a basic mathematical structure that possesses a stable topology and at least a unit of computational capability or behavior.

Definition 28. *Granular computing* is a new computational methodology that models and implements computational structures and functions by a granular system, where each granule in the system carries out a predefined function or behavior by interacting to other granules in the system.

It is recognized that any abstract or concrete granule can be formally modeled by abstract systems in system algebra. On the basis of Definition 13, an abstract granule can be formally described as follows.

Definition 29. A *computing granule* G on the *universal system environment* $\mathfrak{U}$ is a 7-tuple, i.e.:

$$G \triangleq S = (C, R^c, R^i, R^o, B, \Omega, \Theta) \tag{34}$$

where

- C is a finite nonempty set of *cell* or component of the system, $C = \{c_1, c_2, ..., c_n\} \subseteq \text{Þ}\mathcal{C} \sqsubseteq \mathfrak{U}$.
- R is a finite nonempty set of *relations* between pairs of the components in the system, $R = \{r_1, r_2, ..., r_m\} \subseteq C \times C$.
- $R^c = C \times C$ is a set of *internal relations*.
- $R^i \subseteq C_\Theta \times C$ is a set of external *input relations*, $C_\Theta \subseteq \text{Þ}\mathcal{C} \sqsubseteq \mathfrak{U}$.
- $R^o \subseteq C \times C_\Theta$ is a set of external *output relations*.
- B is a set of *behaviors* (or functions), $B = \{b_1, b_2, ..., b_p\} \subseteq \text{Þ}\mathcal{B} \sqsubseteq \mathfrak{U}$.
- Ω is a set of *constraints* on the memberships of components, the conditions of relations, and the scopes of behaviors, $\Omega = \{\omega_1, \omega_2, ..., \omega_q\}$.
- Θ is the *environment* of G with a nonempty set of components C_Θ outside C, i.e., $\Theta = C_\Theta \subseteq \text{Þ}\mathcal{C} \sqsubseteq \mathfrak{U}$.

Definition 30. A *granular system* S_G is a composition of multiple granules in a system where all granules interact with each other for a common goal of system functionality.

Properties of granular systems obey the properties of generic abstract systems as described in Section 3.2 [50]. The set of relational and compositional operations on granules and granular systems towards granular computing are identical as those modeled in system algebra.

5 Conclusions

The abstract, rigorous, and expressive needs in cognitive informatics, computational intelligence, software engineering, and knowledge engineering have led to new forms of mathematics collectively known as denotational mathematics. Denotational mathematics has been introduced as a category of expressive mathematical structures that deals with high-level mathematical entities such as abstract objects, complex relations, behavioral information, concepts, knowledge, processes, and systems. The domain and architecture of denotational mathematics have been described. New mathematical entities, novel mathematical structures, and applications of denotational mathematics have been explored toward the modeling and description of the natural and machine intelligent systems.

Extensions of conventional analytic mathematics onto more complicated mathematical entities beyond numbers and sets lead to the contemporary denotational mathematics. Three paradigms of denotational mathematics, such as concept algebra, system algebra, and Real-Time Process Algebra (RTPA), have been introduced. Within the new forms of denotational mathematics, *concept algebra* has been designed to deal with the new abstract mathematical structure of concepts and their representation and manipulation in knowledge engineering. RTPA has been developed to deal with series of behavioral processes and architectures of software and intelligent systems. *System algebra* has been created to the rigorous treatment of abstract systems and their algebraic relations and operations. Applications of denotational mathematics in cognitive informatics and computational intelligence have been elaborated with a set of case studies and examples. This work has demonstrated that denotational mathematics is an ideal mathematical means for a set of emerging disciplines that deal with concepts, knowledge, behavioral processes, and human/machine intelligence.

Acknowledgement. The author would like to acknowledge the Natural Science and Engineering Council of Canada (NSERC) for its partial support to this work. The author would like to thank the valuable comments and suggestions of the anonymous reviewers.

References

1. Aho, A.V., Sethi, R., Ullman, J.D.: Compilers: Principles, Techniques, and Tools. Addison-Wesley Publication Co., New York (1985)
2. Anderson, J.R.: The Architecture of Cognition. Harvard Univ. Press, Cambridge (1983)
3. Ashby, W.R.: Principles of the Self-Organizing System. In: von Foerster, H., Zopf, G. (eds.) Principles of Self-Organization, pp. 255–278. Pergamon, Oxford (1962)
4. Ashby, W.R.: Requisite Variety and Implications for Control of Complex Systems. Cybernetica 1, 83–99 (1985)
5. Bender, E.A.: Mathematical Methods in Artificial Intelligence. IEEE CS Press, Los Alamitos (1996)
6. Chomski, N.: Three Models for the Description of Languages. I.R.E. Transactions on Information Theory 2(3), 113–124 (1956)

7. Codin, R., Missaoui, R., Alaoui, H.: Incremental Concept Formation Algorithms Based on Galois (Concept) Lattices. Computational Intelligence 11(2), 246–267 (1995)
8. Colins, A.M., Loftus, E.F.: A Spreading-Activation Theory of Semantic Memory. Psychological Review 82, 407–428 (1975)
9. Ellis, D.O., Fred, J.L.: Systems Philosophy. Prentice-Hall, Englewood Cliffs (1962)
10. Ford, J.: Chaos: Solving the Unsolvable. Dynamics and Fractals. Academic Press, London (1986)
11. Ganter, B., Wille, R.: Formal Concept Analysis, pp. 1–5. Springer, Heidelberg (1999)
12. Hampton, J.A.: Psychological Representation of Concepts of Memory, pp. 81–110. Psychology Press, Hove, England (1997)
13. Heylighen, F.: Self-Organization, Emergence and the Architecture of Complexity. In: Proc. 1st European Conf. on System Science (AFCET), Paris, pp. 23–32 (1989)
14. Higman, B.: A Comparative Study of Programming Languages, 2nd edn., MacDonald (1977)
15. Hoare, C.A.R.: Communicating Sequential Processes. Communications of the ACM 21(8), 666–677 (1978)
16. Hoare, C.A.R.: Communicating Sequential Processes. Prentice-Hall International, London (1985)
17. Hurley, P.J.: A Concise Introduction to Logic, 6th edn. Wadsworth Pub. Co., ITP (1997)
18. Klir, G.J.: Facets of Systems Science. Plenum, New York (1992)
19. Lewis, H.R., Papadimitriou, C.H.: Elements of the Theory of Computation, 2nd edn. Prentice-Hall International, Englewood Cliffs (1998)
20. Lin, T.Y.: Granular Computing on Binary Relations (I): Data Mining and Neighborhood Systems. In: Proc. Rough Sets in Knowledge Discovery, pp. 107–120. Physica-Verlag, Heidelberg (1998)
21. Louden, K.C.: Programming Languages: Principles and Practice. PWS-Kent Publishing Co., Boston (1993)
22. Matlin, M.W.: Cognition, 4th edn. Harcourt Brace College Pub., NY (1998)
23. Medin, D.L., Shoben, E.J.: Context and Structure in Conceptual Combination. Cognitive Psychology 20, 158–190 (1988)
24. Milner, R. (ed.): A Calculus of Communication Systems. LNCS, vol. 92. Springer, Heidelberg (1980)
25. Murphy, G.L.: Theories and Concept Formation. In: Mechelen, I.V., et al. (eds.) Categories and Concepts, Theoretical Views and Inductive Data Analysis, pp. 173–200. Academic Press, London (1993)
26. Pavel, M.: Fundamentals of Pattern Recognition, 2nd edn. Addision-Wesley, Reading (1993)
27. Pawlak, Z.: Rough Logic, Bulletin of the Polish Academy of Science. Technical Science 5(6), 253–258 (1987)
28. Pedrycz, W. (ed.): Granular Computing: An Emerging Paradigm. Physica-Verlag, Heidelberg (2001)
29. Pnueli, A.: The Temporal Logic of Programs. In: Proc. 18th IEEE Symposium on Foundations of Computer Science, pp. 46–57. IEEE, Los Alamitos (1977)
30. O'Grady, W., Archibald, J.: Contemporary Linguistic Analysis: An Introduction, 4th edn. Pearson Education Canada Inc., Toronto (2000)
31. Rapoport, A.: Mathematical Aspects of General Systems Theory. General Systems Yearbook 11, 3–11 (1962)
32. Skarda, C.A., Freeman, W.J.: How Brains Make Chaos into Order. Behavioral and Brain Sciences 10 (1987)

33. Skowron, A., Stepaniuk, J.: Information Granules: Towards Foundations of Granular Computing. International Journal of Intelligent Systems 16, 57–85 (2001)
34. Smith, E.E., Medin, D.L.: Categories and Concepts. Harvard Univ. Press, Cambridge (1981)
35. von Bertalanffy, L.: Problems of Life: An Evolution of Modern Biological and Scientific Thought. C.A. Watts, London (1952)
36. von Neumann, J.: The Principles of Large-Scale Computing Machines. Annals of History of Computers 3(3), 263–273 (reprinted, 1946)
37. Wang, Y.: The Real-Time Process Algebra (RTPA). Annals of Software Engineering: A International Journal, USA 14, 235–274 (2002)
38. Wang, Y.: Keynote: On Cognitive Informatics. In: Proc. 1st IEEE International Conference on Cognitive Informatics (ICCI 2002), Calgary, Canada, pp. 34–42. IEEE CS Press, Los Alamitos (2002)
39. Wang, Y.: On Cognitive Informatics. Brain and Mind: A Transdisciplinary Journal of Neuroscience and Neurophilosophy, USA 4(3), 151–167 (2003)
40. Wang, Y.: Using Process Algebra to Describe Human and Software System Behaviors. Brain and Mind 4(2), 199–213 (2003)
41. Wang, Y.: On the Informatics Laws and Deductive Semantics of Software. IEEE Transactions on Systems, Man, and Cybernetics (C) 36(2), 161–171 (2006)
42. Wang, Y.: Cognitive Informatics and Contemporary Mathematics for Knowledge Representation and Manipulation. In: Wang, G.-Y., Peters, J.F., Skowron, A., Yao, Y. (eds.) RSKT 2006. LNCS (LNAI), vol. 4062, pp. 69–78. Springer, Heidelberg (2006)
43. Wang, Y.: Software Engineering Foundations: A Software Science Perspective. CRC Series in Software Engineering, vol. II. Auerbach Publications, NY, USA (2007)
44. Wang, Y.: The Theoretical Framework of Cognitive Informatics. International Journal of Cognitive Informatics and Natural Intelligence 1(1), 1–27 (2007)
45. Wang, Y.: The OAR Model of Neural Informatics for Internal Knowledge Representation in the Brain. International Journal of Cognitive Informatics and Natural Intelligence 1(3), 64–75 (2007)
46. Wang, Y.: Keynote: On Theoretical Foundations of Software Engineering and Denotational Mathematics. In: Proc. 5th Asian Workshop on Foundations of Software, Xiamen, China, pp. 99–102 (2007)
47. Wang, Y.: The Theoretical Framework and Cognitive Process of Learning. In: Proc. 6th International Conference on Cognitive Informatics (ICCI 2007), Lake Tahoe, CA, pp. 470–479. IEEE CS Press, Los Alamitos (2007)
48. Wang, Y.: On the Big-R Notation for Describing Iterative and Recursive Behaviors. International Journal of Cognitive Informatics and Natural Intelligence 2(1), 17–28 (2008)
49. Wang, Y.: On Concept Algebra: A Denotational Mathematical Structure for Knowledge and Software Modeling. International Journal of Cognitive Informatics and Natural Intelligence 2(2), 1–19 (2008)
50. Wang, Y.: On System Algebra: A Denotational Mathematical Structure for Abstract System Modeling. International Journal of Cognitive Informatics and Natural Intelligence 2(2), 20–42 (2008)
51. Wang, Y.: RTPA: A Denotational Mathematics for Manipulating Intelligent and Computational Behaviors. International Journal of Cognitive Informatics and Natural Intelligence 2(2), 44–62 (2008)
52. Wang, Y.: Deductive Semantics of RTPA. International Journal of Cognitive Informatics and Natural Intelligence 2(2), 95–121 (2008)

53. Wang, Y., Wang, Y., Patel, S., Patel, D.: A Layered Reference Model of the Brain (LRMB). IEEE Trans. on Systems, Man, and Cybernetics (C) 36(2), 124–133 (2006)
54. Wang, Y., Lotfi, A.: On the System Algebra Foundation for Granular Computing. International Journal of Software Science and Computational Intelligence 1(1) (January 2009)
55. Wille, R.: Restructuring Lattice Theory: An Approach Based on Hierarchies of Concepts. In: Rival, I. (ed.) Ordered Sets, pp. 445–470. Reidel, Dordrecht (1982)
56. Wilson, L.B., Clark, R.G.: Comparative Programming Language. Addison-Wesley Publishing Co., Reading (1988)
57. Woodcock, J., Davies, J.: Using Z: Specification, Refinement, and Proof. Prentice Hall International, London (1996)
58. Yao, Y.Y.: Information Granulation and Rough Set Approximation. International Journal of Intelligent Systems 16(1), 87–104 (2001)
59. Yao, Y.Y.: A Comparative Study of Formal Concept Analysis and Rough Set Theory in Data Analysis. In: Tsumoto, S., Słowiński, R., Komorowski, J., Grzymała-Busse, J.W. (eds.) RSCTC 2004. LNCS (LNAI), vol. 3066, pp. 59–68. Springer, Heidelberg (2004)
60. Zadeh, L.A.: Fuzzy Sets and Systems. In: Fox, J. (ed.) Systems Theory, pp. 29–37. Polytechnic Press, Brooklyn (1965)
61. Zadeh, L.A.: Outline of a New Approach to Analysis of Complex Systems. IEEE Trans. on Sys. Man and Cyb. 1(1), 28–44 (1973)
62. Zadeh, L.A.: Fuzzy Sets and Information Granularity. In: Gupta, M.M., Ragade, R., Yager, R. (eds.) Advances in Fuzzy Set Theory and Applications, pp. 3–18. North-Holland, Amsterdam (1979)

Mereological Theories of Concepts in Granular Computing

Lech Polkowski

Polish-Japanese Institute of Information Technology
Warsaw, Poland
Department of Mathematics and Computer Science
University of Warmia and Mazury, Olsztyn, Poland
polkow@pjwstk.edu.pl

"...A few ideas that are not new ...." (Cyprian Kamil Norwid)

Abstract. This article is conceived as a homage to mathematicians and computer theorists working on basic concepts concerning knowledge and their usage in application contexts. Due to their work, we now have in our possession very impressive tools for analysis of uncertainty like rough set theory and fuzzy set theory along with hybrid ramifications between the two and with other areas of research in the realm of cognitive technologies in particular a very promising area of cognitive informatics.

In this work, we strive at presenting basic issues in granular theory of knowledge emphasizing formal aspects of our approach. This approach can be seen as a continuation of the line of analysis initiated by Gottlob Frege with his idea of exact and inexact concepts through analysis of the idea of knowledge by Popper, Lesniewski, Łukasiewicz and others, to the implementation of the Fregean idea in the theory of knowledge known as rough set theory initiated by Pawlak and pursued by many followers.

The basic tool in our analysis of the idea of a concept and a fortiori of knowledge is mereological theory of concepts (Lesniewski): we try to convince the reader that ideas of that theory suit well needs of analysis of knowledge and granulation of it.

This work arises from some previous attempts at developing this ideas in a wider context of ontological discussion; the author is indebted to the colloquia series SEFIR at Lateran University in Rome, where he was able to present the basic ideas in a lecture in January 2005; for this opportunity he is grateful to Professors Giandomenico Boffi and Alberto Pettorossi.

The author wishes to dedicate this work to the memory of Professors Helena Rasiowa and Zdzisław Pawlak who influenced very much his research interests in this area.

Keywords: mereological theory of concepts, approximate reasoning, granular computing.

1 Introduction: On the Notions of a Concept, Knowledge and Reasoning

To begin with, let us observe that man was using always symbols, words, or tokens in order to single out from the complex world some of its objects,

M.L. Gavrilova et al. (Eds.): Trans. on Comput. Sci. II, LNCS 5150, pp. 30–45, 2008.

phenomena, or relations deemed important: (see, e.g., Plutarch, *Moralia, The Letter on "E", XVII, 3*: (capital "E" ("Ei: you are" written on the wall of the Delphi temple):: "My own view is that the letter signifies neither number, nor order, nor conjunction, nor any other omitted part of speech; it is a complete and self-operating mode of addressing the God; the word once spoken brings the speaker into apprehension of his power."

The extent of a symbol, term or token understood as the community of entities that fall under this symbol, term or token, has been understood as a concept.

Concepts serve as building blocks for knowledge. The notion of knowledge is many–faceted and difficult to be defined precisely and uniquely and it is used very frequently without any attempt at definition as a notion that explains per se. We follow J. M. Bocheński [4] in claiming that the world is a system of states of things, related to themselves by means of the network of relations; things, their features, relations among them and states are reflected in knowledge: things in objects or notions, features and relations in notions (or, concepts), states of things in sentences. Sentences constitute knowledge. Knowledge allows its possessor to classify new objects, model processes, make predictions.

Processes of reasoning include an effort by means of which sentences are created; various forms of reasoning depend on the chosen system of notions, symbolic representation of notions, forms of manipulating symbols.

Formal reasoning processes may be divided, according to J. M. Bochenski [4], cf., Łukasiewicz [20], Hintikka [13], into two main classes: deductive reasoning and reductive (encompassing so called inductive reasoning) reasoning. Deductive reasoning proceeds formally in systems endowed with implicative assertions, according to the schema:

$$\text{if } A \text{ implies } B \text{ and } A \text{ then } B. \tag{1}$$

On the contrary, reductive reasoning proceeds according to the schema:

$$\text{if } A \text{ implies } B \text{ and } B \text{ then } A. \tag{2}$$

In reasoning with certainty, it is assumed that all assertions are assigned values of truth states, from falsity (0) through intermediate states, to truth (1). Deduction rules assign truth values to the assertion B on the basis of the truth values of the implication $A \Rightarrow B$ and the assertion A. This is especially manifest in axiomatic systems, see, e.g., Hilbert and Ackermann [12] in which one assumes a collection of assertions called axioms, assigned by default to be true, from which by deduction rules other true assertions are derived.

Reductive reasoning poses more problems: in order to infer that A on the basis of $A \Rightarrow B$ and B requires usually evidence for B of satisfactory strength. Problems related to this inference are expressed, e.g., by the well–known Hempel "paradoxes of confirmation" [11], [3].

By these problems, reductive reasoning borders on reasoning under uncertainty (approximate reasoning), as many approximate schemas, e.g., based on probability or Bayesian reasoning, were proposed for it, see, e.g.,[1], [44].

Approximate reasoning in the strict sense seems to be a form of reasoning in which assertions are assigned degrees of uncertainty, i.e., one is aware that the actual value of truth of an assertion is known only to a degree; in this there seems to lie the main distinguishing feature of this form of reasoning. Clearly, it is not possible to draw a clear line between this form of reasoning and some forms of formal deductive reasoning; for instance, the state 1/2 of truth in the 3–valued logic of Łukasiewicz [21] may be interpreted as the state "do not know" i.e., an expression of uncertainty to degree of 1/2. However, in 3–valued logic, the state of truth of 1/2 is known "exactly" and it is not subject to evaluation on the basis of evidence or knowledge.

We will in this exposition adhere to the point of view that approximate reasoning involves clear realization of uncertainty of knowledge which is expressed in its formal shape by some numerical factors that convey the degree of uncertainty and are evaluated on the basis of actual knowledge.

Among known forms of approximate reasoning, one may mention: rough set–based reasoning, see, e.g., [24], [27], in which knowledge is expressed by means of information systems and the underlying logic is the decision logic of descriptors; fuzzy set–based reasoning, see, e.g., [47], [10], in which degrees of uncertainty are expressed by means of membership degrees of an element in a set; various forms of non–monotonic reasoning, see, e.g., [5].

2 Mathematical Approaches to the Notion of a Concept

A view on concepts as non–empty categories of entities, each category having of more than one entity in its scope, was adopted by Aristotle in *Prior Analytics.* A calculus of concepts in which concepts are related one to another by means of inclusion ("all a is b") and intersection ("some a is b", "no a is b") developed by Aristotle has been known as Syllogistics. Primitive terms of Syllogistic were proposed as: Aab ("all a is b"), Iab ("some a is b"), Oab ("some a is not b"), Eab ("no a is b"). Formulas of Syllogistic are of the form $\frac{Xuv, Ywt}{Zef}$ where X, Y, Z are among A, I, O, E. Theorems i.e. true formulas are called syllogisms (moods).

An axiomatic form was given to Syllogistic by Łukasiewicz [22]; it exploits axiomatic schemes: 1.Aaa 2. Iaa 3.$\frac{Amb, Aam}{Aab}$ 4. $\frac{Amb, Ima}{Iab}$, and three derivation rules: a. Modus Ponens $\frac{p,\ if\ p\ then\ q}{q}$; b. Substitution c. Replacement by equivalents: $Oab = NAab, Eab = NIab$, where N is the operator of negation.

This mechanism of inference allows for derivation of all 24 true moods of Syllogistics. As proved by Słupecki [43], the Aristotle Syllogistic is a complete logical calculus of concepts with containment and intersection as relations among them.

With creation of set theory in works by Cantor, Dedekind, Hausdorff and others, the view emerged that concepts can be modeled in the new language of set theory. A theory of concepts was created by Gottlob Frege [6], [7], [8]. Frege created second–order logical calculus in Begriffsschrift and Grundgesetze, introducing variables for entities (like $x, y, z, ...$) as well as variables for functions (like $f, g, h, ...$) with the intent of forming names of entities by means of terms like $f(x)$.

By means of truth values $T(true)$, $F(alse)$, concepts were defined as functions mapping entities to truth values. An entity a falls under a concept f in case $f(a) = T$, in symbols $[f]a$.

The essential role is played in the Frege system by Substitution Rule:

For each formula with only free occurrences of $[f]x$, one is justified in replacing each such occurrence with a formula $\alpha(x)$ in which x is free.

For each concept f, its extension εf was postulated which was required to satisfy the following

Basic law V:

$\varepsilon f = \varepsilon g \Leftrightarrow \forall x.([f]x = [g]x)$, meaning: extensions of concepts are identical if and only if for every object, that object falls under f if and only if it falls under g.

Membership in an extension was defined by Frege as follows: $x \in y \Leftrightarrow \exists f.(y = \varepsilon f \wedge [f]x)$.

Thus, for every concept f: $x \in \varepsilon f \Leftrightarrow [f]x$ (The law of extensions).

It is a consequence to Basic Law V that :

$\forall f.\exists x.(x = \varepsilon f)$,

meaning: to every concept there is an extension.

As it is well known, these intuitively acceptable intuitions about the nature of concepts led to a contradiction: the well–known Russell Paradox. As formulated by Frege (Grundgesetze II, Appendix): consider the concept g corresponding via Principle of Comprehension to the formula: $\exists f.x = \varepsilon f \wedge \neg[f]x$. There exists the extension of g: εg.

Then: $g \in \varepsilon g \Leftrightarrow \neg(g \in \varepsilon g)$, a contradiction.

The Russell Paradox forced a fundamental reconstruction of set theory which resulted in axiomatic systems of Zermelo–Fraenkel or Goedel–Bernays see [14]; in each of them the notion of a set was restricted by reducing possibilities of new set construction. Nevertheless, many statements about sets turned out to be undecidable, e.g., the Continuum Hypothesis, the Souslin Hypothesis etc., see [14]. As a result many additional axioms independent of standard axioms of set theory has emerged leading to a variety of set theories.

"The true reason for the incompleteness inherent in all formal systems of mathematics is that the formation of ever higher types can be continued into the transfinite, while in any formal system at most denumerably many of them are available. (...) The undecidable propositions constructed here become decided whenever appropriate higher types are added (for example, the type ω to the system P [Peano Arithmetic]). An analogous situation prevails for the axiom system of set theory" (Goedel [9]; translated by A. Kanamori in [15]).

Language of set theory has become standard in discussing concepts in many paradigms of reasoning, in spite of foundational difficulties with this theory. For instance, rough set theory, or fuzzy set theory have been formulated predominantly in this language. For theories of concepts in those paradigms, see, e.g., [25], [10]. Theory of concept algebras in the realm of cognitive informatics was presented in [45], [46].

A theory of concepts invoking the aristotelian spirit and removing the existence of the empty concept – a reason for the Russell Paradox – was formulated by Lesniewski [16], [17], as the theory of Ontology. Ontology formally defines the notion of a concept as a distributive entity; the primitive relation is ε ("is").

The Axiom of Ontology (Lesniewski) is formulated as follows:

$$x\varepsilon y \Leftrightarrow (\exists z.z\varepsilon x) \wedge (\forall u, v.u, v\varepsilon x \Rightarrow u = v) \wedge (\forall z.z\varepsilon x \Rightarrow z\varepsilon y).$$

Thus, y is a distributive entity.

The consequences of the Axiom of Ontology are:

1. If there exists y such that $x\varepsilon\ y$ then x is non–void.
2. If there exists y such that $x\varepsilon\ y$ then x is a single entity (an individual).
3. If there exists y such that $x\varepsilon y$ then $x\varepsilon\ x$ for each entity x.
4. If $x\varepsilon\ y$ for some y and $z\varepsilon\ x$ then $z = x$.

An archetypal example, quoted by Lesniewski himself [16] was the sentence "Socrates is a man" in which an individual "Socrates" is stated to fall under the concept of "man".

Duns Scotus, in *Treatise on God as First Principle* [42]: " Aristotle says in the seventh book of Metaphysics: ""If anything were compounded of but one element that one would be the thing itself"" ". Clearly, this view is in contradiction to set theory which discerns between a singleton set and its element but it is valid in Ontology by Lesniewski: each individual x satisfies the formula $x\varepsilon x$.

Ontologuy sets apart the collection of individual entities – subjects in formulas of the form $x\varepsilon y$. On these individuals, a calculus of parts is developed. The notion of a part here is abstracted as a mathematical notion from real examples of complex objects which can be decomposed into parts.

Thus, Mereology is a calculus on individual concepts as specified by Ontology; it is based on the predicate π of part, defined for individual entities, subject to requirements:

P1. $x\pi y \wedge y\pi z \Rightarrow x\pi z$.

P2. $\neg(x\pi x)$.

The relation of a part is therefore a strict order on individual entities; making it into a partial non–strict order follows the standard lines. This new order is the ingredient relation.

The ingredient relation ing_π induced by a part relation π is defined as follows:

$$x\ ing_\pi\ y \Leftrightarrow x = y\ or\ x\ \pi\ y.$$

Clearly, the relation ing_π is a partial non–strict ordering on individual entities.

Let us note that the relation of proper inclusion $\subset$ is a part relation whereas the corresponding ingredient relation is the relation of improper inclusion $\subseteq$.

In Mereology, as in naive set theory, the necessity was felt for having a tool for converting collections of entities into a single representative entity.

Passing from distributive entities (concepts, collections) to individual entities is provided in Mereology by the class operator Cls.

The class $Cls(M)$ of a distributive concept (collection, property) M is defined by requiring the two properties to be satisfied.

C1. $x\varepsilon\ M \Rightarrow x\ ing_{\pi}\ Cls(M)$.

C2. $x\ ing_{\pi}\ Cls(M) \Rightarrow \exists u, v. u\ ing_{\pi}\ x \wedge u\ ing_{\pi}\ v \wedge v\varepsilon\ M$.

Let us note that in case when the relation π is the relation $\subset$, and M is a non–void collection of sets, the class of M is the union $\bigcup M$ of the collection M.

An example [16] of a class operator is the chessboard which is the class of white and black squares. We offer yet another example: assuming the part relation to be the strict order ¡ on the interval [0,1], the associated ingredient relation is the partial order $\leq$ and for a collection $M \subset [0,1]$ of reals, the class of M is $Cls(M) = sup\ M$.

The basic properties of the class operator are, see [16],

Proposition 1. If y is an individual entity, then $Cls(y) = y$.

Proposition 2. For each x, y: if for each z from $z\ ing_{\pi}\ x$ it follows that $t\ ing_{\pi}\ x, t\ ing_{\pi}\ y$ for some t, then $x\ ing_{\pi}\ y$.

Proposition 2 is called the Inference Rule of Mereology (IRM); its usage allows for reasoning about entities on the basis of ingredient relations; for the proof, see [16].

3 Rough Mereology

Rough Mereology, see [28], [29], [37], [39], [32], extends Mereology by considering the relation of a part to a degree; thus, it is suited for needs of Approximate Reasoning, allowing to formally define the graded state of truth.

The basic notion of Rough Mereology is the concept of being a part of a given object to a degree at least equal to a fixed value. It is denoted with the symbol $\mu(y, r)$ where y is an individual object and r is a real number in the interval $[0, 1]$.

The term $x\varepsilon\ \mu(y, r)$ is read as "the individual x is a part of the individual y to a degree at least r".

The concept μ is required to satisfy the following conditions.

RM1. $x\varepsilon\ \mu(y, 1)$ if and only if $x\ ing_{\pi}\ y$.

RM2. If $x\varepsilon\ \mu(y, 1)$ and $z\varepsilon\ \mu(x, r)$ then $z\varepsilon\ \mu(y, r)$.

RM3. If $x\varepsilon\ \mu(y, r)$ and $s < r$ then $x\varepsilon\ \mu(y, s)$.

The concept μ has been called by the author a *rough inclusion*.

4 Granulation of Knowledge

As an application of the mereological theory of concepts, we propose to present the theory of granulation, studied by us in many works both of theoretical as well as applications oriented character, see, e.g., [31], [33]. Granular computing was proposed by Zadeh [48] and it was brought into the realm of relational systems via the notion of neighborhood systems by Lin [18].

The formal theory of granulation presented here is in a sense akin to Lin's neighborhood systems: granules to be defined can be regarded as counterparts in the language of mereology to closed balls in metric spaces with metrics in the latter case replaced with rough inclusions in the former case.

We assume that a universal concept U is given; therefore, each entity u of our interest satisfies $u\varepsilon\ U$. Given a concept C such that each $c\varepsilon\ C$ is a concept under which some entities responding to U fall, and a part relation π, the concept $I(C,\pi)$ of individual entities under C relative to π is defined, according to Ontology Axiom. We assume that a rough inclusion μ, compatible with π in the sense of the requirement RM1, is selected which for any two entities $x, y\varepsilon\ I(C,\pi)$ determines whether the term $x\varepsilon\ \mu(y,r)$ is satisfied.

Given an individual $y\varepsilon\ I(C,\pi)$, and a real number $r \in [0,1]$, we define the *granule about* y *of the radius* r, denoted with the symbol $g(y,r,\mu)$ as,

$$g(y,r,\mu) = Cls(\mu(y,r)). \tag{3}$$

Thus, we define a granule as the class in mereological sense of the concept of being as part to a given degree to a given entity.

From the definition, basic properties of granules follow.

Proposition 3. For entity $x\varepsilon\ I(C)$, if $y\ ing_\pi\ x$ and $z\varepsilon\ \mu(y,r)$, then $z\varepsilon\ \mu(x,r)$.

Proof. Assume that $y\ ing_\pi\ x$ and $z\varepsilon\ \mu(y,r)$. By RM1, $y\varepsilon\ \mu(x,1)$ which along with $z\varepsilon\ \mu(y,r)$ yields $z\varepsilon\ \mu(x,r)$ by RM2.

Thus, belonging to the concept of being part to a degree r is upper–hereditary with respect to the relation of ingredient.

In this very general setting, a few interesting specific and expressive properties of granules can be proved. In particular, the intuitively posing itself idea that granules have a topological character of neighborhoods would require a demonstration.

In order to give examples of rough inclusions as well as to establish some properties of interesting rough inclusions, we discuss a few methods of inducing them.

4.1 Metric–induced Rough Inclusions

Consider a metric ρ on the individuals $I(C,\pi)$ of the universal concept U; a concept $\rho(y,r)$ will mean that the distance ρ to the individual y is not greater than r, i.e., $x\varepsilon\ \rho(y,r)$ if and only if $\rho(x,y) \le r$.

It was pointed first by Henri Poincaré [26] that ρ does induce a *tolerance relation* τ_δ by the relation $x\tau_\delta y$ if and only if $\rho(x,y) < \delta$; this relation is certainly reflexive and symmetric but needs not be transitive.

Following this idea, we define a concept $\mu_\rho(y,r)$ by requiring that $x\varepsilon\ \mu_\rho(y,r)$ if and only if $x\varepsilon\ \rho(y,1-r)$. From the standard properties of a metric function, the following facts about $\mu = \mu_\rho$ will follow.

Proposition 4. (i) μ is symmetric, i.e., from $x\varepsilon\ \mu_\rho(y,r)$ it follows that $y\varepsilon\ \mu_\rho(x,r)$.

(ii) $x\varepsilon\ \mu(y,1)$ if and only if $x = y$.

(iii) If $x\varepsilon\mu(y,1)$ and $z\varepsilon\ \mu(x,r)$ then $z\varepsilon\ \mu(y,r)$.

(iv) If $x\varepsilon\ \mu(y,r)$ and $s < r$ then $x\varepsilon\ \mu(y,s)$.

(v) If $z\varepsilon\ \mu(x,r)$ and $x\varepsilon\ \mu(y,s)$ then $z\varepsilon\ \mu(y,L(r,s))$ where $L(r,s) = max\{0, x + y - 1)$ is the *Łukasiewicz functor (tensor product, t–norm).*

Proof. (i) follows by symmetry of ρ. (ii) follows by the property that $\rho(x,y)=0$ if and only if $x=y$. (iii) follows by definition of μ and (ii). (iv) follows by definition of μ. Finally, (v) is a consequence to the triangle inequality for ρ: $z\varepsilon\ \mu(x,r)$ and $x\varepsilon\ \mu(y,s)$ imply, respectively, that $\rho(z,x)\leq 1-r$ and $\rho(x,y)\leq 1-s$ hence $\rho(z,y)\leq(1-r)+(1-s)$ which implies by definition of μ that $z\varepsilon\ \mu(y,1-[(1-r)+(1-s)])$, i.e., $z\varepsilon\ \mu(y,r+s-1)$. The final form of the thesis in (v) comes after taking into account that values of degrees are non–negative.

The statement (v) is said to express the *transitivity property of* μ_ρ.

The statement (ii) shows that the ingredient relation in this case is identity and the part relation is empty.

4.2 Rough Inclusions Induced by Means of Representations of Entities: The Attribute–Value Formalism

In knowledge related problems in Computer Science, entities in the universal concept U are not discussed per se, as in the above section with metrics, but they are represented by means of a chosen system for knowledge representation. This formalism allows for applying to entity representations various mathematical and logical operators which would be inapplicable to entities directly.

In the case of our discussion, it will be most suitable to choose the *attribute–value formalism.*

This formalism consists in selecting for entities in U of a finitely many *attributes* forming a set A; each $a\in A$ is conceived as a mapping on entities valued in a set V_a of $a--values$. We assume for simplicity that attribute values are real numbers.

Each pair (a,v) where v is a value of the attribute a defines a concept $[a,v]$ by means of $x\varepsilon\ [a,v]$ if and only if $a(u)=v$.

In consequence of this adopted formalism, each entity u is represented by the *information concept* $Inf(u)=\{[a,a(u)]:a\in A\}$.

On information concepts, one can perform standard operations, irrespective of the true nature of entities in U. In particular, one is able to define rough inclusions on entities by acting on their representations. We demonstrate a few basic ideas on this aspect.

4.3 Archimedean t–Norms in Inducing Rough Inclusions

The Łukasiewicz t–norm $L(r,s)=max\{0,r+s-1\}$ is an example of an Archimedean t–norm t, see Hájek [10] or Polkowski [27] which is characterized by the property that $t(x,x)=x$ only for $x=0,1$. It is known, see, e.g., [10], that the only two up to an isomorphism such t–norms are L and the *product t–norm* $P(r,s)=r\cdot s$. Thus, a general form of an Archimedean t–norm is either $\phi^{-1}\circ L\circ\phi$ or $\phi^{-1}\circ P\circ\phi$ where ϕ is a chosen authomorphism of the structure $([0,1],\leq)$.

Given an Archimedean t–norm t, a representation holds, see Ling [19],

$$t(x,y)=g(f(x)+f(y)), \tag{4}$$

where f is a decreasing continuous mapping of $[0, 1]$ into itself, and g is the pseudo–inverse to f, see Ling [19].

For a given Archimedean t–norm t, with f, g as in (4), and $g^{-1}(1) = 0$, we define a concept $\mu_t(y, r)$,

$$x\varepsilon\ \mu_t(y, r) \text{ iff } g(\frac{|A \setminus\ IND(x, y)|}{|A|}) \geq r, \tag{5}$$

where $[a, v]\varepsilon\ IND(x, y)$ if and only if $a(x) = v = a(y)$.

Proposition 5. μ_t is a rough inclusion.

Proof. For RM1, $x\varepsilon\ \mu_t(y, 1)$ if and only if $Inf(x) = IND(x, y) = Inf(y)$, i.e., when x, y are *indiscernible*; thus the ingredient relation is in this case the indiscernibility relation and part relation is indiscernibility of distinct entities.

Assuming that $x\varepsilon\ \mu_t(y, 1)$ and $z\varepsilon\ \mu_t(x, r)$, we have $Inf(x) = IND(x, y) = Inf(y)$ and $g\frac{|A\setminus\ IND(x,z)|}{|A|}) \geq r$; thus, $g\frac{|A\setminus\ IND(z,y)|}{|A|}) \geq r$ as well and $z\varepsilon\ \mu_t(y, r)$ verifying RM2.

RM3 follows obviously.

Example: The case of the Łukasiewicz t–norm L In this case, $f(x) = 1 - x = g(x)$, see Ling [19], and thus,

$$x\varepsilon\ \mu_L(y, r) \text{ iff } \frac{|IND(x, y)|}{|A|} \geq r. \tag{6}$$

Let us note that μ_L coincides with the rough inclusion μ_H induced according to sect.4.1 from the reduced Hamming distance $H(x, y) = \frac{|A\setminus\ IND(x,y)|}{|A|}$.

The transitivity property established in sect.4.1 for metric induced rough inclusions, holds for rough inclusions induced from Archimedean t–norms in a more general form.

Proposition 6. For an Archimedean rough inclusion t and the rough inclusion μ_t induced from t according to (5), the transitivity property takes place: if $z\varepsilon\ \mu_t(x, r)$ and $x\varepsilon\ \mu_t(y, s)$ then $z\varepsilon\ \mu_t(y, t(r, s))$.

Proof. We denote the term $A \setminus IND(x, y)$ with the symbol $DIS(x, y)$. We have by assumptions: $g(\frac{|DIS(z,x)|}{|A|}) \geq r$ and $g(\frac{|DIS(x,y)|}{|A|}) \geq r$. Hence, passing to the inverse f: $\frac{|DIS(z,x)|}{|A|} \leq f(r)$, $\frac{|DIS(x,y)|}{|A|} \leq f(s)$. Clearly, $|DIS(z, y)| \leq |DIS(z, x)| + |DIS(x, y)|$ and thus $\frac{|DIS(z,y)|}{|A|} \leq f(r) + f(s)$; passing again to the inverse g we obtain $g(\frac{|DIS(z,y)|}{|A|}) \geq g(f(r) + f(s))$. As by (4), $g(f(r) + f(s)) = t(r, s)$, we obtain finally $z\varepsilon\ \mu_t(y, t(r, s))$.

4.4 The Case of Residual Implication–Induced Rough Inclusions

In order to extend the analysis in sect. 4.3 to continuous t–norms, we make use of *residual implications* induced by continuous t–norms. For a continuous t–norm t, one defines the residual implication (residuum) $\Rightarrow_t$ of t as,

$$z \leq x \Rightarrow_t y \text{ iff } t(z, x) \leq y. \tag{7}$$

Assume a mereological universe (U, π), with the associated ingredient relation ing and let $\phi : (U, ing) \to ([0,1], \leq)$ be a morphism, i.e., the equivalence $u\ ing\ v\ textrmiff\ \phi(u) \leq \phi(v)$ holds. Then,

Proposition 7. The concept $\mu_\phi(v, r)$ defined as $u\varepsilon\ \mu_\phi(v, r)$ if and only if $\phi(u) \Rightarrow_t \phi(v) \geq r$ is a rough inclusion.

Proof. It suffices to check that conditions RM1–RM3 are fulfilled. By (7), $x \Rightarrow_y = 1$ if and only if $x \leq y$; thus, $u\varepsilon\ \mu_\phi(v, 1)$ if and only if $\phi(u) \leq \phi(v)$ if and only if $u\ ing\ v$. This verifies RM1. For RM2, assume that $w\varepsilon\ \mu_\phi(u, r)$ and $u\varepsilon\ \mu_\phi(v, 1)$. Thus, $\phi(w) \Rightarrow_t \phi(u) \geq r$ and $\phi(u) \leq \phi(v)$. By (7), $t(r, \phi(w)) \leq \phi(u)$ hence $t(r, \phi(w)) \leq \phi(v)$ and, again by (7), $\phi(w) \Rightarrow_t \phi(v) \geq r$, i.e., $w\varepsilon\ \mu_\phi(v, r)$ proving RM2. RM3 holds obviously.

For various forms of the morphism ϕ, one obtains by Proposition 7, corresponding rough inclusions.

Rough inclusions obtained from residual implications also have the transitivity property.

Proposition 8. For any rough inclusion of the form μ_ϕ transitivity property holds in the form: If $w\varepsilon\ \mu_\phi(u, r)$ and $u\varepsilon\ \mu_\phi(v, s)$ then $w\varepsilon\ \mu_\phi(v, t(r, s))$.

Proof. Assume that $w\varepsilon\ \mu_\phi(u, r)$ and $u\varepsilon\ \mu_\phi(v, s)$, i.e., by (7), $t(r, \phi(w)) \leq \phi(u)$ and $t(s, \phi(u)) \leq \phi(v)$. As t is monotone coordinate–wise, $t(s, t(r, \phi(w))\leq \phi(v)$. By symmetry and associativity of t, the last inequality is $t(t(r, s), \phi(w)) \leq \phi(v)$, i.e., $\phi(w) \Rightarrow_t \phi(v) \geq t(r, s)$. Thus, $w\varepsilon\ \mu_\phi(v, t(r, s))$.

Let us observe that μ_{phi} is not symmetric, contrary to μ_t with t Archimedean.

Symmetrization of μ_ϕ can be achieved along standard lines: let $u\varepsilon\ \mu_t^\sigma(v, r)$ if and only if $u\varepsilon\ \mu_t(v, r)$ and $v\varepsilon\ \mu_t(u, r)$.

Clearly: μ_t^σ is a rough inclusion; let us observe that $\mu_t^\sigma(v, 1)$ is the identity.

4.5 Granular Topologies

Our intuition about granules as neighborhoods of entities can be put into a formal statement. We denote by μ_t a rough inclusion induced either by an Archimedean t–norm or by a residual implication of a continuous t–norm.

We define classes of the form $N(v, r) = Cls(M_t(v, r))$, where

$$u\varepsilon M_t(v, r)\ tetxrmiff\ \exists s > r. u\varepsilon\ \mu_t(v, s). \tag{8}$$

We expect that the system $\{N(v, r) : v\varepsilon U; r \in [0, 1]\}$ will turn out a neighborhood basis for a topology τ_{μ_t}. This is justified by,

Proposition 9. The following hold,

1. If $u\ ing\ N(v, r)$ then $\exists \delta > 0. N(u, \delta)\ ing\ N(v, r)$.
2. If $s > r$ then $N(v, s)\ ing\ N(v, r)$.
3. If $w\ ing\ N(v, r)$ and $w\ ing\ N(u, s)$ then $\exists \delta > 0$vvsuch that $N(v, \delta)\ ing\ N(v, r)$ and $N(w, \delta)\ ing\ N(u, s)$.

Proof. For Property 1. $u\ ing\ N(v,r)$ implies that there exists an s¿r such that $u\varepsilon\mu_t(v,s)$. Let $\delta < 1$ be such that $t(x,s) > r$ whenever $x > \delta$; δ exists by continuity of t and the identity $t(1,s) = s$. Thus, if $w\ ing\ N(u,\delta)$, then $w\varepsilon\mu_t(u,\eta)$ with $\eta > \delta$ and, by transitivity property, $w\varepsilon\mu_t(v,t(\eta,s))$; as $t(\eta,s) \geq r$, by RM3, $w\varepsilon\mu_t(v,r)$. By Inference Rule (IRM), $N(u,\delta)\ ing\ N(v,r)$.

Property 2. follows by the Inference Rule (IRM) directly from definitions. Property 3. is a consequence to Properties 1. and 2.

Specific properties of topologies τ_{μ_t} depend on the chosen parameter t. For instance, in case of μ_L induced by the Archimedean t–norm L of Łukasiewicz, the topology τ_{μ_L} is a metric topology induced by the distance function (metric) H acting on attribute–value representation of entities: $H(Inf(u),Inf(v)) = \frac{|Inf(u)\cap Inf(v)|}{|A|}$.

We conclude this discussion of granulation with a characterization of granules induced by Archimedean t–norms as well as granules induced by residual implications of continuous t–norms.

Proposition 10. For any rough inclusion μ: either μ_t induced by an Archimedean t–norm or any symmetrized form μ_t^σ of a rough inclusion obtained from a residual implication of a continuous t–norm in a manner indicated above, the following equivalence holds: $w\ ing\ g(v,r,\mu)$ if and only if $w\varepsilon\mu(v,r)$.

Proof. Assume that $w\ ing\ g(v,r,\mu)$. Recall that the granule $g(v,r,\mu)$ is the class of the property $\mu(v,r)$. By the class definition and our assumption, there are entities q,p such that $q\ ing\ w$, $q\ ing\ p$, $p\varepsilon\mu(v,r)$. Thus: $q\varepsilon\ \mu(p,1)$ and by transitivity property $q\varepsilon\ \mu(v,t(r,1))$, i.e., $q\varepsilon\ \mu(v,r)$. As $w\varepsilon\ \mu(q,1)$ (by symmetry and RM1 because of $q\ ing\ w$), again by transitivity property, $w\varepsilon\ \mu(v,r)$.

4.6 Rough Inclusions on Finite Sets

For our further purpose, it will be convenient to discuss a particular case of rough inclusions on a universal concept of finite sets Fin. Close relations of containment of sets to ingredient relations suggest that a natural rough inclusion can be defined as,

$$x\varepsilon\ \nu_3(y,r) \text{ iff } \begin{cases} x \subseteq y \text{ and} r = 1 \\ x\cap y = \emptyset \text{ and } r = 0 \\ r = \frac{1}{2} \text{ otherwise} \end{cases} \tag{9}$$

One can also mimic recipes for rough inclusions worked out in the general case above: we restrict ourselves to the rough inclusion induced by the Archimedean t–norm L of Łukasiewicz,

$$x\varepsilon\ \nu_L(y,r) \text{ iff } g(\frac{|x\setminus y|}{|x|}) \geq r. \tag{10}$$

As $g(u) = 1 - u$, the formula for ν_L can be rewritten in the form:

$$x\varepsilon\ \nu_L(y,r) \text{ iff } \frac{|x\cap y|}{|x|} \geq r.$$

5 Approximate Reasoning: A Granular Rough Mereological Logic GRML

Granulation makes it possible to construct intensional logics, see [2] for this notion, in which possible worlds at which extensions of formulas are evaluated are granules of individuals and states of truth are reals in the interval $[0, 1]$, cf. e.g., [37].

We assume a universal concept U of entities with a part relation π, the corresponding ingredient relation *ing*, a compatible rough inclusion μ, and a rough inclusion ν on finite concepts (sets); granules $g(v, r, \mu)$ are defined as in (3). We consider predicates ϕ interpreted in individual entities in U, i.e., the meaning $[\phi]$ is a concept under which some individuals in U fall. We assume finiteness of U. As logical connectives, we choose the implication C (if ... then) and the negation N (it is not true that ...). Semantic interpretation of connectives C and N will be chosen as the following:

$$[N\phi] = U \setminus [\phi], \tag{11}$$

and,

$$[C\phi\psi]] = (U \setminus [\phi]) \cup [\psi]. \tag{12}$$

The *intension* $I(\phi, \mu, \nu)$ of ϕ is a mapping from the concept of granules about entities in U into real numbers; the *extension* of ϕ at a granule g is the value $E(g, \phi, \mu, \nu) = I(\phi, mu, \nu)(g)$ of truth state of ϕ at the granule g.

This value will be defined as,

$$E(g, \phi, \mu, \nu) \geq r \text{ iff } g\varepsilon\ \nu([\phi], r), \tag{13}$$

i.e., the value of the extension of a predicate at a granule is defined by the value of degree of partiality of the granule in the meaning of the predicate.

We call a meaningful formula ϕ a *theorem* if and only if $E(g, \phi, \mu, \nu) = 1$ for each granule g.

5.1 The Case of the Archimedean t–Norm L of ŁUkasiewicz

For illustration of ways in which granular logics work, we choose the t–norm L and a fortiori the rough inclusion μ_L for granule computing and the rough inclusion ν_L on finite concepts (sets). We assume obviously that knowledge is represented in the attribute–value formalism.

With respect to negation the extension operator behaves according to the formula,

$$E(g, \phi, \mu_L, \nu_L) \geq r \text{ iff } E(g, N\phi, \mu_L, \nu_L) \leq 1 - r, \tag{14}$$

which follows from (11) and the form of ν_L (10). One can simplify this result to the statement that $E(g, N\phi, \mu_L, \nu_L) = 1 - E(g, \phi, \mu_L, \nu_L)$ by taking the maximal values of degrees of partiality in both sides.

The behavior of the extension operator with respect to the implication functor follows similarly from (12) and (10),

$$E(g, C\phi\psi, \mu_L, \nu_L) \leq [(1 - E(g, \phi, \mu_L, \nu_L)) + E(g, \phi, \mu_L, \nu_L)]. \tag{15}$$

so finally,

The formula on the right hand side of inequality (15) is the value of the Łukasiewicz implication of many–valued logic [23], [41].

Proposition 11. If a formula ϕ is a theorem of granular logic, then the sentential collapse ϕ^c of ϕ is a theorem of the Łukasiewicz many – valued logic.

Indeed, by (14) and (15) it follows that if the value of extension of ϕ is 1 then ϕ regarded as a sentence in the Łukasiewicz logic has value of truth 1.

An analysis of modal statements. Modalities related to knowledge concern operators "certainly ϕ...", "possibly ϕ...". In attribute–value formalism, these operators are formally defined within rough set theory.

The idea is to introduce a concept IND of indiscernibility: entities u, v are *indiscernible*, $(u, v)\varepsilon\ IND$ if and only if $Inf(u) = Inf(v)$. The concept of indiscernibility class $[u]_{IND}$ is next: $v\varepsilon\ [u]_{IND}$ if and only if $(u, v)\varepsilon\ IND$. A concept E is *IND–exact* if and only if there exists a property (concept) F on indiscernibility classes such that $E = Cls(F)$.

Otherwise, a concept is *in–exact.* For any concept X, one can form the class $l(X)$ of all indiscernibility classes which are ingredients of X; it is an exact concept which is an ingredient of X and moreover, any exact concept which is an ingredient of X is an ingredient of $l(X)$.

Dually, there exists an exact concept $u(X)$ such that X is an ingredient of $u(X)$ and $u(X)$ is an ingredient of any exact concept of which X is an ingredient. $l(X)$ is the *lower approximation* to X and $u(X)$ is the *upper approximation* to X.

We define functors L of necessity and M of possibility (the formula $L\phi$ is read "certainly ϕ" and the formula $M\phi$ is read: "possibly ϕ"),

$$E(g, L\phi, \mu_L, \nu_L) \geq r \text{ iff } g\varepsilon\ \nu_L(l([\phi]), r), \tag{16}$$

and,

$$E(g, M\phi, \mu_L, \nu_L) \geq r \text{ iff } g\varepsilon\ \nu_L(u([\phi]), r). \tag{17}$$

We now present using the granular logic GRML that defined by us modalities L, M obey the rules of modal logic S5.

Proposition 12. The following formulas of modal logic are theorems of GRML,

1. (K) $CL(C\phi\psi)CL\phi L\psi$.
2. (T) $CL\phi\phi$.
3. (S4) $CL\phi LL\phi$.
4. (S5) $CM\phi LM\phi$.

Proof. We verify that the formula (K) is a theorem of GRML. Other formulas are theorems by virtue of duality. We have, $[CL(C\phi\psi)CL\phi L\psi]$=$U \setminus lU \setminus [\phi] \cup [\psi] \cup (U \setminus l([\phi])) \cup l([\psi]))$.

Assuming that u is such that u does not fall into $U \setminus l([\phi]) \cup l([\psi])$, we have that

(i) $[u]_{IND}\ ing\ [\phi]$; (ii) $[u]_{IND} \cap [\psi] = \emptyset$.

It follows by (i,ii) that u does not fall into $l(U \setminus [\phi]) \cup [\psi]$ (as if were u in $l(U \setminus [\phi]) \cup [\psi]$ it would mean that $[u]_{IND}\ ing\ (U \setminus [\phi]) \cup [\psi]$ hence by (ii) one would have $[u]_{IND}\ ing\ U \setminus [\phi]$, contradicting (i), i.e., the meaning of (K) is U.

6 On Applications of Rough Inclusions and Granulation in Computer Science

We finally point to some applications of the presented theory of granulation and reasoning, described in the literature. Among them are:

1. Fusion of knowledge. The problem is in merging a finite number of information systems. In [29], it was shown that a simple merging scheme in which a Cartesian product ag is formed of two information systems ag_1, ag_2 results in granule synthesizer $S(G)$ which from granules $g_1(v_1, r_1, \mu_L)$ at ag_1 and $g_2(v_2, r_2, \mu_L)$ at ag_2 forms a granule $g((v_1, v_2), L(r_1, r_2), \mu_L)$ at ag; similarly, the logic synthesizer $L(G)$ forms from extensions $E(g_1(v_1, r_1), \phi_1, \mu_L, \nu_L)$ at ag_1 and $E(g_2(v_2, r_2), \phi_2, \mu_L, \nu_L)$ at ag_2 an extension at ag for the granule $g((v_1, v_2), L(r_1, r_2), \mu_L)$ of the form

$$E(g_2(v_2, r_2), \phi_2, \mu_L, \nu_L) \cdot E(g_2(v_2, r_2), \phi_2, \mu_L, \nu_L),$$

i.e., uncertainty is fused according to the product t–norm. Details can be found also in [34], [35].

2. Neural computing. A perceptron–like neuron is constructed whose activation function is a rough inclusion built from the Archimedean product t–norm $P(r, s) = r \cdot s$, see [30].

3. Calculus of vague statements. Posed by L. Zadeh paradigm of Computing with Words, or, Perception Calculus, requires a formalism in which vague statements like "John is old" acquire degrees of truth allowing for computing on semantic level with degrees of truth of syntactically composed statements. An application of reasoning based on granulation to this problem was presented in [38].

4. Classification problems. The problem here is to classify entities in a test universe Tst into classes on the basis of rules learned from the training universe Trn. An approach based on granulation see, e.g., [36] consists in forming a granular training universe $GTrn$ consisting of a chosen according to some strategy covering of Trn by granules of a fixed radius, and then factoring attributes through granules by a chosen strategy, e.g., majority voting, averaging etc. etc. The obtained new system is subject to inducing rules which are then applied to the test Tst universe of entities. Results are very promising.

References

1. Adams, E. W.: Probability and the logic of conditionals. In: [13]
2. van Bentham, J.: A Manual of Intensional Logic. CSLI Stanford University (1988)
3. Black, M.: Notes on the paradoxes of confirmation. In: [13]
4. Bochenski, I.M.: Die zeitg önossichen Denkmethoden. A. Francke AG Verlag, Bern (1954) (9th edn. 1986)
5. Bochman, A.: A Logical Theory of Nonmonotonic Inference and Belief Change. Springer, Berlin (2001)
6. Frege, G.: Begriffschrift, eine der arithmetischen nachgebildete Formelsprache des reinen Denken. Louis Nebert, Halle a. S (1879)

7. Frege, G.: Grundgesetze der Arithmetik I. Verlag Hermann Pohle, Jena (1893)
8. Frege, G.: Grundgesetze der Arithmetik II. Verlag Hermann Pohle, Jena (1903)
9. Gödel, K.: Über Formal Unentscheidbare Sätze der Principia Mathematica. Monatshefte 38 (1931)
10. Hájek, P.: Metamathematics of Fuzzy Sets. Kluwer, Dordrecht (1998)
11. Hempel, C.G.: Studies in the logic of confirmation. Mind 54 (1945)
12. Hilbert, D., Ackermann, W.: Grundzüge der theoretischen Logik, Berlin (1938)
13. Hintikka, J., Suppes, P. (eds.): Aspects of Inductive Logic. North–Holland, Amsterdam (1966)
14. Jech, T.: Set Theory: The Third Millennium Edition, Revised and Expanded. Springer, Berlin (2003)
15. Kanamori, A.: The mathematical development of set theory from Cantor to Cohen. The Bulletin of Symbolic Logic 2(1) (1996)
16. Lesniewski, S.: Podstawy ogolnej teoryi mnogosci (On the Foundations of General Set Theory) (in Polish). The Polish Scientific Circle, Moscow (1916); see also a later digest in: Topoi 2, 7–52 (1982); engl. transl.: Foundations of the General Theory of Sets. I. In: Surma, S. J., Srzednicki, J., Barnett, D. I., Rickey, V. F. (eds.): Stanislaw Lesniewski: Collected Works, vol. 1. Kluwer, Dordrecht (1992)
17. Srzednicki, J.T., Rickey, V.F., Czelakowski, J. (eds.): Leśniewski's Systems. Ontology and Mereology. Nijhoff and Ossolineum, The Hague and Wrocław (1984)
18. Lin, T.Y.: Neighborhood systems and relational database (abstract). In: Proceedings of CSC 1988, p. 725 (1988)
19. Ling, C.-H.: Representation of associative functions. Publ. Math. Debrecen 12, 189–212 (1965)
20. Łukasiewicz, J.: Concerning the reversibility of the relation ratio–consequence (in Polish). Przegląd Filozoficzny 16 (1913)
21. Łukasiewicz, J.: Farewell lecture by Professor Jan Łukasiewicz. Warsaw University Hall, March 7 (1918). In: [23]
22. Łukasiewicz, J.: Aristotle's Syllogistic. Clarendon Press, Oxford (1957)
23. Borkowski, L. (ed.): Jan Lukasiewicz. Selected Works. North–Holland and PWN–Polish Scientific Publishers, Amsterdam and Warsaw (1970)
24. Pawlak, Z.: Rough Sets: Theoretical Aspects of Reasoning about Data. Kluwer, Dordrecht (1991)
25. Pawlak, Z.: Rough sets. Int. J. Computer and Information Sci. 11, 341–356 (1982)
26. Poincaré, H.: Science and Hypothesis, London (1905)
27. Polkowski, L.: Rough Sets. Mathematical Foundations. Physica Verlag, Heidelberg (2002)
28. Polkowski, L.: A rough set paradigm for unifying rough set theory and fuzzy set theory (a plenary lecture). In: Wang, G., Liu, Q., Yao, Y., Skowron, A. (eds.) RSFDGrC 2003. LNCS (LNAI), vol. 2639, pp. 70–78. Springer, Heidelberg (2003); cf. Fundamenta Informaticae 54, 67–88 (2003)
29. Polkowski, L.: Toward rough set foundations. Mereological approach (a plenary lecture). In: Tsumoto, S., Słowiński, R., Komorowski, J., Grzymała-Busse, J.W. (eds.) RSCTC 2004. LNCS (LNAI), vol. 3066, pp. 8–25. Springer, Heidelberg (2004)
30. Polkowski, L.: Rough–fuzzy–neurocomputing based on rough mereological calculus of granules. Intern. J. Hybrid Intell. Systems 2, 91–108 (2005)
31. Polkowski, L.: Formal granular calculi based on rough inclusions (a feature talk). In: [40], pp. 57–62 (2005)

32. Polkowski, L.: Rough mereological reasoning in rough set theory. A survey of results and problems (a plenary lecture). In: Wang, G.-Y., Peters, J.F., Skowron, A., Yao, Y. (eds.) RSKT 2006. LNCS (LNAI), vol. 4062, pp. 79–92. Springer, Heidelberg (2006)
33. Polkowski, L.: The paradigm of granular rough computing. In: Proceedings ICCI 2007. 6th IEEE Intern. Conf. on Cognitive Informatics, pp. 145–163. IEEE Computer Society, Los Alamitos (2007)
34. Polkowski, L.: An approach to granulation of knowledge and granular computing based on rough mereology: A survey. In: Kreinovich, V., Pedrycz, W., Skowron, A. (eds.) Handbook of Granular Computing. John Wiley, New York (to appear, 2008)
35. Polkowski, L.: Granulation of knowledge: Similarity based approach in information and decision systems. In: Lin, T. (ed.) Encyclopedia of Systems and Complexity. Springer, Berlin (to appear)
36. Polkowski, L., Artiemjew, P.: On Granular Rough Computing: Factoring Classifiers Through Granulated Decision Systems. In: Kryszkiewicz, M., Peters, J.F., Rybinski, H., Skowron, A. (eds.) RSEISP 2007. LNCS (LNAI), vol. 4585, pp. 280–289. Springer, Heidelberg (2007)
37. Polkowski, L., Semeniuk–Polkowska, M.: On rough set logics based on similarity relations. Fundamenta Informaticae 64, 379–390 (2005)
38. Polkowski, L., Semeniuk–Polkowska, M.: A formal approach to Perception Calculus of Zadeh by means of rough mereological logic. In: Proceedings IPMU 2006, Paris (2006)
39. Polkowski, L., Semeniuk–Polkowska, M.: Mereology in approximate reasoning about concepts. In: Valore, P. (ed.) Formal Ontology and Mereology, Polimetrica Intern. Publishers, Monza (2006)
40. Proceedings 2005 IEEE Intern. Conf. Granular Computing, Tsinghua Univ., Beijing, China. IEEE Press, Los Alamitos (2005)
41. Rosser, J.B., Turquette, A.R.: Many–Valued Logics. North Holland, Amsterdam (1958)
42. Scotus, Duns: Treatise on God as the First Principle, `http://www.ewtn.com/library/THEOLOGY/GODASFIR.HTM`
43. Słupecki, J.: S. Leśniewski's calculus of names. Studia Logica 3, 7–72 (1955); Also In: [17], pp. 59–122
44. Suppes, P.: A Bayesian approach to the paradoxes of confirmation. In: [13]
45. Wang, Y.: On concept algebra and knowledge representation. In: IEEE ICCI, pp. 320–331 (2006)
46. Yao, Y.T.: Concept formation and learning: A cognitive informatics perspective. In: IEEE ICCI, pp. 42–51 (2004)
47. Zadeh, L.A.: Fuzzy sets. Information and Control 8, 338–353 (1965)
48. Zadeh, L.A.: Fuzzy sets and information granularity. In: Gupta, M., Ragade, R., Yaeger, R.R. (eds.) Advances in Fuzzy Set Theory and Applications, pp. 3–18. North–Holland, Amsterdam (1979)

On Mathematical Laws of Software

Yingxu Wang

Theoretical and Empirical Software Engineering Research Centre (TESERC)
International Center for Cognitive Informatics (ICfCI)
Dept. of Electrical and Computer Engineering
Schulich School of Engineering, University of Calgary
2500 University Drive, NW, Calgary, Alberta, Canada T2N 1N4
Tel.: (403) 220 6141; Fax: (403) 282 6855
yingxu@ucalgary.ca

Abstract. Recent studies on the laws and mathematical constraints of software have resulted in fundamental discoveries in computing and software engineering toward exploring the nature of software. It was recognized that software is not constrained by any physical laws discovered in the natural world. However, software obeys the laws of mathematics, cognitive informatics, system science, and formal linguistics. This paper investigates into the mathematical laws of software and computing behaviors. A generic mathematical model of programs is created that reveals the nature of software as abstract processes and its uniqueness beyond other mathematical entities such as sets, relations, functions, and abstract concepts. A comprehensive set of mathematical laws for software and its behaviors is established based on the generic mathematical model of programs and the fundamental computing behaviors elicited in Real-Time Process Algebra (RTPA). A set of 95 algebraic laws of software behaviors is systematically derived, which encompasses the laws of meta-processes, process relations, and system compositions. The comprehensive set of mathematical laws of software lays a theoretical foundation for analyzing and modeling software behaviors and software system architectures, as well as for guiding rigorous practice in programming. They are also widely applicable for the rigorous modeling and manipulation of human cognitive processes and computational intelligent behaviors.

Keywords: Software science, software engineering, denotational mathematics, software, programs, computational intelligence, modeling, analysis, mathematical models, generic model of software, process models, algebraic laws, laws of meta-processes, laws of process relations, laws of process compositions, RTPA.

1 Introduction

The wander on laws of software science can be traced back to George Boole in his work on "*The Laws of Thought* [3]." Two fundamental discoveries in computing and software engineering, the *constrain laws* of software [9], [16] and the *process metaphor* [14], [15], [20], have fundamentally influenced the theoretical and empirical

M.L. Gavrilova et al. (Eds.): Trans. on Comput. Sci. II, LNCS 5150, pp. 46–83, 2008.

research on exploring the nature of software. The former reveals that, although software is not constrained by any physical laws discovered in the nature world, it do obey the laws of mathematics [16], [21], [38], [39], [41], as well as cognitive informatics [33], [34], [36], [37], [40], system science [39], and formal linguistics and semantics [6], [7], [11], [13], [26], [31], [36], [39], [43]. The latter indicates that any software system and its behaviors can be modeled and described by a set of processes [14], which can be treated as a new mathematical entity beyond sets, relations, functions, and abstract concepts [38], [43], [44], [45], [46], [48].

In the exploration of the generic laws of programming, Hoare and his colleagues proposed a set of mathematic laws of programming [16]. This work covered a set of algebraic laws of programs for Dijkstra's *nondeterministic sequential programming language* [9]. Wang investigated the cognitive informatics laws of software in [36] and organizational laws of software engineering in [47], which enhanced the understanding toward the fundamental computing behaviors and their composing rules [2], [4], [8], [14], [15], [17], [20], [25], [27].

The latest reveal of the *generic program model* (GPM) in Wang (2007) [39] on the basis of Real-Time Process Algebra (RTPA) [32], [35], [39], [43], [46] have led to the establishment of a comprehensive set of 95 mathematical laws of fundamental software behaviors in the categories of meta-processes and process relations. The mathematical laws of software provide a solid foundation underlying software architectural and behavioral modeling and analyses. They can be used to reveal equivalency between process expressions, to simplify complicated process behaviors, to express interactive process relations, and to prove the correctness of software behaviors.

This paper investigates into the mathematical laws of software and computing behaviors. Emphases will be put on the laws of the most complicated real-time computing requirements and behaviors, such as laws for system dispatching, interrupt, timing, addressing, dynamic memory allocation, and Boolean/numeric evaluations. An algebraic treatment of fundamental behaviors of software systems is presented in Section 2, with the introduction of GPM and RTPA. Algebraic laws of software for the most fundamental and elementary computing behaviors are described in Section 3 known as the laws of meta-processes of software. Then, algebraic laws of software for constructing and composing complex process behaviors and architectures are described in Section 4 known as the laws of algebraic operations of software. The two categories of mathematical laws of software cover a comprehensive set of laws for software behaviors, architectures, and their interactions, which form a foundation for rigorous reasoning and modeling of software systems and computing mechanisms.

2 The Algebraic Treatment of Fundamental Behaviors of Software Systems in RTPA

On the basis of the process metaphor of software systems, abstract processes can be rigorously treated as a mathematical entity beyond sets, relations, functions, and abstract concepts. Real-Time Process Algebra (RTPA) is a denotational mathematical structure for denoting and manipulating system behavioral processes [32], [35], [36],

[39], [43], [46]. RTPA is designed as a coherent algebraic system for intelligent and software system modeling, specification, refinement, and implementation. RTPA encompasses 17 meta- processes and 17 relational process operations. RTPA can be used to describe both logical and physical models of software and intelligent systems. Logic views of system architectures and their physical platforms can be described using the same set of notations. When a system architecture is formally modeled, the static and dynamic behaviors performed on the architectural model can be specified by a three-level refinement scheme at the system, class, and object levels in a top-down approach. RTPA has been successfully applied in real-world system modeling and code generation for software systems, human cognitive processes, and intelligent systems.

Definition 1. *RTPA* is a denotational mathematical structure for algebraically denoting and manipulating system behavioural processes and their attributes by a triple, i.e.:

$$RTPA \triangleq (\mathfrak{T}, \mathfrak{P}, \mathfrak{N}) \tag{1}$$

where

- $\mathfrak{T}$ is a set of 17 primitive types for modeling system architectures and data objects;
- $\mathfrak{P}$ a set of 17 meta-processes for modeling fundamental system behaviors;
- $\mathfrak{R}$ a set of 17 relational process operations for constructing complex system behaviors.

Detailed descriptions of $\mathfrak{T}$, $\mathfrak{P}$, and $\mathfrak{N}$ in RTPA will be extended in the following subsections.

2.1 The Generic Mathematical Model of Programs and Software Systems

Program modeling is on coordination of computational behaviors with given data objects. On the basis of RTPA, a generic program model can be described by a formal treatment of statements, processes, and complex processes from the bottom-up in the program hierarchy.

Definition 2. A *process P* is the basic unit of an applied computational behavior that is composed by a set of statements p_i, $1 \le i \le n$-1, with left-associated cumulative relations, r_{ij}, i.e.:

$$\begin{aligned} P &= \mathop{R}_{i=1}^{n-1} (p_i \; r_{ij} \; p_j), j = i+1 \\ &= (...(((p_1) \; r_{12} \; p_2) \; r_{23} \; p_3) \; ... \; r_{n-1,n} \; p_n) \end{aligned} \tag{2}$$

where $p_i \in \mathfrak{P}$ and $r_{ij} \in \mathfrak{R}$.

With the formal process model as defined above, the generic mathematical model of programs can be derived as follows.

Definition 3. A *program* $\wp$ is a composition of a finite set of m processes according to the time-, event-, and interrupt-based process dispatching rules of RTPA, i.e.:

$$\wp = \mathop{R}_{k=1}^{m}(@\, e_k \hookrightarrow P_k) \tag{3}$$

Definitions 2 and 3 indicate that a program is an *embedded relational entity*, where a statement in a program is an instantiation of a meta-instruction of a programming language that executes a basic unit of coherent function and leads to a predictable behavior.

Theorem 1. The *Generic Program Model* (GPM) states that a software system or a program $\wp$ is an algebraic structure with a set of embedded relational processes, in which all previous processes of a given process form the context of the current process, i.e.:

$$\begin{aligned} \wp &= \mathop{R}_{k=1}^{m}(@\, e_k \hookrightarrow P_k) \\ &= \mathop{R}_{k=1}^{m}[@\, e_k \hookrightarrow \mathop{R}_{i=1}^{n-1}(p_i(k)\, r_{ij}(k)\, p_j(k))],\ j = i+1,\ p_i, p_j \in \mathfrak{P},\ r_{ij} \in \mathfrak{R} \end{aligned} \tag{4}$$

Proof. Theorem 1 can be directly proved on the basis of Definitions 2 and 3. Substituting P_k in Definition 3 with Eq. 2, a generic program $\wp$ obtains the form as a series of embedded relational processes as presented in Theorem 1.

The GPM model given in Theorem 1 reveals that a program is a finite and nonempty set of embedded binary relations between a current statement and all previous ones that formed the *semantic context* or environment of computing. Theorem 1 provides a unified software model, which is a formalization of the well accepted but informal process metaphor for software systems in computing.

2.2 The Type System of RTPA

A type is a set in which all member data objects share a common logical property or attribute. The maximum range of values that a set of variables can assume is the domain of a type, and a type is always associated with a set of predefined or allowable operations in computing. A type can be classified as *primitive* and *derived* (complex) types. The former are the most elemental types that cannot further divided into simpler ones; the latter are a compound form of multiple primitive types based on given type rules. In computing, most primitive types are provided by programming languages; while most user defined types are derived ones.

Definition 4. A *type system* specifies data object modeling and manipulation rules in computing.

A set of 17 primitive types of RTPA in computing and human cognitive process modeling is elicited from works in [5], [19], [22], [28], [32], [35], [39], [43], [46], which is summarized in Table 1. In Table 1, the first 11 primitive types are for mathematical and logical manipulation of data objects, and the remaining 6 are for system architectural modeling.

It is noteworthy that although a generic computing behavior is constrained by the *mathematical domain* D_m of types, an executable program is constrained by the *language-defined domain* D_l, and at most time, it is further restricted by the *user-defined domain* D_u, where $D_u \subseteq D_l \subseteq D_m$.

Table 1. RTPA Primitive Types and their Domains

No.	Type	Syntax	D_m	D_l
1	Natural number	**N**	[0, +∞]	[0, 65535]
2	Integer	**Z**	[-∞, +∞]	[-32768, +32767]
3	Real	**R**	[-∞, +∞]	[-2147483648, 2147483647]
4	String	**S**	[0, +∞]	[0, 255]
5	Boolean	**BL**	[**T**, **F**]	[**T**, **F**]
6	Byte	**B**	[0, 255]	[0, 255]
7	Hexadecimal	**H**	[0, +∞]	[0, max]
8	Pointer	**P**	[0, +∞]	[0, max]
9	Time	**TI = hh:mm:ss:ms**	**hh**: [0, 23] **mm**: [0, 59] **ss**: [0, 59] **ms**: [0, 999]	**hh**: [0, 23] **mm**: [0, 59] **ss**: [0, 59] **ms**: [0, 999]
10	Date	**D = yy:MM:dd**	**yy**: [0, 99] **MM**: [1, 12] **dd**: [1, 31]	**yy**: [0, 99] **MM**: [1, 12] **dd**: [1, 31]
11	Date/Time	**DT = yyyy:MM:dd: hh:mm:ss:ms**	**yyyy**: [0, 9999] **MM**:[1, 12] **dd**: [1, 31] **hh**: [0, 23] **mm**: [0, 59] **ss**: [0, 59] **ms**: [0, 999]	**Yyyy**: [0, 9999] **MM**:[1, 12] **dd**: [1, 31] **hh**: [0, 23] **mm**: [0, 59] **ss**: [0, 59] **ms**: [0, 999]
12	Run-time determinable type	**RT**	–	–
13	System architectural type	**ST**	–	–
14	Random event	**@*e*S**	[0, +∞]	[0, 255]
15	Time event	**@*t*TM**	[0**ms**, 9999 **yyyy**]	[0**ms**, 9999 **yyyy**]
16	Interrupt event	**@*int*⊙**	[0, 1023]	[0, 1023]
17	Status	**ⓢ*s*BL**	[**T**, **F**]	[**T**, **F**]

Lemma 1. The *primary types of computational objects* state that the *RTPA type system* $\mathfrak{T}$ encompasses 17 primitive types elicited from fundamental computing needs, i.e.:

$$\mathfrak{T} \triangleq \{\mathbf{N}, \mathbf{Z}, \mathbf{R}, \mathbf{S}, \mathbf{BL}, \mathbf{B}, \mathbf{H}, \mathbf{P}, \mathbf{TI}, \mathbf{D}, \mathbf{DT}, \mathbf{RT}, \mathbf{ST}, @e\mathbf{S}, @t\mathbf{TM}, @int\odot, ⓢs\mathbf{BL}\} \quad (5)$$

where the primitive types stand for *natural number, integer, real, string, Boolean, byte, hexadecimal, pointer, time, date, date/time, run-time determinable type, system architectural type, random event, time event, interrupt event,* and *system status.*

RTPA provides a coherent notation system and a rigorous mathematical structure for modeling both software and intelligent systems. RTPA can be used to describe both *logical* and *physical* models of systems, where logic views of the architecture of a software system and its operational platform can be described using the same set of notations. When the system architecture is formally modelled, the static and dynamic behaviors that perform on the system architectural model, can be specified by a three-level refinement scheme at the system, class, and object levels in a top-down approach. Although CSP [14], [15], the timed-CSP [4], [10], [23], and other process algebra treated any computational operation as a process, RTPA distinguishes the concepts of meta-processes from those of complex and derived processes, which are composed by relational process operations on the meta-processes.

2.3 The Meta-processes of Software Behaviors in RTPA

RTPA adopts the foundationalism in order to elicit the most primitive computational processes known as the *meta-processes*. In this approach, complex processes are treated as derived processes from these meta-processes based on a set of algebraic process composition rules known as the *process relations.*

Definition 5. A *meta-process* in RTPA is a primitive computational operation that cannot be broken down to further individual actions or behaviors.

A meta-process is an elementary process that serves as a basic building block for modeling software behaviors. In RTPA, a set of 17 meta-processes has been elicited as shown in Table 2, from essential and primary computational operations commonly identified in existing formal methods and modern programming languages [1], [12], [16], [18], [52], [53]. Mathematical notations and syntaxes of the meta-processes are formally described in Table 2, while formal semantics of the meta-processes of RTPA may be referred to [36], [39], [43].

Lemma 2. The *RTPA meta-process system* $\mathfrak{P}$ encompasses 17 fundamental computational operations as defined in Table 2, i.e.:

$$\mathfrak{P} = \{:=, \blacklozenge, \Rightarrow, \Leftarrow, \not\Leftarrow, \succ, \prec, |\succ, |\prec, @, \triangleq, \uparrow, \downarrow, !, \otimes, \boxtimes, \S\} \quad (6)$$

As shown in Lemma 2 and Table 2, each meta-process is a basic operation on one or more operands such as variables, memory elements, or I/O ports. Structures of the operands and their allowable operations are constrained by their types as described in the preceding subsection.

It is noteworthy that not all generally important and fundamental computational operations as shown in Table 2 had been explicitly identified in conventional formal methods. For instances, the evaluation, addressing, memory allocation/release, timing/duration, and the system processes. However, all these are found necessary and essential in modeling system architectures and behaviors [39].

Table 2. RTPA Meta-Processes

No.	Meta Process	Notation	Syntax
1	Assignment	:=	$y\mathbb{T} := x\mathbb{T}$
2	Evaluation	◆	$◆_{\mathbb{T}} exp\mathbb{T} \rightarrow \mathbb{T}$
3	Addressing	⇒	$id\mathbb{T} \Rightarrow \text{MEM}[ptr\mathbf{P}]\ \mathbb{T}$
4	Memory allocation	⇐	$id\mathbb{T} \Leftarrow \text{MEM}[ptr\mathbf{P}]\ \mathbb{T}$
5	Memory release	⇍	$id\mathbb{T} \nLeftarrow \text{MEM}[\perp]\mathbb{T}$
6	Read	⋗	$\text{MEM}[ptr\mathbf{P}]\mathbb{T} \gtrdot x\mathbb{T}$
7	Write	⋖	$x\mathbb{T} \lessdot \text{MEM}[ptr\mathbf{P}]\mathbb{T}$
8	Input	|⋗	$\text{PORT}[ptr\mathbf{P}]\mathbb{T} \mid\gtrdot x\mathbb{T}$
9	Output	|⋖	$x\mathbb{T} \mid\lessdot \text{PORT}[ptr\mathbf{P}]\mathbb{T}$
10	Timing	$\underline{\underline{@}}$	$@t\mathbf{TM}\ \underline{\underline{@}}\ \S t\mathbf{TM}$ $\mathbf{TM} = \mathbf{yy{:}MM{:}dd}$ $\mid \mathbf{hh{:}mm{:}ss{:}ms}$ $\mid \mathbf{yy{:}MM{:}dd{:}hh{:}mm{:}ss{:}ms}$
11	Duration	$\underline{\underline{\Delta}}$	$@t_n\mathbf{TM}\ \underline{\underline{\Delta}}\ \S t_n\mathbf{TM} + \Delta n\mathbf{TM}$
12	Increase	↑	$\uparrow(n\mathbb{T})$
13	Decrease	↓	$\downarrow(n\mathbb{T})$
14	Exception detection	!	$!\ (@e\mathbf{S})$
15	Skip	⊗	⊗
16	Stop	⊠	⊠
17	System	§	$\S(SysID\mathbf{ST})$

2.4 Process Operations of RTPA

Definition 6. A *process relation* in RTPA is an algebraic operation and a compositional rule between two or more meta-processes in order to construct a complex process.

A set of 17 fundamental process relations has been elicited from fundamental algebraic and relational operations in computing in order to build and compose complex processes in the context of real-time software systems. Syntaxes and usages of the 17

RTPA process relations are formally described in Table 3. Deductive semantics of these process relations may be referred to [36], [39], [43], [46].

Lemma 3. The software composing rules state that the *RTPA process relation system* $\mathfrak{R}$ encompasses 17 fundamental algebraic and relational operations elicited from basic computing needs as defined in Table 3, i.e.:

$$\mathfrak{R} = \{\rightarrow, \curvearrowright, |, |\ldots|\ldots, R^*, R^+, R^i, \circlearrowleft, \rightarrowtail, ||, ∯, |||, \gg, ↯, \hookrightarrow_t, \hookrightarrow_e, \hookrightarrow_i\} \quad (7)$$

Table 3. RTPA Process Relations and Algebraic Operations

No.	Process Relation	Notation	Syntax
1	Sequence	$\rightarrow$	$P \rightarrow Q$
2	Jump	$\curvearrowright$	$P \curvearrowright Q$
3	Branch	$\mid$	$♦exp\mathbf{BL} = \mathbf{T} \rightarrow P$ $\mid ♦\sim \rightarrow Q$
4	Switch	$\mid \ldots \mid$	$♦exp\mathbb{T} =$ $i \rightarrow P_i$ $\mid \sim \rightarrow \varnothing$ where $\mathbb{T} \in \{\mathbf{N}, \mathbf{Z}, \mathbf{B}, \mathbf{S}\}$
5	While-loop	R^*	$\underset{exp\mathbf{BL}=\mathbf{T}}{\overset{\mathbf{F}}{R}} P$
6	Repeat-loop	R^+	$P \rightarrow \underset{exp\mathbf{BL}=\mathbf{T}}{\overset{\mathbf{F}}{R}} P$
7	For-loop	R^i	$\underset{i\mathbf{N}=1}{\overset{n\mathbf{N}}{R}} P(i\mathbf{N})$
8	Recursion	$\circlearrowleft$	$\underset{i\mathbf{N}=n\mathbf{N}}{\overset{0}{R}} P^{i\mathbf{N}} \circlearrowleft P^{i\mathbf{N}-1}$
9	Function call	$\rightarrowtail$	$P \rightarrowtail F$
10	Parallel	$\|\|$	$P \|\| Q$
11	Concurrence	∯	P ∯ Q
12	Interleave	$\|\|\|$	$P \|\|\| Q$
13	Pipeline	$\gg$	$P \gg Q$
14	Interrupt	↯	P ↯ Q
15	Time-driven dispatch	$\hookrightarrow_t$	$@t_i\mathbf{TM} \hookrightarrow_t P_i$
16	Event-driven dispatch	$\hookrightarrow_e$	$@e_i\mathbf{S} \hookrightarrow_e P_i$
17	Interrupt-driven dispatch	$\hookrightarrow_i$	$@int_i⦿ \hookrightarrow_i P_i$

3 Laws of Meta-processes of Software Behaviors

A meta-process in RTPA is the most fundamental and elementary process that cannot be broken up further. Specific laws that constrain RTPA meta-processes in software engineering and computing are explored in this section. As a summary, the algebraic laws of the 17 meta-processes are listed in Table 4.

3.1 Laws of Assignments

Hoare wrote in 1969: "Assignment is undoubtedly the most characteristic feature of programming a digital computer, and one that most clearly distinguishes it from other branches of mathematics. It is surprising therefore that the axiom governing our reasoning about assignment is quite as simple as any to be found in elementary logic [13]."

Table 4. Algebraic Laws of Meta-Processes of Software Behaviors

No.	Process	Notation	Specific laws
1	Assignment	:=	**L1**(Selectivity), **L2**(Transitivity), **L3**(Most recent effectiveness), **L4**(Compositionality), **L5**(Decompositionality)
2	Evaluation	◆	**L6**(Boolean evaluation), **L7**(Ordinal evaluation), **L8**(Ordinal power set evaluation), **L9**(Numerical evaluation), **L10** (Logical expressive equivalence), **L11**(Shortcut of conjunctive evaluation), **L12**(Shortcut of disjunctive evaluation)
3	Addressing	⇒	**L13**(Definite memory addressing), **L14**(Power set memory addressing)
4	Memory allocation	⇐	**L15**(Memory allocation),
5	Memory release	$\not\Leftarrow$	**L16**(Memory release)
6	Read	>	**L17**(Memory read)
7	Write	<	**L18**(Memory write)
8	Input	|>	**L19**(Port input)
9	Output	|<	**L20**(Port output)
10	Timing	$\underline{\underline{@}}$	**L21**(System clock), **L22**(Absolute timing), **L23**(Relative timing)
11	Duration	≜	**L24**(Duration timing)
12	Increase	↑	-
13	Decrease	↓	-
14	Exception detection	!	-
15	Skip	⊗	**L25**(Jump equivalence), **L26**(Skip absorption), **L27**(Skew skip absorption)
16	Stop	⊠	-
17	System	§	**L33**(Serial architecture representation), **L46**(Recurrent CLM representation), **L56**(Nested architecture representation), **L63**(Parallel architecture representation), **L77**(Coupled architecture representation)

The assignment process is transitive, and it constrained by the following laws of software.

Law 1. The law of *assignment selectivity* states that the assignment operation, :=, is selective on variables and/or values that share the same or equivalent type $\mathbb{T}$, i.e.:

$$
\begin{aligned}
y\mathbb{T} := x\mathbb{T} \triangleq\ & \blacklozenge(T(y) = T(x) \vee T(y) \simeq T(x)) \\
& \to y\mathbb{T} = V(x)\mathbb{T} \\
\mid\ & \blacklozenge\sim \\
& \to !(@\text{TypeError}\mathbf{S}) \\
& \to \varnothing
\end{aligned}
\tag{8}
$$

where, $\mathbb{T}$ is a predefined primitive type in RTPA, $\mathbb{T} \in \mathfrak{T}$, and T(*z*) and V(*z*) are a type and value evaluation function, respectively, on a given variable *z*.

Law 2. The law of *assignment transitivity* states that an assignment operation is transitive among multiple related variables that share the same or equivalent type $\mathbb{T}$, i.e.:

$$
(y\mathbb{T} := z\mathbb{T}) \to (x\mathbb{T} := y\mathbb{T}) \triangleq x\mathbb{T} := z\mathbb{T} \tag{9}
$$

Law 2 can be expressed in a more general form as assignment compositeness in Law 4.

Law 3. The law of *most recent effectiveness of assignments* states that *n* sequential assignment operations on the same variable are mutually exclusive, where only the last assignment is effective in the series of assignments, i.e.:

$$
\mathop{R}_{i=1}^{n} y\mathbb{T} := x_i\mathbb{T} \triangleq (y\mathbb{T} := x_n\mathbb{T}) \tag{10}
$$

Law 3 can be used to eliminate redundant or unnecessary assignments in programming or system specifications, particularly in a distributed environment.

Law 4. The law of *assignment compositionality* states that an assignment operation is compositional through multiple functions or expressions, i.e.:

$$
\begin{aligned}
& (y\mathbb{T} := g(z\mathbb{T})\mathbb{T}) \to (x\mathbb{T} := f(y\mathbb{T})\mathbb{T}) \\
& \triangleq (x\mathbb{T} := f \circ g(z\mathbb{T})\mathbb{T})\mathbb{T} \\
& = (x\mathbb{T} := f(g(z\mathbb{T})\mathbb{T})\mathbb{T}
\end{aligned}
\tag{11}
$$

Law 4 is the foundation of programming and component-based system composition in software engineering. It is in line with the generic program model as presented in Section 2.1. It is obvious that Law 3 is a special case of Law 4 where both functions *f* and *g* are assignments.

An inverse expression of Law 4 forms the law of decompositionality for assignments.

Law 5. The law of *assignment decompositionality* states that an assignment operation is decompositional through multiple functions or expressions, i.e.:

$$
\begin{aligned}
x\mathbb{T} &:= f \circ g(z\mathbb{T})\mathbb{T})\mathbb{T} \\
&\triangleq x\mathbb{T} := f(g(z\mathbb{T})\mathbb{T})\mathbb{T} \\
&= (y\mathbb{T} := g(z\mathbb{T})\mathbb{T}) \rightarrow (x\mathbb{T} := f(y\mathbb{T})\mathbb{T})
\end{aligned} \tag{12}
$$

Law 5 forms the foundation of programming and component-based system specification in software engineering.

3.2 Laws of Evaluations

The evaluation processes of expressions in computing can be classified as Boolean, ordinal, and numerical. The evaluation processes are constrained by the following laws of software.

Law 6. The law of *Boolean evaluation* states that a given Boolean expression *exp***BL** can be evaluated exclusively by $\blacklozenge_{BL}$, which results in one of the Boolean constants **T** or **F**, i.e.:

$$\blacklozenge_{BL}(exp\mathbf{BL})\mathbf{BL} \triangleq \blacklozenge_{BL}: exp\mathbf{BL} \rightarrow \{\mathbf{T}, \mathbf{F}\} \tag{13}$$

Typical branch constructs obey Law 6 where each branch is selected by a Boolean constant **T** or **F**.

Law 7. The law of *ordinal evaluation* states that a given natural number expression *exp***N** can be evaluated ordinally by $\blacklozenge_{N}$, which results in a unique ordinal number, i.e.:

$$\blacklozenge_{N}(exp\mathbf{N})\mathbf{N} \triangleq \blacklozenge_{N}: exp\mathbf{N} \rightarrow \mathbf{N} \quad \mathrm{M} \tag{14}$$

Typical switch constructs obey Law 7 where each branch is selected by an ordinal number $n \in \mathbf{N}$.

Law 8. The law of *ordinal power set evaluation* states that a given natural number expression *exp***N** can be evaluated ordinally by $\blacklozenge_{\text{ÞN}}$, which results in a subset of natural numbers, i.e.:

$$\blacklozenge_{\text{ÞN}}(exp\mathbf{N})\text{Þ}\mathbf{N} \triangleq \blacklozenge_{\text{ÞN}}: exp\mathbf{N} \rightarrow \text{Þ}\mathbf{N} \tag{15}$$

A more general and flexible switch constructs obey Law 8 where each branch is selected by a subset of numbers $\text{Þ}\mathbf{N} \subseteq \mathbf{N}$.

Law 9. The law of *numerical evaluation* states that a given real expression *exp***R** or integer expression *exp***Z** can be evaluated numerically by $\blacklozenge_{R}$, or $\blacklozenge_{Z}$, which results in a real number or an integer, respectively, i.e.:

$$\blacklozenge_{R}(exp\mathbf{R})\mathbf{R} \triangleq \blacklozenge_{R}: exp\mathbf{R} \rightarrow \mathbf{R} \tag{16a}$$

$$\blacklozenge_{Z}(exp\mathbf{Z})\mathbf{Z} \triangleq \blacklozenge_{Z}: exp\mathbf{Z} \rightarrow \mathbf{Z} \tag{16b}$$

Law 9 is the mathematical foundation of measurement theories and software engineering measurement.

Law 10. The law of *logical expressive equivalence* states that a pair of Boolean expressions are equivalent, denoted by $\simeq$, *iff*: (a) Both expressions share the same set of variables; and either (b) The symbolic expressions can be transformed into the same form, or (c) The truth tables of both expressions are identical, i.e.:

$$\begin{aligned} & exp_1(x_1\mathbf{BL}, x_2\mathbf{BL}, ..., x_n\mathbf{BL})\mathbf{BL} \simeq exp_2(x_1\mathbf{BL}, x_2\mathbf{BL}, ..., x_n\mathbf{BL})\mathbf{BL} \\ & \triangleq \overset{n}{\underset{i=1}{R}} \forall x_i \in \mathbf{BL}, \blacklozenge_{\mathrm{BL}}(exp_1\mathbf{BL})\mathbf{BL} = \blacklozenge_{\mathrm{BL}}(exp_2\mathbf{BL})\mathbf{BL} \end{aligned} \tag{17}$$

According to Law 10, equivalent evaluation of Boolean expressions can be determined either by identical truth tables or by transformation of one expression into the other.

Law 11. The law of *shortcut of conjunctive evaluation* states that a conjunctive Boolean expression with multiple Boolean variables, $exp\mathbf{BL} = x_1\mathbf{BL} \wedge x_2\mathbf{BL} \wedge \ldots \wedge x_n\mathbf{BL}$, can be evaluated directly as $\mathbf{F}$ whenever at most one of them is false, i.e.:

$$\blacklozenge_{\mathrm{BL}}(x_1\mathbf{BL} \wedge x_2\mathbf{BL} \wedge ... \wedge x_n\mathbf{BL})\mathbf{BL} \triangleq \mathbf{F},\ \exists x_i\mathbf{BL} = \mathbf{F}, 0 \le i \le n \tag{18}$$

Law 12. The law of *shortcut of disjunctive evaluation* states that a disjunctive Boolean expression with multiple Boolean variables, $exp\mathbf{BL} = x_1\mathbf{BL} \vee x_2\mathbf{BL} \vee \ldots \vee x_n\mathbf{BL}$, can be evaluated directly as $\mathbf{T}$ whenever at most one of them is true, i.e.:

$$\blacklozenge_{\mathrm{BL}}(x_1\mathbf{BL} \vee x_2\mathbf{BL} \vee ... \vee x_n\mathbf{BL})\mathbf{BL} \triangleq \mathbf{T},\ \exists x_i\mathbf{BL} = \mathbf{T}, 0 \le i \le n \tag{19}$$

3.3 Laws of Addressing

Addressing is a fundamental computational process that maps a variable to a memory location or block. The addressing processes are constrained by the following laws of software.

Law 13. The law of *definite memory addressing* $\Rightarrow$ states that a declared logical identifier $id\mathbf{B}$ in type byte can be associated to a unique physical memory address, i.e.:

$$\begin{aligned} & id\mathbf{BL} \Rightarrow \mathrm{MEM}[ptr\mathbf{P}]\mathbf{B} \triangleq \\ & \pi : id\mathbf{B} \rightarrow ptr\mathbf{P} \rightarrow \mathrm{MEM}[ptr\mathbf{P}]\mathbf{B} \end{aligned} \tag{20}$$

More generally, when the logical representation of an arbitrary typed identifier $id\mathbb{T}$ occupies more than one memory elements, the following laws can be introduced.

Law 14. The law of *power set memory addressing* states that a declared logical identifier $id\mathbb{T}$ can be mapped into a unique block of n continuous logical memory elements for representing a variable in type $\mathbb{T}$, i.e.:

$$\begin{aligned} & id\mathbb{T} \Rightarrow \mathrm{MEM}[ptr\mathbf{P}]\mathbb{T} \triangleq \\ & \pi : id\mathbb{T} \rightarrow ptr\mathbf{P} \rightarrow \mathrm{MEM}[ptr\mathbf{P}]\mathbb{T} \end{aligned} \tag{21}$$

where the power set of addressing results is determined by $ptr\mathbf{P} = [ptr\mathbf{P},\ ptr\mathbf{P} + n\mathbf{N} - 1]$, and n is language implementation-specific.

Law 15. The law of *memory allocation* $\Leftarrow$ states that a unique block of n continuous physical memory elements can be allocated to a declared logical identifier $id\mathbb{T}$, i.e.:

$$\begin{aligned} id\mathbb{T} &\Leftarrow \text{MEM}[ptr\mathbf{P}]\mathbb{T} \triangleq \\ &\pi^{-1}\text{: MEM}[ptr\mathbf{P}]\mathbb{T} \to id\mathbb{T} \end{aligned} \tag{22}$$

where the power set of allocation pointer is $ptr\mathbf{P} = [ptr\mathbf{P},\ ptr\mathbf{P} + n\mathbf{N} - 1]$, and n is language implementation-specific.

Law 16. The law of *memory release* $\not\Leftarrow$ states that a unique block of n continuous logical memory elements can be dissociated from a declared logical identifier $id\mathbb{T}$ by the following sequential processes, i.e.:

$$\begin{aligned} id\mathbb{T} \not\Leftarrow \text{MEM}[ptr\mathbf{P}\mathbf{P}]\mathbb{T} \triangleq & \\ (id\mathbb{T} \Rightarrow & \text{MEM}[ptr\mathbf{P}\mathbf{P}]\mathbb{T} \\ \to & \text{MEM}[ptr\mathbf{P},\ ptr\mathbf{P} + n\mathbf{N} - 1]\mathbb{T} := \perp \\ \to & ptr\mathbf{P} := \perp \\ \to & id\mathbb{T} := \perp \\ ) & \end{aligned} \tag{23}$$

where the first step denotes a readdressing process in order to establish the association between the identifier, the memory block, and the pointers, and $\perp$ denotes a value undefined.

3.4 Laws of I/O Manipulations

Input/output (I/O) processes are important computational operations for modeling interactive behaviors of systems. The I/O manipulation processes are constrained by the following laws of software.

Law 17. The law of *memory read* states that the read operation on memory, $\gtrdot$, is equivalent to an assignment operation, where the allocated memory element pointed by $ptr\mathbf{P}$ is treated as a logical variable, i.e.:

$$\text{MEM}[ptr\mathbf{P}]\mathbb{T} \gtrdot x\mathbb{T} \triangleq x\mathbb{T} := \text{MEM}[ptr\mathbf{P}]\mathbb{T} \tag{24}$$

Law 18. The law of *memory write* states that the write operation on memory, $\lessdot$, is equivalent to an assignment operation, where the allocated memory element pointed by $ptr\mathbf{P}$ is treated as a logical variable, i.e.:

$$x\mathbb{T} \lessdot \text{MEM}[ptr\mathbf{P}]\mathbb{T} \triangleq \text{MEM}[ptr\mathbf{P}]\mathbb{T} := x\mathbb{T} \tag{25}$$

Law 19. The law of *port input* states that the input from a port, $|\!>$, is equivalent to an assignment operation, where the allocated port buffer pointed by $ptr\mathbf{P}$ is treated as a logical variable, i.e.:

$$\text{PORT}[ptr\mathbf{P}]\mathbb{T} \,|\!> x\mathbb{T} \triangleq x\mathbb{T} := \text{PORT}[ptr\mathbf{P}]\mathbb{T} \tag{26}$$

Law 20. The law of *port output* states that the output to a port, $|\!<$, is equivalent to an assignment operation, where the allocated port buffer pointed by $ptr\mathbf{P}$ is treated as a logical variable, i.e.:

$$x\mathbb{T} \,|\!< \text{PORT}[ptr\mathbf{P}]\mathbb{T} \triangleq \text{PORT}[ptr\mathbf{P}]\mathbb{T} := x\mathbb{T} \tag{27}$$

3.5 Laws of Time Manipulations

Time manipulation is a necessary dimension in computing that is supplemental to the logic and space dimensions for modeling interactive system behaviors. The time manipulation processes are constrained by the following laws of software.

Law 21. The law of *system clock* states that a software system, in particular a real-time system, needs to maintain two clocks at the system level known as the *absolute clock* $\S t\mathbf{TM}$ and the *relative clock* $\S t_n\mathbf{N}$, as follows:

$$\text{a)}\quad \S t\mathbf{TM} \triangleq [0{:}1{:}1{:}0{:}0{:}0{:}0\mathbf{yyyy{:}MM{:}dd{:}hh{:}mm{:}ss{:}ms},\ 9999{:}12{:}31{:}23{:}59{:}59{:}999\mathbf{yyyy{:}MM{:}dd{:}hh{:}mm{:}ss{:}ms}] \tag{28a}$$

where a subset of the date/time range **TM** may be implemented for nonreal-time or noncontinuous systems.

$$\text{b)}\quad \S \tau\mathbf{N} \triangleq [0\mathbf{ms},\ 1\times 10^8\mathbf{ms}] \tag{28b}$$

where the range of the relative clock is determined by $\S \tau\mathbf{N} = 24\mathbf{hh} \times 60\mathbf{mm} \times 60\mathbf{ss} \times 1000\mathbf{ms} = 8.64 \times 10^7\mathbf{ms}$. The relative clock $\S \tau\mathbf{N}$ may be reset to zero at midnight every day in system modeling.

Law 22. The law of *absolute timing* states that the absolute system clock $\S t\mathbf{TM}$ can be used to set, $\underline{\underline{@}}$, a timing event or the execution of a process $@t\mathbf{TM}$ at a precise point of calendar time, i.e.:

$$@t\mathbf{TM} \ \underline{\underline{@}}\ \S t\mathbf{TM} \tag{29}$$

where $\mathbf{TM} = \mathbf{yy{:}MM{:}dd} \mid \mathbf{hh{:}mm{:}ss{:}ms} \mid \mathbf{yyyy{:}MM{:}dd{:}hh{:}mm{:}ss{:}ms}$.

Law 23. The law of *relative timing* states that the relative clock $\S t\mathbf{N}$ can be used to set, $\underline{\underline{@}}$, a timing event or the execution of a process $@t\mathbf{TM}$ at a certain relative time point, i.e.:

$$@t\mathbf{N} \ \underline{\underline{@}}\ \S \tau\mathbf{ms} \tag{30}$$

where $\S t\mathbf{N} \in [0,\ 1 \times 10^8\mathbf{ms}]$.

Law 24. The law of *duration timing* states that an absolute timing event $@t\mathbf{TM}$ on the system clock $\S t\mathbf{TM}$, or a relative timing event $@t\mathbf{N}$ on the system clock $\S t\mathbf{N}$, can be set, $\triangleq$, with and a given duration $\Delta n\mathbf{TM}$ or $\Delta n\mathbf{ms}$ for a timing event or the execution of a process, i.e.:

$$@t\mathbf{TM} \triangleq \S t\mathbf{TM} + \Delta n\mathbf{TM} \tag{31a}$$

or

$$@t\mathbf{N} \triangleq \S \tau\mathbf{N} + \Delta n\mathbf{ms} \tag{31b}$$

3.6 Laws of Skip

Skip is special meta-process in programming that has no semantic effect on the current process P^k at a given embedded layer k in a program, such as a branch, loop, or function. However, it redirects the system to jump to execute an upper-layer process P^{k-1} in the embedded hierarchy. Therefore, skip is also known as *exit* or *break* in programming. The skip processes are constrained by the following laws of software.

Law 25. The law of *jump equivalence* of skip processes states that a skip does nothing functionally but adjusts the internal control to jump to a predefined process outside the current executing layer, i.e.:

$$P \to \otimes \to Q = P \curvearrowright Q \tag{32}$$

where the *jump* process operation will be explained in Section 4.2.

Law 26. The law of *skip absorption* states that skips can be replaced by jump process relations in a branch, parallel, concurrent, or pipeline process structure, i.e.:

$$\begin{aligned} ((P \to \otimes)\ \mathbb{R}\ (Q \to \otimes)) \to S &= (P \curvearrowright S)\ \mathbb{R}\ (Q \curvearrowright S) \\ &= (P\ \mathbb{R}\ Q) \curvearrowright S \end{aligned} \tag{33}$$

where $\mathbb{R} = \{|, ||, ∯, \gg\}$.

Law 27. The law of *skew skip absorption* states that a skip can be replaced by a jump process in a skew branch, parallel, concurrent, or pipeline process structure, i.e.:

$$((P \to T)\ \mathbb{R}\ (Q \to \otimes)) \to S = (P \to T \to S)\ \mathbb{R}\ (Q \curvearrowright S) \tag{34}$$

where $\mathbb{R} = \{|, ||, ∯, \gg\}$.

According to Table 4, the top-level meta-process, *system*, obeys a set of architectural representation laws, which will be described in Section 4, such as Law 33 for serial architecture representation, Law 46 for recurrent CLM representation, Law 56 for nested architecture representation, Law 63 for parallel architecture representation, and Law 77 for coupled architecture representation.

4 Laws of Algebraic Operations of Software Behaviors

The relational operations of process behaviors defined in RTPA provides a set of 17 process composing rules for constructing complex processes and manipulating advanced computing behaviors on the basis of the RTPA meta-processes. A comprehensive set of algebraic laws for relational process operations will be established in this section.

Let *P, Q, S* be meta or complex processes *P, Q, S* $\in \mathfrak{R}$, and $\mathbb{R}$, $\mathbb{R}'$ be different relational operators, $\mathbb{R}, \mathbb{R}' \in \mathfrak{R}$, then a set of algebraic laws of software process compositions can be elicited as defined in Table 5, where the simplest process can be a single event.

Table 5 provides a set of generic algebraic laws of software process relations and compositional rules. Observing the table it can be seen that the 12 relational laws, $\mathcal{L}1$ – $\mathcal{L}12$, can be classified into six pairs, i.e., associative/dissociative, reflexive/irreflexive, symmetric/antisymmetric, transitive/intransitive, distributive/ nondistributive, and elicitive/nonelicitive. It is noteworthy that each pair of laws is exclusive, i.e., a specific process relational operation $\mathbb{R} \in \mathfrak{R}$ only obeys one of the laws in a certain pair as shown in Table 5.

Table 5. Generic Algebraic Laws of Software Processes

No.	Law	Description
Law $\mathcal{L}1$	Associative	$\mathbb{R}_1 \circ (\mathbb{R}_2 \circ \mathbb{R}_3) \Rightarrow (\mathbb{R}_1 \circ \mathbb{R}_2) \circ \mathbb{R}_3$
Law $\mathcal{L}2$	Dissociative	$\mathbb{R}_1 \circ (\mathbb{R}_2 \circ \mathbb{R}_3) \not\Rightarrow (\mathbb{R}_1 \circ \mathbb{R}_2) \circ \mathbb{R}_3$
Law $\mathcal{L}3$	Reflexive	$P \mathbb{R} Q \Rightarrow P = Q$
Law $\mathcal{L}4$	Irreflexive	$P \mathbb{R} Q \Rightarrow P \neq Q$
Law $\mathcal{L}5$	Symmetric	$P \mathbb{R} Q \Rightarrow Q \mathbb{R} P$
Law $\mathcal{L}6$	Asymmetric	$P \mathbb{R} Q \not\Rightarrow Q \mathbb{R} P$
Law $\mathcal{L}7$	Transitive	$P \mathbb{R} S \wedge S \mathbb{R} Q \Rightarrow P \mathbb{R} Q$
Law $\mathcal{L}8$	Intransitive	$P \mathbb{R} S \wedge S \mathbb{R} Q \not\Rightarrow P \mathbb{R} Q$
Law $\mathcal{L}9$	Distributive	$P \mathbb{R} (Q \mathbb{R}' S) \Rightarrow (P \mathbb{R} Q)\ \mathbb{R}'\ (P \mathbb{R} S)$
Law $\mathcal{L}10$	Nondistributive	$P \mathbb{R} (Q \mathbb{R}' S) \not\Rightarrow (P \mathbb{R} Q)\ \mathbb{R}'\ (P \mathbb{R} S)$
Law $\mathcal{L}11$	Elicitive	$(P \mathbb{R} Q)\ \mathbb{R}'\ (P\ R\ S) \Rightarrow P \mathbb{R} (Q \mathbb{R}' S)$
Law $\mathcal{L}12$	Nonelicitive	$(P \mathbb{R} Q)\ \mathbb{R}'\ (P\ R\ S) \not\Rightarrow P \mathbb{R} (Q \mathbb{R}' S)$

A mapping of the 12 generic laws into each of the relational operators is summarized in Table 6. The laws and properties may be used to enhance the understanding of the mechanisms and behaviors of the fundamental processes in software engineering and computational intelligence. In addition to the generic laws for software processes and behaviors, there are special laws for most of the relational operations as identified in the right-most column of Table 6, which will be described individually in the corresponding laws.

The following subsections describe the algebraic laws for each of the 17 relational process operations $\mathbb{R} \in \mathfrak{R}$, plus any additional special law that a particular process operation must obey.

4.1 Laws of Sequential Processes

The sequential operation of processes is associative, reflective, distributive, and elicitive. However, it is asymmetric and intransitive. The sequential process operations are constrained by the following laws of software.

Table 6. Algebraic Laws of Relational Process Operations

No.	Process relation	Notation	Generic algebraic laws												Special algebraic laws
			$\mathcal{L}$ 1	$\mathcal{L}$ 2	$\mathcal{L}$ 3	$\mathcal{L}$ 4	$\mathcal{L}$ 5	$\mathcal{L}$ 6	$\mathcal{L}$ 7	$\mathcal{L}$ 8	$\mathcal{L}$ 9	$\mathcal{L}$ 10	$\mathcal{L}$ 11	$\mathcal{L}$ 12	
1	Sequence	$\rightarrow$	✓		✓			✓		✓	✓		✓		L32, L33
2	Jump	$\curvearrowright$	✓		✓			✓		✓	✓		✓		-
3	Branch	$\mid$		✓		✓		✓		✓	✓		✓		L41
4	Switch	$\mid \ldots \mid$		✓		✓		✓		✓		✓	✓		L43
5	While	R^*	-	-	-	-	-	-	-	-	-	-	-	-	L44
6	Repeat	R^+	-	-	-	-	-	-	-	-	-	-	-	-	L44
7	For-do	R^i	-	-	-	-	-	-	-	-	-	-	-	-	L44, L45, L46
8	Recursion	$\circlearrowleft$	-	-	-	-	-	-	-	-	-	-	-	-	L47, L48, L49
9	Function call	$\rightarrowtail$	✓			✓		✓		✓	✓		✓		L56
10	Parallel	$\parallel$	✓			✓	✓		✓			✓	✓		L62, L63
11	Concurrence	∯	✓			✓	✓		✓			✓	✓		L69
12	Interleave	$\parallel\mid$	✓			✓	✓		✓			✓	✓		-
13	Pipeline	»	✓			✓		✓		✓		✓	✓		L76, L77
14	Interrupt	↯		✓		✓		✓		✓		✓		✓	L78, L79, L80
15	Time-dispatch	$\hookrightarrow_t$		✓		✓	✓			✓		✓	✓		L84, L85
16	Even-dispatch	$\hookrightarrow_e$		✓		✓	✓			✓		✓	✓		L89, L90
17	Interrupt-dispatch	$\hookrightarrow_i$		✓		✓	✓			✓		✓	✓		L94, L95

a) Associativity of Sequential Processes

Law 28. The law of *associativity* of sequential processes states that a list of sequential processes can be arbitrarily associated whenever their original order of sequence is preserved, i.e.:

$$P \to Q \to S = (P \to Q) \to S = P \to (Q \to S) \tag{35}$$

Law 28, sequential associativity, is the foundation of system modularization, decomposition, and integration in software engineering.

b) Reflective of Sequential Processes

Law 29. The law of *reflectivity* of sequential processes states that a list of sequential processes is reflective, *iff* they are identical, i.e.:

$$P \to Q = Q \to P \Rightarrow P = Q \tag{36}$$

c) Distributivity of Sequential Processes

Law 30. The law of *distributivity* of sequential processes states that a linear process *S* can be distributed into a pair of disjunctive conditional branch processes $P \mid Q$, i.e.:

$$S \to (P \mid Q) = (S \to P) \mid (S \to Q) \tag{37a}$$

or

$$(P \mid Q) \to S = (P \to S) \mid (Q \to S) \tag{37b}$$

d) Elicitivity of Common Statements

Law 31. The law of *elicitivity of common statements* states that common processes in a pair of disjunctive conditional branch processes can be elicited as shared processes, i.e.:

$$(S \to P) \mid (S \to Q) = S \to (P \mid Q) \tag{38a}$$

or

$$(P \to S) \mid (Q \to S) = (P \mid Q) \to S \tag{38b}$$

Law 31 is an inverse statement of Law 30, which is frequently used to elicit the common component from a branch structure.

In addition, sequential processes also obey the following special laws.

e) Procedural Representation of Compound Statements

Law 32. The law of *procedural representation of compound statements* states that an arbitrary subset of sequential processes *P, Q, S* can be associated into a procedure or function *F*, i.e.:

$$\begin{aligned} &P \to (Q_1 \to Q_2 \to \ldots \to Q_n) \to S = \\ &P \to F \to S, F = (Q_1 \to Q_2 \to \ldots \to Q_n) \end{aligned} \tag{39}$$

where *F* is the *procedure* elicited from the list of processes.

Law 32 is an extension of Law 28, sequential associativity, which provides a powerful means to derive structural programs that implements the principles of encapsulation, abstraction, and information hiding in software engineering.

f) Representation of Serial Architectures

Law 33. The law of *serial architecture representation* states that the sequential processes can be used as an abstract representation of recurring serial architectures (SAs) in system modeling, i.e.:

$$\mathbf{SAST} \triangleq (P_1 \rightarrow P_2 \rightarrow ... \rightarrow P_n) \tag{40}$$

where each process P_i, $1 \le i \le n$, denotes a component in the system.

Law 33 shows that system architecture modeling share the same properties as those of behavioral processes. Both of them can be modeled by processes and their relational operations.

4.2 Laws of Jump Processes

The jump operation of processes is associative, reflective, distributive, and elicitive, but it is asymmetric and intransitive. The jump process operations are constrained by the following laws of software.

a) Associativity of Jump Processes

Law 34. The law of *associativity* of jump processes states that multiple jumps between a list of sequential processes can be arbitrarily associated whenever their original order of sequence is preserved, i.e.:

$$P \curvearrowright Q \curvearrowright S = (P \curvearrowright Q) \curvearrowright S = P \curvearrowright (Q \curvearrowright S) \tag{41}$$

b) Reflective of Jump Processes

Law 35. The law of *reflectivity* of jump processes states that the jump operation between two processes is reflective, *iff* they are identical, i.e.:

$$P \curvearrowright Q = Q \curvearrowright P \Rightarrow P = Q \tag{42}$$

c) Distributivity of Jump Processes

Law 36. The law of *distributivity* of jump processes states that a process S can be distributed into a pair of disjunctive conditional branch processes $P \mid Q$ by the jump operation, i.e.:

$$S \curvearrowright (P \mid Q) = (S \curvearrowright P) \mid (S \curvearrowright Q) \tag{43a}$$

or

$$(P \mid Q) \curvearrowright S = (P \curvearrowright S) \mid (Q \curvearrowright S) \tag{43b}$$

d) Elicitivity of Jump Processes

Law 37. The law of *elicitivity of jump statements* states that a common process in a pair of branch processes can be elicited as a shared process, i.e.:

$$(S \curvearrowright P) \mid (S \curvearrowright Q) = S \curvearrowright (P \mid Q) \tag{44a}$$

or

$$(P \curvearrowright S) \mid (Q \curvearrowright S) = (P \mid Q) \curvearrowright S \tag{44b}$$

The elicitivity of jump operations is an inversed operation of jump distributivity as given in Law 36.

4.3 Laws of Branch Processes

The branch operation of processes is distributive and elicitive, but it is dissociative, irreflexive, asymmetric, and intransitive. The branch process operations are constrained by the following laws of software.

a) Distributive of Branch Processes

Law 38. The law of *distributivity of branch* states that a process S can be distributed into a pair of disjunctive conditional branch processes P | Q by a relational operation $\mathbb{R}$, i.e.:

$$S \; \mathbb{R} \; (P \mid Q) = (S \; \mathbb{R} \; P) \mid (S \; \mathbb{R} \; Q) \tag{45a}$$

or

$$(P \mid Q) \; \mathbb{R} \; S = (P \; \mathbb{R} \; S) \mid (Q \; \mathbb{R} \; S) \tag{45b}$$

where $\mathbb{R} = \{ \parallel, \oiint, \curvearrowright, \gg, \mid\mid\mid \}$.

b) Elicitivity of Invariant Process from Branch Structures

Law 39. The law of *elicitivity* of sequential processes states that a common sequential process *S* within a disjunctive conditional process can be elicited and executed outside the conditional construct, i.e.:

$$\begin{array}{l} \blacklozenge exp\mathbf{BL} = \mathbf{T} \\ \quad \to S \\ \quad \to P \\ \mid \blacklozenge \sim \\ \quad \to S \\ \quad \to Q \\ = S \to (\; \blacklozenge exp\mathbf{BL} = \mathbf{T} \\ \qquad\qquad \to P \\ \qquad\quad \mid \blacklozenge \sim \\ \qquad\qquad \to Q \\ \quad) \end{array} \tag{46a}$$

where $exp\mathbf{BL}$ is a Boolean expression, and $\blacklozenge \sim$ denotes otherwise.

Law 39 can be similarly expressed in other form as follows:

$$
\begin{array}{l}
\quad \blacklozenge exp\mathbf{BL} = \mathbf{T} \\
\quad\quad \rightarrow P \\
\quad\quad \rightarrow S \\
\mid \blacklozenge \sim \\
\quad\quad \rightarrow Q \\
\quad\quad \rightarrow S \\
= (\quad \blacklozenge exp\mathbf{BL} = \mathbf{T} \\
\quad\quad\quad \rightarrow P \\
\quad\quad \mid \blacklozenge \sim \\
\quad\quad\quad \rightarrow Q \\
\quad) \\
\quad \rightarrow S
\end{array}
\qquad (46b)
$$

where S is independent from the Boolean expression *exp***BL** or the execution of S will not affect the Boolean value of exp**BL**.

Law 39, sequential elicitivity, is the theoretical foundation of common function or object elicitation, improvement of programming efficiency, and well structured programming in software engineering.

c) Skew Symmetry of Branch Processes

Law 40. The law of *skew symmetry of branch* states that a branch or conditional choice is commutative on the true and false branches, i.e.:

$$
\begin{array}{l}
\quad \blacklozenge exp\mathbf{BL} = \mathbf{T} \\
\quad\quad \rightarrow P \\
\mid \blacklozenge \sim \\
\quad\quad \rightarrow Q \\
= \quad \blacklozenge exp\mathbf{BL} = \mathbf{F} \\
\quad\quad \rightarrow Q \\
\mid \blacklozenge \sim \\
\quad\quad \rightarrow P
\end{array}
\qquad (47)
$$

d) Embedded Branch Processes

Law 41. The law of *embedded branch* states that multiple branches can be nested on the else branches in order to form a multi-layer branch structure, i.e.:

$$
\begin{array}{l}
\quad \blacklozenge exp_0\mathbf{BL} = \mathbf{T} \\
\qquad \to P_0 \\
\quad \mid \blacklozenge\sim \\
\qquad \to \blacklozenge exp_1\mathbf{BL} = \mathbf{T} \\
\qquad\qquad \to P_1 \\
\qquad\quad \mid \blacklozenge\sim \\
\qquad\qquad \ldots \\
\qquad\qquad\quad \to \blacklozenge exp_n\mathbf{BL} = \mathbf{T} \\
\qquad\qquad\qquad\quad \to P_n \\
\qquad\qquad\qquad \mid \blacklozenge\sim \\
\qquad\qquad\qquad\quad \to \otimes \\
= \blacklozenge exp\mathbf{N} = \; 0 \to P_0 \\
\qquad\qquad\quad \mid 1 \to P_1 \\
\qquad\qquad\quad \mid \ldots \\
\qquad\qquad\quad \mid n \to P_n \\
\qquad\qquad\quad \mid \sim \to \otimes
\end{array}
\tag{48}
$$

4.4 Laws of Switch Processes

The switch operation of processes is elicitive, but it is dissociative, irreflexive, asymmetric, intransitive, and nondistributive. Therefore, Law 38 for branch structures can be extended to the switch structure as follows.

a) Elicitivity of Sequential Processes

Law 42. The law of *elicitivity* of sequential processes states that a common sequential process Q within a switch process can be elicited and executed outside the switch construct, i.e.:

$$
\begin{array}{l}
\quad \blacklozenge exp\mathbf{N} = \; 0 \to (P_0 \to Q) \\
\qquad\qquad\quad \mid 1 \to (P_1 \to Q) \\
\qquad\qquad\quad \mid \ldots \\
\qquad\qquad\quad \mid n \to (P_n \to Q) \\
\qquad\qquad\quad \mid \sim \to Q \\
= \blacklozenge exp\mathbf{N} = \; 0 \to P_0 \\
\qquad\qquad\quad \mid 1 \to P_1 \\
\qquad\qquad\quad \mid \ldots \\
\qquad\qquad\quad \mid n \to P_n \\
\qquad\qquad\quad \mid \sim \to \varnothing \\
\quad \to Q
\end{array}
\tag{49a}
$$

or

$$\begin{aligned}\blacklozenge exp\mathbf{N} = \ & 0 \rightarrow (Q \rightarrow P_0) \\ & \mid 1 \rightarrow (Q \rightarrow P_1) \\ & \mid \ldots \\ & \mid n \rightarrow (Q \rightarrow P_n) \\ & \mid \sim \rightarrow Q \\ = Q \rightarrow (\blacklozenge exp\mathbf{N} = \ & 0 \rightarrow P_0 \\ & \mid 1 \rightarrow P_1 \\ & \mid \ldots \\ & \mid n \rightarrow P_n \\ & \mid \sim \rightarrow \varnothing \\ )\end{aligned} \tag{49b}$$

The sequential elicitivity is the theoretical foundation of common function or object elicitation, improvement of programming efficiency, and well structured programming in software engineering.

b) Equivalent Embedded Branch Processes

Law 43. The law of *equivalent embedded branch* states that the switch operation is equivalent to the multiple embedded structures in computing, i.e.:

$$\begin{aligned}\blacklozenge exp\mathbf{N} = \ & 0 \rightarrow P_0 \\ & \mid 1 \rightarrow P_1 \\ & \mid \ldots \\ & \mid n \rightarrow P_n \\ & \mid \sim \rightarrow \otimes \\ = \ & \blacklozenge exp_0\mathbf{BL} = \mathbf{T} \\ & \quad \rightarrow P_0 \\ & \mid \blacklozenge \sim \\ & \quad \rightarrow \blacklozenge exp_1\mathbf{BL} = \mathbf{T} \\ & \qquad \rightarrow P_1 \\ & \quad \mid \blacklozenge \sim \\ & \qquad \ldots \\ & \qquad \rightarrow \blacklozenge exp_n\mathbf{BL} = \mathbf{T} \\ & \qquad\quad \rightarrow P_n \\ & \qquad \mid \blacklozenge \sim \\ & \qquad\quad \rightarrow \otimes\end{aligned} \tag{50}$$

4.5 Laws of Iterative Processes

Although the iterative operations of processes do not obey any of the generic algebraic laws as given in Table 5, they are constrained by the following special laws of software.

a) Equivalence between Different Forms of Iterations

Law 44. The law of *equivalence between different forms of iterations* states that all forms of iterative constructs, such as while-do R^*, repeat-do R^+, and for-do R^i, are equivalent, i.e.:

$$\text{a)}\quad R^* P = \underset{exp\mathbf{BL}=\mathbf{T}}{\overset{\mathbf{F}}{R}} P \qquad (51a)$$

$$\text{b)}\quad \begin{aligned} R^+ P &= P \to R^* P \\ &= P \to \underset{exp\mathbf{BL}=\mathbf{T}}{\overset{\mathbf{F}}{R}} P \end{aligned} \qquad (51b)$$

$$\text{c)}\quad \begin{aligned} R^i P(i\mathbf{N}) &= \underset{i\mathbf{N}=1}{\overset{n}{R}} P(i\mathbf{N}) \\ &= (i\mathbf{N} := 1 \\ &\quad \to exp\mathbf{BL} = (i\mathbf{N} \le n\mathbf{N}) \\ &\quad \to \underset{exp\mathbf{BL}=\mathbf{T}}{\overset{\mathbf{F}}{R}} P(i\mathbf{N}) \\ &\quad \to \uparrow (i\mathbf{N}) \\ &\quad) \end{aligned} \qquad (51c)$$

where R^*, R^+, and R^i are known as the big-R notation of iterative behaviors and operations in software engineering [32], [39], [42].

b) Cumulativeness of Iterations

Law 45. The law of *cumulativeness of iterations* states that two of sequential iterations of identical process P can be concatenated, i.e.:

$$\underset{i\mathbf{N}=0}{\overset{n-1}{R}} P(i\mathbf{N}) \to \underset{i\mathbf{N}=n}{\overset{n+m-1}{R}} P(i\mathbf{N}) = \underset{i\mathbf{N}=0}{\overset{n+m-1}{R}} P(i\mathbf{N}) \qquad (52)$$

c) Recurrent Denoting of Logic Architectures

Law 46. The law of *recurrent Component Logical Model* (CLMs) *representation* [32], [39], [42] states that the iterative process relations can be used as an abstract model to denote repetitive architectural patterns in computing, i.e.:

$$CLM\mathbf{ST} \triangleq \mathop{R}_{i\mathbf{N}=0}^{n-1} CLM[i\mathbf{N}] \tag{53}$$

Law 46 can be illustrated by the following example.

Example 1. The port architecture of a computer, *PA*, with 1024 ports in various types $\mathbb{T}_i$, can be denoted as follows:

$$PA\mathbf{ST} \triangleq \mathop{R}_{i\mathbf{N}=0}^{1023} \text{PORT}[i\mathbf{N}]\mathbb{T}_i \tag{54}$$

where $\mathbb{T}_i \in \{\mathbf{B}, \mathbf{H}, \mathbf{N}, \mathbf{Z}, \mathbf{R}, \mathbf{S}\} \subset \mathfrak{T}$.

4.6 Laws of Recursive Processes

The recursive operation of processes is constrained by the following special laws of software.

a) Two-Phase Recursions

Law 47. The law of *two-phase recursion* states that a recursion is carried out by a series of deductive embedding processes (denoted by $\circlearrowleft$) and then followed by an inversed series of inductive de-embedding processes (denoted by $\circlearrowright$), i.e.:

$$\begin{aligned} &\mathop{R}_{i\mathbf{N}=n\mathbf{N}}^{0} P^{i\mathbf{N}} \circlearrowleft P^{i\mathbf{N}-1} \rightarrow \mathop{R}_{i\mathbf{N}=0}^{n\mathbf{N}} P^{i\mathbf{N}} \circlearrowright P^{i\mathbf{N}+1} \\ &= P^n \circlearrowleft P^{n-1} \circlearrowleft \ldots \circlearrowleft P^1 \circlearrowleft \boldsymbol{P^0} \circlearrowright P^1 \circlearrowright \ldots \circlearrowright P^{n-1} \circlearrowright P^n \end{aligned} \tag{55}$$

where in the first phase of *embedding*, a given layer of nested process is deduced to a lower layer till it is embodied to a known value P^0. In the second phase of *de-embedding*, the value of a higher layer process is deduced by the lower layer starting from the base layer P^0, where its value has already been known at the end of the preceding phase.

b) Terminable Recursions

Law 48. The *law of terminable recursion* states that a recursive function is terminable or non circular, *iff*: (a) A *base value* P^0 exists for certain arguments for which the function does not refer to itself; and (b) In each recursion, the argument of the function must be closer to the base value, i.e.:

$$\mathop{R}_{i\mathbf{N}=n\mathbf{N}}^{0} P^{i\mathbf{N}} \circlearrowleft P^{i\mathbf{N}-1} \tag{56}$$

c) Equivalence between Recursions and Iterations

Law 49. The law of *equivalence between recursive and iteration* states that a recursive structure can always be represented by an equivalent iterative structure, i.e.:

$$\overset{0}{\underset{i\mathbf{N}=n\mathbf{N}}{R}} P^{i\mathbf{N}} \circlearrowleft P^{i\mathbf{N}-1} = \overset{0}{\underset{i\mathbf{N}=n\mathbf{N}}{R}} (\blacklozenge i\mathbf{N}>0 \\ \rightarrow P^{i\mathbf{N}} := P^{i\mathbf{N}-1} \\ \mid \blacklozenge\sim \\ \rightarrow P^{0} \\ ) \tag{57}$$

where a process P at layer i of embedment, P^i, calls itself at an inner layer i-1, P^{i-1}, $0 \leq i \leq n$, and n is the *depth of recursion* or embedment that is determined by an explicitly specified conditional expression $exp\mathbf{BL} = \mathbf{T}$ inside the body of P.

4.7 Laws of Function Calls

The function call operation of processes is associative, distributive, and elicitive, but it is irreflexive, asymmetric, and intransitive. The function-call process operations are constrained by the following laws of software.

a) Function Elicitation from Recurring Patterns

It is a good practice in programming if the common portion of program R in both paths of branch structures can be elicited as a shared process or a list of sequential statements. The law of elicitity is in line with the principle of information hiding [24].

Law 50. Recurring patterns, processes, algorithms, and methods of classes in programming can be elicited and predefined as a procedure or function.

Some typical recurring algorithms and processes are provided below, such as the *unit increment/decrement* function and the *modular* function.

Example 2. A pair of unit increment and decrement functions can be introduced to simplify frequently used expressions in programming.

$$x\mathbb{T} := x\mathbb{T} + 1 \\ \triangleq \uparrow (x\mathbb{T}), \mathbb{T} \in \{\mathbf{N}, \mathbf{Z}, \mathbf{B}, \mathbf{H}, \mathbf{P}, \mathbf{TI}, \mathbf{D}, \mathbf{DT}\} \subseteq \mathfrak{T} \tag{58a}$$

$$x\mathbb{T} := x\mathbb{T} - 1 \\ \triangleq \downarrow (x\mathbb{T}), \mathbb{T} \in \{\mathbf{N}, \mathbf{Z}, \mathbf{B}, \mathbf{H}, \mathbf{P}, \mathbf{TI}, \mathbf{D}, \mathbf{DT}\} \subseteq \mathfrak{T} \tag{58b}$$

where the definitions of the specific types in the set of abstract type $\mathbb{T}$ may be referred to [32], [39].

Example 3. Let $x\mathbf{Z}$ be an arbitrary integer, and $M\mathbf{N}$ be a positive integer, a modular function *mod* yields the integer remainder $r\mathbf{N}$, i.e.:

$$x\mathbf{Z} \bmod M\mathbf{N} = r\mathbf{N}, \quad 0 \leq r\mathbf{N} < M\mathbf{N} \\ = \begin{cases} x\mathbf{Z} - k\mathrm{M}\mathbf{N}, & x\mathbf{Z} > 0 \\ kM\mathbf{N} - x\mathbf{Z}, & x\mathbf{Z} < 0 \end{cases} \tag{59}$$

For instances, according to Eq. 59, 18 *mod* 12 = 18 – 1 • 12 = 6, and -26 *mod* 7 = 4 • 7 – 26 = 2.

Programmers may define their own functions as a basic system construction mechanism in computing.

b) Associativity of Function Calls

Law 51. The law of *associativity of function calls* states that a sequence of linear procedure calls can be arbitrarily associated or grouped, i.e.:

$$P \rightarrowtail Q \rightarrowtail S = (P \rightarrowtail Q) \rightarrowtail S = P \rightarrowtail (Q \rightarrowtail S) \tag{60}$$

c) Distributivity of Function Calls

Law 52. The law of *distributivity of function calls* states that a sequential process S can be distributed into a pair of disjunctive conditional processes $P \mid Q$, i.e.:

$$(P \mid Q) \rightarrowtail S = (P \rightarrowtail S) \mid (Q \rightarrowtail S) \tag{61}$$

d) Function Elicitation from Branch Structures

Law 53. The law of *function elicitation from branch structures* states that a common pattern S in both branches of a conditional structure can be elicited and separately encapsulated as a predefined defined procedure, i.e.:

$$\begin{array}{l} \blacklozenge exp\mathbf{BL} = \mathbf{T} \\ \quad \to P \\ \quad \rightarrowtail S \\ \mid \blacklozenge \sim \\ \quad \to Q \\ \quad \rightarrowtail S \\ = (\ \blacklozenge exp\mathbf{BL} = \mathbf{T} \\ \qquad \to P \\ \quad \mid \blacklozenge \sim \\ \qquad \to Q \\ \quad) \\ \quad \rightarrowtail S \end{array} \tag{62}$$

e) Function Elicitation from Switch Structures

Law 54. The law of *function elicitation from switch structures* states that a common pattern S in both branches of a conditional structure can be elicited and separately encapsulated as a predefined procedure, i.e.:

$$\begin{aligned}
\blacklozenge exp\mathbf{N} = \; & 0 \rightarrow (P_0 \rightarrowtail S) \\
& | \; 1 \rightarrow (P_1 \rightarrowtail S) \\
& | \; ... \\
& | \; n \rightarrow (P_n \rightarrowtail S) \\
& | \; \sim \rightarrowtail S \\
= \; \blacklozenge exp\mathbf{N} = \; & 0 \rightarrow P_0 \\
& | \; 1 \rightarrow P_1 \\
& | \; ... \\
& | \; n \rightarrow P_n \\
& | \; \sim \rightarrow \varnothing \\
\rightarrowtail S &
\end{aligned} \tag{63}$$

f) Function Elicitation from Parallel Structures

Law 55. The law of *function elicitation from parallel structures* states that a common pattern *S* in both sides of a parallel structure can be elicited and separately encapsulated as a predefined procedure, i.e.:

$$(P \rightarrowtail S) \,||\, (Q \rightarrowtail S) = (P \,||\, Q) \rightarrowtail S \tag{64}$$

The laws of procedure elicitation from various constructs and process relations are the theoretical foundation of common function or object elicitation, improvement of programming efficiency, and structured programming in software engineering.

g) Representation of Embedded Architectures

Law 56. The law of *nested architecture representation* states that the function-call processes are an abstract representation of recurring nested architectures, *NAs*, of systems, i.e.:

$$NA\mathbf{ST} \triangleq P_1 \rightarrowtail P_2 \rightarrowtail ... \rightarrowtail P_n \tag{65}$$

where each process P_i, $1 \le i \le n$, denotes a nested component in the system.

4.8 Laws of Parallel Processes

The parallel operation of processes is associative, symmetric, transitive, and elicitive, but it is irreflexive and nondistributive. The parallel process operations are constrained by the following laws of software, assuming that all processes belong to and are synchronized in the same system in the parallel structure.

a) Associativity of Parallel Processes

Law 57. The law of *associativity of parallel processes* states that a list of parallel processes can be arbitrarily associated or grouped, i.e.:

$$P \,||\, Q \,||\, S = (P \,||\, Q) \,||\, S = P \,||\, (Q \,||\, S) = P \,||\, (S \,||\, Q) \tag{66}$$

b) Symmetry of Parallel Processes

Law 58. The law of *symmetry of parallel processes* states that parallel process relations are commutative, i.e.:

$$P \parallel Q = Q \parallel P \tag{67}$$

c) Transitivity of Parallel Processes

Law 59. The law of *transitivity of parallel processes* states that parallel process relations are transitive between each other, i.e.:

$$(P \parallel Q) \parallel (Q \parallel S) = P \parallel Q \parallel S \tag{68}$$

d) Elicitivity of Parallel Processes

Law 60. The law of *elicitivity of parallel processes* states that the common process in two groups of parallel processes can be elicited, i.e.:

$$(P \parallel S) \to (Q \parallel S) = (P \to Q) \parallel S = S \parallel (P \to Q) \tag{69}$$

e) Elicitivity of Event

Law 61. The law of *elicitivity of event* in parallel processes states that a common event which triggers different parallel processes is extractive, i.e.:

$$@\, e\mathbf{S} \hookrightarrow P \parallel @\, e\mathbf{S} \hookrightarrow Q = @\, e\mathbf{S} \hookrightarrow (P \parallel Q) \tag{70}$$

f) Idempotency of Identical Parallel Processes

Law 62. The law of *idempotency of identical parallel processes* states that parallel process relation between the same process is idempotent, i.e.:

$$P \parallel P = P \tag{71}$$

g) Representation of Parallel Architectures

Law 63. The law of *parallel architecture representation* states that the parallel processes can be used as an abstract model of recurring parallel architectures, *PAs*, of systems, i.e.:

$$PA\mathbf{ST} \triangleq P_1 \parallel P_2 \parallel \ldots \parallel P_n \tag{72}$$

where each process P_i, $1 \le i \le n$, denotes a parallel component in the system.

4.9 Laws of Concurrent Processes

The concurrent operation of processes is associative, symmetric, transitive, and elicitive, but it is irreflexive and nondistributive. The concurrent process operations are constrained by the following laws of software. The differences between concurrent and parallel processes are that the former are implemented and executed on separated machines or they are asynchronized in a distributed environment.

a) Associativity of Concurrent Processes

Law 64. The law of *associativity of* concurrent *processes* states that a list of concurrent processes can be arbitrarily associated or grouped, i.e.:

$$P \oiint Q \oiint S = (P \oiint Q) \oiint S = P \oiint (Q \oiint S) = (P \oiint S) \oiint Q \tag{73}$$

b) Symmetry of Concurrent Processes

Law 65. The law of *symmetry of concurrent processes* states that concurrent process relations are commutative, i.e.:

$$P \oiint Q = Q \oiint P \tag{74}$$

c) Transitivity of Concurrent Processes

Law 66. The law of *transitivity of concurrent processes* states that concurrent process relations are transitive between each other, i.e.:

$$(P \oiint Q) \oiint (Q \oiint S) = P \oiint Q \oiint S \tag{75}$$

d) Elicitivity of Concurrent Processes

Law 67. The law of *elicitivity of concurrent processes* states that the common process in two groups of concurrent processes can be elicited, i.e.:

$$(P \oiint S) \to (Q \oiint S) = (P \to Q) \oiint S = S \oiint (P \to Q) \tag{76}$$

e) Elicitivity of Event

Law 68. The law of *elicitivity of event* in concurrent processes states that a common event which triggers different concurrent processes is extractive, i.e.:

$$@e\mathbf{S} \hookrightarrow P \oiint @e\mathbf{S} \hookrightarrow Q = @e\mathbf{S} \hookrightarrow (P \oiint Q) \tag{77}$$

f) Idempotency of Identical Concurrent Processes

Law 69. The law of *idempotency of identical concurrent processes* states that concurrent process relation between the same process is idempotent, i.e.:

$$P \oiint P = P \tag{78}$$

4.10 Laws of Interleave Processes

The interleave operation of processes is associative, symmetric, transitive, and elicitive, but it is irreflexive and nondistributive. The interleave process operations are constrained by the following laws of software.

a) Associativity of Interleave Processes

Law 70. The law of *associativity of interleave processes* states that a list of interleaved processes can be arbitrarily associated or grouped, i.e.:

$$P \mid\mid\mid Q \mid\mid\mid S = (P \mid\mid\mid Q) \mid\mid\mid S = P \mid\mid\mid (Q \mid\mid\mid S) = P \mid\mid\mid (S \mid\mid\mid Q) \tag{79}$$

b) Symmetry of Interleave Processes

Law 71. The law of *symmetry of interleave processes* states that interleaved process relations are commutative, i.e.:

$$P \mid\mid\mid Q = Q \mid\mid\mid P \tag{80}$$

c) Transitivity of Interleave Processes

Law 72. The law of *transitivity of interleave processes* states that interleaved process relations are transitive between each other, i.e.:

$$(P \mid\mid\mid Q) \mid\mid\mid (Q \mid\mid\mid S) = P \mid\mid\mid Q \mid\mid\mid S \tag{81}$$

d) Elicitivity of Interleave Processes

Law 73. The law of *elicitivity of interleave processes* states that the common process in two groups of interleaved processes can be elicited, i.e.:

$$(P \mid\mid\mid S) \to (Q \mid\mid\mid S) = (P \to Q) \mid\mid\mid S = S \mid\mid\mid (P \to Q) \tag{82}$$

4.1.1 Laws of Pipeline Processes

The pipeline operation of processes is associative and elicitive, but it is irreflexive, asymmetric, intransitive, and nondistributive. The pipeline process operations are constrained by the following laws of software.

a) Associativity of Pipeline Processes

Law 74. The law of *associativity of pipeline processes* states that a list of interlinked processes can be arbitrarily associated or grouped, i.e.:

$$P \gg Q \gg S = (P \gg Q) \gg S = P \gg (Q \gg S) = P \gg (S \gg Q) \tag{83}$$

b) Elicitivity of Pipeline Processes

Law 75. The law of *elicitivity of interleave processes* states that the common process in two groups of interleaved processes can be elicited, i.e.:

$$(S \to P) \gg (S \to Q) = S \to (P \gg Q) \tag{84}$$

or

$$(P \to S) \gg (Q \to S) = (P \gg Q) \to S \tag{85}$$

c) Pairwise Coupling of Pipeline Processes

Law 76. The law of *pairwise coupling of pipeline processes* states that each of the outputs of process P, O_{Pi}, $1 \le i \le n$, is connected to the counterpart process Q's input, I_{Qi}, i.e.:

$$\underset{i\mathbf{N}=1}{\overset{n\mathbf{N}}{R}} (O_{Pi} = I_{Qi}) \tag{86}$$

where O_P and I_Q are the outputs and inputs of processes P and Q, and $\#O_P = \#I_Q$.

d) Representation of Coupled Architectures

Law 77. The law of *coupled architecture representation* states that the pipeline processes can be used as an abstract model of recurring pairwise coupled architectures, *CAs*, of systems, i.e.:

$$\textit{CA}\mathbf{ST} \triangleq P_1 \gg P_2 \gg \ldots \gg P_n \tag{87}$$

where each process P_i , $1 \le i \le n$, denotes a component in the system.

It is noteworthy that Laws 33, 46, 56, 63, and 77 provide a set of five laws for basic system architectures in system modeling known as the serial, recurrent (CLM), nested, parallel, and coupled structures, respectively. These laws also indicate that not only system behaviors but also system architectures can be modeled by the RTPA process relational operations [32], [39], [42].

4.1.2 Laws of Interrupt Processes

The interrupt operation of processes is dissociative, irreflexive, asymmetric, intransitive, nondistributive, and nonelicitive. However, interrupt process operations are constrained by the following special laws.

a) Parallel Mechanism of Interrupt Processes

Law 78. The law of *parallel mechanism of interrupt processes* states that an interrupt service process Q is parallel to the main process P, which is triggered by the ith interrupt event $@int_i$⊙, i.e.:

$$P \not\!\!\!\rightarrow Q \triangleq P \,||\, @int_i\odot \nearrow Q_i \searrow \odot \tag{88}$$

where ↗ and ↘ denote an interrupt service and an interrupt return, respectively.

b) Hierarchy of Interrupt Priorities

Law 79. The law of *hierarchy of interrupt priorities* states that multiple interrupt resources interacting with a computing system can be configured at different levels of priorities l, i.e.:

$$\mathop{R}_{l\mathbf{N}=1}^{n\mathbf{N}} int_l \hookrightarrow Q_l \tag{89}$$

c) Maximum Duration of Interrupt Services

Law 80. The law of *the maximum duration of interrupt service* states that the duration of an interrupt process in $P \not\!\!\!\rightarrow Q$ should not exceed the basic time slice of system dispatching $\S t_d$, i.e.:

$$t_{\text{int}} = t_Q < \S t_{\text{d}} \tag{90}$$

4.1.3 Laws of Time Dispatch Processes

The time dispatch operation of processes at the system level is symmetric and elicitive, but it is dissociative, ireflexive, intransitive, and nondistributive. The time-dispatch process operations are constrained by the following laws of software.

a) Elicitivity of Timing Event

Law 81. The law of *elicitivity of timing events* in system dispatch states that a common timing event which triggers two different dispatches can be elicited and the two relational processes can be joined, i.e.:

$$(@\, t\mathbf{TM} \hookrightarrow P)\ \mathbb{R}\ (@\, t\mathbf{TM} \hookrightarrow Q) = @\, t\mathbf{TM} \hookrightarrow (P \mathbb{R} Q) \tag{91}$$

where $\mathbb{R} \in \{\rightarrow, \|, ∯, |||, »\}$.

b) Elicitivity of Process in Timing Dispatches

Law 82. The law of *elicitivity of timing dispatch processes* states that a common process S can be elicited from a pair of time-driven dispatch processes, i.e.:

$$\begin{aligned} & S \rightarrow (@\, t_1\mathbf{TM} \hookrightarrow P) \mid S \rightarrow (@\, t_2\mathbf{TM} \hookrightarrow Q) \\ & = S \rightarrow (@\, t_1\mathbf{TM} \hookrightarrow P \mid @\, t_2\mathbf{TM} \hookrightarrow Q) \end{aligned} \tag{92}$$

$$\begin{aligned} & (@\, t_1\mathbf{TM} \hookrightarrow P) \rightarrow S \mid (@\, t_2\mathbf{TM} \hookrightarrow Q) \rightarrow S \\ & = (@\, t_1\mathbf{TM} \hookrightarrow P \mid @\, t_2\mathbf{TM} \hookrightarrow Q) \rightarrow S \end{aligned} \tag{93}$$

c) Symmetry of Timing Dispatches

Law 83. The law of *symmetry of timing dispatching processes* states that a time-driven dispatch is symmetric or commutative, i.e.:

$$@\, t_1\mathbf{TM} \hookrightarrow P \mid @\, t_2\mathbf{TM} \hookrightarrow Q = @\, t_2\mathbf{TM} \hookrightarrow Q \mid @\, t_1\mathbf{TM} \hookrightarrow P \tag{94}$$

d) Skew Symmetry of Timing Dispatches

Law 84. The law of *skew symmetric of timing dispatch* states that a time-driven dispatch is symmetric or commutative on a pair of complemented events, i.e.:

$$@\, t\mathbf{TM} \hookrightarrow P \mid @\, \overline{t\mathbf{TM}} \hookrightarrow Q = @\, \overline{t\mathbf{TM}} \hookrightarrow Q \mid @\, t\mathbf{TM} \hookrightarrow P \tag{95}$$

f) Idempotency of Timing Dispatches

Law 85. The law of *idempotency of timing dispatch* states that a time-driven dispatch can be omitted if it is unconditional, i.e.:

$$@\, t\mathbf{TM} \hookrightarrow P \mid @\, \overline{t\mathbf{TM}} \hookrightarrow P = P \tag{96}$$

4.1.4 Laws of Event Dispatch Processes

The event dispatch operation of processes at the system level is symmetric and elicitive, but it is dissociative, ireflexive, intransitive, and nondistributive. The event-dispatch process operations are constrained by the following laws of software.

(a) Elicitivity of Operating Event

Law 86. The law of *elicitivity of operating events* in system dispatch states that a common event which triggers two different dispatches can be elicited and the two relational processes can be joined, i.e.:

$$(@\, e\mathbf{\$} \hookrightarrow P)\ \mathbb{R}\ (@\, e\mathbf{\$} \hookrightarrow Q) = @\, e\mathbf{\$} \hookrightarrow (P \mathbb{R} Q) \tag{97}$$

where $\mathbb{R} \in \{\rightarrow, \|, \oiint, |||, \gg\}$.

b) Elicitivity of Process in Event Dispatches

Law 87. The law of *elicitivity of event dispatch* states that a common process S can be elicited from a pair of event dispatch processes, i.e.:

$$\begin{aligned} & S \rightarrow (@\, e_1\mathbf{\$} \hookrightarrow P) \mid S \rightarrow (@\, e_2\mathbf{\$} \hookrightarrow Q) \\ & = S \rightarrow (@\, e_1\mathbf{\$} \hookrightarrow P \mid @\, e_2\mathbf{\$} \hookrightarrow Q) \end{aligned} \tag{98}$$

$$\begin{aligned} & (@\, e_1\mathbf{\$} \hookrightarrow P) \rightarrow S \mid (@\, e_2\mathbf{\$} \hookrightarrow Q) \rightarrow S \\ & = (@\, e_1\mathbf{\$} \hookrightarrow P \mid @\, e_2\mathbf{\$} \hookrightarrow Q) \rightarrow S \end{aligned} \tag{99}$$

c) Symmetry of Event Dispatches

Law 88. The law of *symmetry of event dispatching processes* states that an event-driven dispatch is symmetric or commutative, i.e.:

$$@\, e_1\mathbf{\$} \hookrightarrow P \mid @\, e_2\mathbf{\$} \hookrightarrow Q = @\, e_2\mathbf{\$} \hookrightarrow Q \mid @\, e_1\mathbf{\$} \hookrightarrow P \tag{100}$$

d) Skew Symmetry of Event Dispatches

Law 89. The law of *skew symmetric of event dispatch* states that an event-driven dispatch is symmetric or commutative on a pair of complemented events, i.e.:

$$@\, e\mathbf{\$} \hookrightarrow P \mid @\, \overline{e}\mathbf{\$} \hookrightarrow Q = @\, \overline{e}\mathbf{\$} \hookrightarrow Q \mid @\, e\mathbf{\$} \hookrightarrow P \tag{101}$$

f) Idempotency of Event Dispatches

Law 90. The law of *idempotency of event dispatch* states that an event-driven dispatch can be omitted if it is unconditional, i.e.:

$$@\, e\mathbf{\$} \hookrightarrow P \mid @\, \overline{e}\mathbf{\$} \hookrightarrow P = P \tag{102}$$

4.1.5 Laws of Interrupt Dispatch Processes

The interrupt dispatch operation of processes at the system level is symmetric and elicitive, but it is dissociative, ireflexive, intransitive, and nondistributive. The interrupt-dispatch process operations are constrained by the following laws of software.

a) Elicitivity of Interrupt Event

Law 91. The law of *elicitivity of interrupt events* in system dispatch states that a common interrupt which triggers two different dispatches can be elicited and the two relational processes can be joined, i.e.:

$$(@\, int\odot \hookrightarrow P)\ \mathbb{R}\ (@\, int\odot \hookrightarrow Q) = @\, int\odot \hookrightarrow (P \mathbb{R} Q) \tag{103}$$

where $\mathbb{R} \in \{\rightarrow, \|, \oiint, |||, \gg\}$.

b) Elicitivity of Process in Interrupt Dispatches

Law 92. The law of *elicitivity of interrupt dispatch processes* states that a common process *S* can be elicited from a pair of interrupt-driven dispatch processes, i.e.:

$$\begin{aligned} &S \rightarrow (@\, int_1 \odot \hookrightarrow P) \mid S \rightarrow (@\, int_2 \odot \hookrightarrow Q) \\ &= S \rightarrow (@\, int_1 \odot \hookrightarrow P \mid @\, int_2 \odot \hookrightarrow Q) \end{aligned} \tag{104a}$$

$$\begin{aligned} &(@\, int_1 \odot \hookrightarrow P) \rightarrow S \mid (@\, int_2 \odot \hookrightarrow Q) \rightarrow S \\ &= (@\, int_1 \odot \hookrightarrow P \mid @\, int_2 \odot \hookrightarrow Q) \rightarrow S \end{aligned} \tag{104b}$$

c) Symmetry of Interrupt Dispatches

Law 93. The law of *symmetry of interrupt dispatching processes* states that an interrupt-driven dispatch is symmetric or commutative, i.e.:

$$@\, int_1 \odot \hookrightarrow P \mid @\, int_2 \odot \hookrightarrow Q = @\, int_2 \odot \hookrightarrow Q \mid @\, int_1 \odot \hookrightarrow P \tag{105}$$

d) Skew Symmetry of Interrupt Dispatches

Law 94. The law of *skew symmetric of interrupt dispatch* states that an interrupt-driven dispatch is symmetric or commutative on a pair of complemented events, i.e.:

$$@\, int \odot \hookrightarrow P \mid @\, \overline{int} \odot \hookrightarrow Q = @\, \overline{int} \odot \hookrightarrow Q \mid @\, int \odot \hookrightarrow P \tag{106}$$

e) Idempotency of Interrupt Dispatches

Law 95. The law of *idempotency of interrupt dispatch* states that an interrupt-driven dispatch can be omitted if it is unconditional, i.e.:

$$@\, int \odot \hookrightarrow P \mid @\, \overline{int} \odot \hookrightarrow P = P \tag{107}$$

5 Conclusions

The exploration on the nature of software and its fundamental behaviors constrained by the laws of mathematics, cognitive informatics, system science, and formal linguistics are a profound effort in computing and software engineering. This paper has presented the mathematical laws of software and fundamental computing behaviors on the basis of the generic mathematical model of programs and RTPA. A comprehensive set of 95 algebraic laws in the categories of meta-processes, process relations, and system compositions has been systematically established, which lays a theoretical foundation for analyzing and modeling software behaviors and software system architectures. Supplementary to the algebraic laws of processes and process relations as presented in this paper, the cognitive informatics laws [33], [34], [36], [40], system laws [39], formal linguistic and semantic laws [6], [36], [39], [43] of software may be referred to given literature.

The applications of the mathematical laws of software and the generic mathematical model of programs have provided new perspectives on software engineering foundations and practices. An RTPA type checker and an RTPA code generator have been implemented based on the algebraic laws, which automatically generates code in C++ or Java based on a formal system model in RTPA [29], [30]. A number of real-world software systems have been formally modeled in RTPA based on the algebraic laws, such as the telephone switching system [35], [39], the lift dispatching system [49], the real-time operating system [50], and the ATM system [51]. The mathematical laws of software and RTPA are not only useful for rigorously modeling and manipulating software systems, but also widely applied in human cognitive process modeling and computational intelligence [35], [39].

Acknowledgements. The author would like to acknowledge the Natural Science and Engineering Council of Canada (NSERC) for its partial support to this work. The author would like to thank the valuable comments and suggestions of the anonymous reviewers.

References

1. Aho, A.V., Sethi, R., Ullman, J.D.: Compilers: Principles, Techniques, and Tools, New York. Addison-Wesley Publication Co, Reading (1985)
2. Baeten, J.C.M., Bergstra, J.A.: Real Time Process Algebra. Formal Aspects of Computing 3, 142–188 (1991)
3. Boole, G.: The Laws of Thought, Prometheus Books, NY (1854) (reprint, 2003)
4. Boucher, A., Gerth, R.: A Timed Model for Extended Communicating Sequential Processes. In: Ottmann, T. (ed.) ICALP 1987. LNCS, vol. 267. Springer, Heidelberg (1987)
5. Cardelli, L., Wegner, P.: On Understanding Types, Data Abstraction and Polymorphism. ACM Computing Surveys 17(4), 471–522 (1985)
6. Chomsky, N.: Three Models for the Description of Languages. I.R.E. Transactions on Information Theory 2(3), 113–124 (1956)
7. Chomsky, N.: On Certain Formal Properties of Grammars. Information and Control 2, 137–167 (1959)
8. Dierks, H.: A Process Algebra for Real-Time Programs. In: Maibaum, T.S.E. (ed.) ETAPS 2000 and FASE 2000. LNCS, vol. 1783, pp. 66–76. Springer, Heidelberg (2000)
9. Dijkstra, E.W.: A Discipline of Programming. Prentice Hall, Englewood Cliffs (1976)
10. Fecher, H.: A Real-Time Process Algebra with Open Intervals and Maximal Progress. Nordic Journal of Computing 8(3), 346–360 (2001)
11. Goguen, J.A., Thatcher, J.W., Wagner, E.G., Wright, J.B.: Initial Algebra Semantics and Continuous Algebras. Journal of the ACM 24(1), 59–68 (1977)
12. Higman, B.: A Comparative Study of Programming Languages, 2nd edn. MacDonald (1977)
13. Hoare, C.A.R.: An Axiomatic Basis for Computer Programming. Communications of the ACM 12(10), 576–580 (1969)
14. Hoare, C.A.R.: Communicating Sequential Processes. Communications of the ACM 21(8), 666–677 (1978)
15. Hoare, C.A.R.: Communicating Sequential Processes. Prentice-Hall International, London (1985)
16. Hoare, C.A.R., Hayes, I.J., He, J., Morgan, C.C., Roscoe, A.W., Sanders, J.W., Sorensen, I.H., Spivey, J.M., Sufrin, B.A.: Laws of Programming. Communications of he ACM 30(8), 672–686 (1987)

17. Klusener, A.S.: Abstraction in Real Time Process Algebra. In: Huizing, C., de Bakker, J.W., Rozenberg, G., de Roever, W.-P. (eds.) REX 1991. LNCS, vol. 600, pp. 325–352. Springer, Heidelberg (1992)
18. Louden, K.C.: Programming Languages: Principles and Practice. PWS-Kent Publishing Co., Boston (1993)
19. Martin-Lof, P.: An Intuitionistic Theory of Types: Predicative Part. In: Rose, H., Shepherdson, J.C. (eds.) Logic Colloquium 1973. North-Holland, Amsterdam (1975)
20. Milner, R.: A Calculus of Communicating Systems. LNCS, vol. 92. Springer, Heidelberg (1980)
21. McDermid, J.A. (ed.): Software Engineer's Reference Book. Butterworth-Heinemann Ltd., Oxford (1991)
22. Mitchell, J.C.: Type systems for programming languages. In: van Leeuwen, J. (ed.) Handbook of Theoretical Computer Science, pp. 365–458. North Holland, Amsterdam (1990)
23. Nicollin, X., Sifakis, J.: An Overview and Synthesis on Timed Process Algebras. In: Larsen, K.G., Skou, A. (eds.) CAV 1991. LNCS, vol. 575, pp. 376–398. Springer, Heidelberg (1992)
24. Parnas, D.L., Clements, P.C.: A Rational Design Process: How and Why to Fake It. IEEE Trans. on Software Engineering 12(2), 251–257 (1986)
25. Reed, G.M., Roscoe, A.W.: A Timed model for Communicating Sequential Processes. In: Kott, L. (ed.) ICALP 1986. LNCS, vol. 226. Springer, Heidelberg (1986)
26. Scott, D.S., Strachey, C.: Towards a Mathematical Semantics for Computer Languages, Programming Research Group Technical Report PRG-1-6, Oxford University (1971)
27. Schneider, S.A.: An Operational Semantics for Timed CSP, Programming Research Group Technical Report TR-1-91, Oxford University (1991)
28. Stubbs, D.F., Webre, N.W.: Data Structures with Abstract Data Types and Pascal. Brooks/Cole Publishing Co., Monterey (1985)
29. Tan, X., Wang, Y., Ngolah, C.F.: A Novel Type Checker for Software System Specifications in RTPA. In: Proc. 17th Canadian Conference on Electrical and Computer Engineering (CCECE 2004), Niagara Falls, ON, Canada, pp. 1549–1552. IEEE CS Press, Los Alamitos (2004)
30. Tan, X., Wang, Y., Ngolah, C.F.: Design and Implementation of an Automatic RTPA Code Generator. In: Proc. 19th Canadian Conference on Electrical and Computer Engineering (CCECE 2006), Ottawa, ON, Canada, pp. 1605–1608 (May 2006)
31. Tarski, A.: The Semantic Conception of Truth. Philosophic Phenomenological Research 4, 13–47 (1944)
32. Wang, Y.: The Real-Time Process Algebra (RTPA). Annals of Software Engineering: A International Journal 14, 235–274 (2002)
33. Wang, Y.: On Cognitive Informatics (Keynote Speech). In: Proc. 1st IEEE International Conference on Cognitive Informatics (ICCI 2002), Calgary, Canada, pp. 34–42. IEEE CS Press, Los Alamitos (2002)
34. Wang, Y.: On Cognitive Informatics. Brain and Mind: A Transdisciplinary Journal of Neuroscience and Neurophilosophy, USA 4(3), 151–167 (2003)
35. Wang, Y.: Using Process Algebra to Describe Human and Software System Behaviors. Brain and Mind 4(2), 199–213 (2003)
36. Wang, Y.: On the Informatics Laws and Deductive Semantics of Software. IEEE Transactions on Systems, Man, and Cybernetics (C) 36(2), 161–171 (2006)
37. Wang, Y.: Keynote: Cognitive Informatics - Towards the Future Generation Computers that Think and Feel. In: Proc. 5th IEEE International Conference on Cognitive Informatics (ICCI 2006), Beijing, China, pp. 3–7. IEEE CS Press, Los Alamitos (2006)

38. Wang, Y.: Cognitive Informatics and Contemporary Mathematics for Knowledge Representation and Manipulation (Invited Plenary Talk). In: Wang, G.-Y., Peters, J.F., Skowron, A., Yao, Y. (eds.) RSKT 2006. LNCS (LNAI), vol. 4062, pp. 69–78. Springer, Heidelberg (2006)
39. Wang, Y.: Software Engineering Foundations: A Software Science Perspective. CRC Series in Software Engineering, vol. II. Auerbach Publications, NY, USA (2007)
40. Wang, Y.: The Theoretical Framework of Cognitive Informatics. International Journal of Cognitive Informatics and Natural Intelligence 1(1), 1–27 (2007)
41. Wang, Y.: Keynote: On Theoretical Foundations of Software Engineering and Denotational Mathematics. In: Proc. 5th Asian Workshop on Foundations of Software, Xiamen, China, pp. 99–102 (2007)
42. Wang, Y.: On the Big-R Notation for Describing Iterative and Recursive Behaviors. International Journal of Cognitive Informatics and Natural Intelligence 2(1), 17–28 (2008)
43. Wang, Y.: Deductive Semantics of RTPA. International Journal of Cognitive Informatics and Natural Intelligence 2(2), 95–121 (2008)
44. Wang, Y.: On Concept Algebra: A Denotational Mathematical Structure for Knowledge and Software Modeling. International Journal of Cognitive Informatics and Natural Intelligence 2(2), 1–19 (2008)
45. Wang, Y.: On System Algebra: A Denotational Mathematical Structure for Abstract System modeling. International Journal of Cognitive Informatics and Natural Intelligence 2(2), 20–42 (2008)
46. Wang, Y.: RTPA: A Denotational Mathematics for Manipulating Intelligent and Computational Behaviors. International Journal of Cognitive Informatics and Natural Intelligence 2(2), 44–62 (2008)
47. Wang, Y.: On Laws of Work Organization in Human Cooperation. International Journal of Cognitive Informatics and Natural Intelligence 1(2), 1–15 (2008)
48. Wang, Y.: On Contemporary Denotational Mathematics for Computational Intelligence. In: Gavrilova, M.L., et al. (eds.) Transactions on Computational Science, II. LNCS, vol. 5150, pp. 6–29. Springer, Heidelberg (2008)
49. Wang, Y., Noglah, C.F.: Formal Specification of a Real-Time Lift Dispatching System. In: Proc. 2002 IEEE Canadian Conference on Electrical and Computer Engineering (CCECE 2002), Winnipeg, Manitoba, Canada, pp. 669–674 (May 2002)
50. Wang, Y., Noglah, C.F.: Formal Description of Real-Time Operating Systems using RTPA. In: Proc. 2003 Canadian Conference on Electrical and Computer Engineering (CCECE 2003), Montreal, Canada, pp. 1247–1250. IEEE CS Press, Los Alamitos (2003)
51. Wang, Y., Zhang, Y.: Formal Description of an ATM System by RTPA. In: Proc. 16th Canadian Conference on Electrical and Computer Engineering (CCECE 2003), Montreal, Canada, pp. 1255–1258. IEEE CS Press, Los Alamitos (2003)
52. Wilson, L.B., Clark, R.G.: Comparative Programming Language. Addison-Wesley Publishing Co, Reading (1988)
53. Woodcock, J., Davies, J.: Using Z: Specification, Refinement, and Proof. Prentice Hall International, London (1996)

Rough Logic and Its Reasoning

Qing Liu and Lan Liu

Department of Computer Science
Nanchang University, Nanchang 330031, China
qliu_ncu@yahoo.com.cn

Abstract. In this article, Rough Logic is defined as a nonstandard logic on a given information system $IS = (U, A)$. Atomic formulae of the logic are defined as $a = v$ or a_v. It is interpreted as $a(x) = v$, where $a \in A$ is an attribute in A, x is an individual variable on U, and v is an attribute value. The compound formula consist of the atomic formulae and logical connectives. Semantics of the logic is discussed. Truth value of the rough logic is defined as a ratio of the number of elements satisfying the logical formula to the total of elements on U. Deductive reasoning and resolution reasoning are also studied. The rough logic will offer a new idea for the applications to classical logic and other nonstandard logic.

Keywords: Atomic formula, Well-formed formula, Deductive reasoning, Resolution reasoning.

1 Introduction

Since Pawlak proposed Rough Sets in 1982, many computer scientists and logician tried to create a Rough Logical Theory, to implement approximate reasoning and problem solving in artificial intelligence with it. Pawlak proposed Rough Logic and Decision Logic in 1987 and in 1991 respectively[1,2]. The former set up five truth values: true, false, roughly true, roughly false and roughly inconsistent; The latter is based on the information table, it is essentially a special case in two-valued logic. Orlowska proposed a Logic of Indiscernibility Relation[3] in 1985, that is, to add an indiscernibility relation predicate in classical logic, to quote new concept for classical logic. In 1993, Charaborty proposed a Rough Logic with Rough Quantifier[4], and to create the logical tools of approximate reasoning. In 1994, Nakamura proposed information logic, a rough logic of incomplete knowledge and grade modal logic[5−7]. Author in this article bend oneself to study the rough logic defined on a given information system for long time, and to discuss approximate reasoning based on the logic, proposed the resolution principle, OI-resolution strategy of the logic, and resolution reasoning of rough proposition logic with lower and upper approximate operators[8−14]. Based on previous work, we study further the rough logic on a given information system, discuss multi-valued and λ-resolution strategy of the logic. The rough logic is different from decision logic by Pawlak. The former is multi-valued logical system defined in a given information system, and the latter is two-valued logical

M.L. Gavrilova et al. (Eds.): Trans. on Comput. Sci. II, LNCS 5150, pp. 84–99, 2008.

system defined in a given information system. The rough logic is also different from other multi-valued logic, the latter is no limited to define in a given information system. So, rough logic in this article is a narrow logical system, but to be fitted to applications in practice. Final, we research approximate reasoning of the rough logic, and it is also illustrated with the real examples defined on information systems.

The significance of studying the logic is to offer a new idea for the applications in classical logic, to stride forward a step for research of "Approximate Proof" by logician Hao Wang forty years ago, to offer new theoretical tools for describing and processing to nonstandard knowledge.

In the article, basic concepts of the rough logic is described in Section 2. Related properties of the rough logic are given in Section 3. Clause forms of the logic are also discussed in Section 4. Deductive reasoning of the logic is studied in Section 5. Resolution principles and strategies of the logic are studied in Section 6. Applications for problem resolving in AI to the logic are presented in Section 7. Final Section is the conclusion of perspective of studying to the logic.

2 Basic Concepts of Rough Logic

2.1 Well-Formed Formulae in the Rough Logic

Rough logic is defined as a nonstandard logic on a given information system $IS = (U, A)$. Atomic formulae of the logic are defined as $a = v$ or a_v. It is interpreted as $a(x) = v$, where $a \in A$ is an attribute in A, x is an individual variable on U, and v is an attribute value. The compound formulae consist of the atomic formulae and usual logical connectives. The rough logic is abbreviated as RL_{IS}. Its truth value is multi-valued on IS.

Definition 1. Well-formed formulas (wffs) of the rough logic are recursively defined as follows[8−20]:

1. All atoms of form a_v are wffs in RL_{IS}, where $a \in A$, $v \in V_a$;
2. Let F_1, F_2 be wffs in RL_{IS}, then $\neg F_1$, $F_1 \vee F_2$, $F_1 \wedge F_2$, $F_1 \rightarrow F_2$ and $F_1 \leftrightarrow F_2$ are also wffs in RL_{IS};
3. The obtainable formulae what (1) – (2) are quoted finite times are wffs in RL_{IS}.

2.2 Interpretation and Assignment in the Rough Logic

Let $IS = (U, A)$ be a given information system. I is an interpretation to individual constant, function and predicate to occur in the formula, and u is an assignment to individual variable occurring in the formula. T_{I_u} is a united assignment symbol to formula F in interpretation I and assignment u. We have the following rules:

1. If $\bullet$ is an individual constant in formula $F(\bullet)$, then $T_{I_u}(\bullet) = I(c) = e$, where c is an individual constant symbol in U, e is an entity on U;

2. If $\bullet$ is an individual variable in formula $F(\bullet)$, then $T_{I_u}(\bullet) = u(x) = e$, where x is an individual variable symbol on U, e is an entity on U;
3. If $\bullet$ is a n-place function of form $\pi(\bullet_1, \cdots, \bullet_n)$, then $I(\pi) = f$, $I(\bullet_i) = x_i$, and $T_{I_u}(\bullet) = f(x_1, \cdots, x_n)$, where f is a mapping symbol from U^n to U;
4. If $\bullet$ is a n-place predicate of form $\rho(\bullet_1, \cdots, \bullet_n)$, then $I(\rho) = P$, $I(\bullet_i) = x_i$, and $T_{I_u}(\bullet) = P(x_1, \cdots, x_n)$, where P is a relation symbol on U.

Noteworthiness, the predicates to occur in rough logical formulae are the attribute in a given information system. Therefore, $a_5 \rightarrow c_0$ in the following table 1 is a rough logical formula defined in the given information system, where predicate symbol a and c are the attribute in A on IS [8,13,20–22]

Table 1. Information Table

U	a	b	c	d	e
1	5	4	0	1	1
2	3	4	0	2	1
3	3	4	0	2	2
4	0	2	0	1	2
5	3	2	1	2	2
6	5	2	1	1	0

This information table in the above look to be very simple, because the attribute values in this table are numeric. Despite an information system from in practice is far more complicated than this table, but we transform always it into the simple information table of attribute value being number[27,46]. "Everything is a matter of numbers", that is, all existing things can be viewed as relationships of numbers in the final analysis. This famous axiom was proposed by Pythagoras, a philosopher and mathematician of ancient Greece[27]. So, we could transform a complicated information system into simple information table according to Pythagoras's assertion, as shown in example 4 in Section 7.

2.3 Truth Values of Rough Logical Formulae

Semantics of the rough logical formula $F(\bullet)$ defined in a given information system $IS = (U, A)$ is a subset on U, denoted by

$$m(F(\bullet)) = \{x \in U : x \models_{IS} F(\bullet)\}$$

where $\models_{IS}$ is a satisfiable symbol on IS. Truth values of the rough logical formulae are defined as follows:

Definition 2. *Let $F_1, F_2 \in RL_{IS}$ be rough logical formula defined in a given information system IS. I and u are an interpretation and an assignment to the formula defined in IS. Thus the truth values are computed by the following rules*

1. *Let T_{I_u} be a united assignment symbol to rough logical formula. It is defined as follows:*

$$T_{I_u}(\bullet) = \lambda$$

where $\lambda \in [0, 1]$ is a real number. I is an interpretation symbol, u is an assignment symbol.

2. $T_{I_u}(F_1) = K(m(F_1))/K(U)$, *$K(\bullet)$ denotes the base number of set $\bullet$. $m(\bullet)$ is a set from the propositional variable $\bullet$ or the formula $\bullet$ to a subset on U;*
3. $T_{I_u}(\neg F_1) = 1 - T_{I_u}(F_1)$;
4. $T_{I_u}(F_1 \vee F_2) = max\{T_{I_u}(F_1), T_{I_u}(F_2)\}$;
5. $T_{I_u}(F_1 \wedge F_2) = min\{T_{I_u}(F_1), T_{I_u}(F_2)\}$;
6. $T_{I_u}((\forall x)F_1(x)) = min\{T_{I_u}(F_1(e_1)), \cdots, T_{I_u}(F_1(e_n))\}$, *where x is an individual variable on U. e_i is a value of x on U. It is an entity(concrete studying object) on U.*

2.4 Semantics Model of Rough Logic

Semantics model of the rough logic is denoted by 6-tuple

$$M = (U, A, T_{I_u}, I, u, m)$$

- U is a universe of discourse objects on $IS = (U, A)$.
- A is a set of attributes on IS.
- T_{I_u} is a united assignment symbol to individual constant, individual variable, function and predicate to occur in the formula.
- I is an interpretation to individual constant, function and predicate to occur in the formula.
- u is an assignment to individual variable to occur in the formula.
- m is a mapping from the formula defined on IS to the subset on U.

Given a model M, formula $F(x) \in RL_{IS}$ is satisfiable in the model M, denoted by

$$M, u \models_{IS} F(x)$$

Definition 3. *Meaning of rough logical formulae $F_1, F_2 \in RL_{IS}$ with respect to logical connectives $\neg, \vee, \wedge, \rightarrow, \leftrightarrow$ and quantifier $\forall$ are defined recursively as follows:*

1. $M, u \models_{IS_\lambda} P(x_1, \cdots, x_n) = T_{I_u}(P(u(x_1), \cdots, P(u(x_n))$,
   where $\lambda > 0.5$, $P(x_1, \cdots, x_n)$ is satisfiable to degree at least λ on IS:
2. $M, u \models_{IS_\lambda} \neg F_1 = 1 - M, u \models_{IS_\lambda} F_1$;
3. $M, u \models_{IS_\lambda} (F_1 \vee F_2) = M, u \models_{IS_\lambda} F_1 \vee M, u \models_{IS_\lambda} F_2$;
4. $M, u \models_{IS_\lambda} (F_1 \wedge F_2) = M, u \models_{IS_\lambda} F_1 \wedge M, u \models_{IS_\lambda} F_2$;
5. $M, u \models_{IS_\lambda} (F_1 \rightarrow F_2) = M, u \models_{IS_\lambda} \neg F_1 \vee M, u \models_{IS_\lambda} F_2$;
6. $M, u \models_{IS_\lambda} (F_1 \leftrightarrow F_2) = (M, u \models_{IS_\lambda} \neg F_1 \vee M, u \models_{IS_\lambda} F_2) \wedge (M, u \models_{IS_\lambda} \neg F_2 \vee M, u \models_{IS_\lambda} F_1)$;
7. $M, u \models_{IS_\lambda} (\forall x)F_1(x) = M, u \models_{IS_\lambda} F_1(x_1) \wedge \cdots \wedge M, u \models_{IS_\lambda} F_1(x_n)$.

3 Related Properties of Rough Logic

The rough logical formulae have the following properties[23,29]:

1. Let $F_1, F_2 \in RL_{IS}$ be rough logical formula. Operations of the meaning $m(F_1)$ and $m(F_2)$ corresponding to them on usual connectives $\neg, \vee, \wedge, \rightarrow, \leftrightarrow$ are denoted by[2,21]:
   - $m(\neg F_1) = U - m(F_1)$;
   - $m(F_1 \vee F_2) = m(F_1) \cup m(F_2)$;
   - $m(F_1 \wedge F_2) = m(F_1) \cap m(F_2)$;
   - $m(F_1 \rightarrow F_2) = m(\neg F_1) \cup m(F_2)$;
   - $m(F_1 \leftrightarrow F_2) = (m(\neg F_1) \cup m(F_2)) \cap (m(\neg F_2) \cup m(F_1))$.
2. $\forall F \in RL_{IS}$, if F is true to degree at least λ on IS, then it is denoted by
$$\models_{IS_\lambda} F$$
3. $\forall F \in RL_{IS}$, α and β are individual constant, or individual variable, or function or predicate or rough logical formula on IS, then the substitution is held. Formally we have
$$\vdash_{IS_\lambda} (\alpha \leftrightarrow \beta) \rightarrow (F(\alpha) \leftrightarrow F(\beta))$$
4. $\forall F_1, F_2 \in RL_{IS}$, we have
$$\vdash_{IS_\lambda} F_1 \leftrightarrow F_2 \text{ iff } \vdash_{IS_\lambda} F_2 \leftrightarrow F_1$$
5. $\forall F \in RL_{IS}$. The identity of F is held. Formally we have
$$\vdash_{IS_\lambda} F \leftrightarrow F$$
6. $\forall F_1, F_2 \in RL_{IS}$, the symmetry of F_1 and F_2 is held. Formally we have
$$\vdash_{IS_\lambda} (F_1 \leftrightarrow F_2) \rightarrow (F_2 \leftrightarrow F_1).$$
7. $\forall F_1, F_2, F_3 \in RL_{IS}$, the absorbance laws are held, denoted by
$$\vdash_{IS_\lambda} (F_1 \vee (F_1 \wedge F_2) \leftrightarrow F_1)$$
   and
$$\vdash_{IS_\lambda} (F_1 \wedge (F_1 \vee F_2) \leftrightarrow F_1)$$
8. Some special properties of the rough logical formulae defined on IS are as follows:
   (i) $a_v \wedge a_u = \perp$, where $a \in A$ is an attribute in A. $u, v \in V_a$ are attribute value in set of attribute values, and $u \neq v$. $\perp$ is a false symbol.
   (ii) $\vee_{v \in V_a} a_v = T$, where $a \in A$ is an arbitrary attribute in A. T is a true symbol.
   (iii) $\neg a_u = \vee_{v \in V_a} a_v$, for each $a \in A$, $u \neq v$.
9. $\forall F_1, F_2 \in RL_{IS}$, if F_1 is the equivalence to F_2, then the complement $\neg F_1$ is also the equivalence to the complement $\neg F_2$, denoted by
$$\vdash_{IS_\lambda} (F_1 \leftrightarrow F_2) \rightarrow (\neg F_1 \leftrightarrow \neg F_2)$$

Proof of the properties is straight from the definitions and properties of the logical formulae on IS.

4 Clause Forms of Rough Logical Formulae

The normal forms are similar to classical logic formally[21,18,36], but propositional variables and predicates in the rough logic are the attributes in a given information systems.

(1) Disjunction Normal Form
Disjunction normal form of rough logical formula in a given information system is similar to classical logic, denoted by

$$F = (A_{1_1} \wedge \cdots \wedge A_{1_r}) \vee \cdots \vee (A_{n_1} \wedge \cdots \wedge A_{n_t})$$

(2) Conjunction Normal Form
Conjunction normal form of rough logical formula in a given information system is similar to classical logic, denoted by

$$F = (A_{1_1} \vee \cdots \vee A_{1_r}) \wedge \cdots \wedge (A_{n_1} \vee \cdots \vee A_{n_t})$$

where A_{i_j} is propositional variable, predicate or their negation. They are the attribute on information systems.

(3) **Skolem Clause Form of the Rough Logic**
In first-order logic, all quantifiers are moved into the front of formula, and each existent quantifier is eliminated from the prefixal form by using Skolem's method. Thus, original logical formula is equally transformed into the prefix form of containing only full quantifiers. To eliminate all full quantifiers, the prefix form is equally transformed into the Skolem clause form. Let F be a rough logical formula in a given information system, which could equally be transformed into following Skolem normal form according to Skolem's ways,

$$F = C_1 \wedge \cdots \wedge C_m$$

where each C_i is a disjunction of atoms or their negation. We call it Skolem clause form.

5 Deductive Reasoning of the Rough Logical Formulae

The Rough Logic defined in this article is based on Rough Set Theory proposed by Pawlak. So, proposed rough logic here is defined in a given information system. The truth values of rough logical formulae are limited to take multi-valued in the given information system. Specially, the rough logical formulae with operators H and L is based on upper and lower approximations in Rough Set Theory. H and L are similar as two operators in Modal Logic respectively. $H\varphi \in RL_{IS}$ is viewed as to take truth values in rough upper approximation and $L\varphi \in RL_{IS}$ is viewed as to take truth values in rough lower approximation[2,4,5,8–10,12,14,16,17,28].

For any rough logical formula $\varphi \in RL_{IS}$, it will be proved by deductive reasoning. Here we will prove that some related properties of the rough logical formulas with operators H and L. For example, $LH\varphi$ is roughly equal to $H\varphi$[2,12,22,23,26,29], namely

$$LH\varphi =_R H\varphi$$

where L and H is lower and upper approximate operators respectively[2,12]. Rough equality may also be called as true to degree at least λ, thus we need only to prove:

$$LH\varphi =_\lambda H\varphi$$

which is equivalent to prove following two forms:

$$B_*LH\varphi =_\lambda B_*H\varphi$$

and

$$B^*H\varphi =_\lambda B^*LH\varphi$$

where $LH\varphi$ and $H\varphi$ are operator rough logical formulae[2,12,18,36], $B_*(\bullet)$ and $B^*(\bullet)$ denote the lower and upper approximations of indiscernibility relation B with respect to $\bullet$ respectively. The form of lower approximation is proved as follows:

Proof

1. $L\neg\varphi \to_\lambda \neg\varphi$	Definition of operator L[2,12,18]
2. $\neg\neg\varphi \to_\lambda \neg L\neg\varphi$	Exchange position of logical formulas in (1)[36]
3. $\varphi \to_\lambda H\varphi$	The properties of $\neg$ and duad of L and H[2,12] in (2)
4. $B_*\varphi \to_\lambda B_*H\varphi$	The properties of rough sets[2,12] in (3)
5. $B_*L\varphi \to_\lambda B_*HL\varphi$	Replace φ by $L\varphi$ in (4)
6. $B_*HL\varphi \to_\lambda B_*L\varphi$	Definition of L and H and properties of rough sets[2,12] in (5)
7. $B_*L\varphi \leftrightarrow_\lambda B_*HL\varphi$	(5) and (6) and λ-equivalence definition
8. $B_*LH\varphi \leftrightarrow_\lambda B_*HLH\varphi$	Replace φ by $H\varphi$ in (7)
9. $B_*LH\varphi \leftrightarrow_\lambda B_*HH\varphi$	Properties of rough sets[2,12]
10. $B_*HH\varphi \leftrightarrow_\lambda B_*H\varphi$	Properties of rough sets[2,12]
11. $B_*LH\varphi \leftrightarrow_\lambda B_*H\varphi$	The "hypothetical syllogism" in (9) and (10)[2,12]

where $\to_\lambda$ and $\leftrightarrow_\lambda$ are called as λ-complication and λ-equivalence respectively. Similarly, we may prove the form of upper approximation.

6 Resolution Reasoning of the Rough Logical Formulae

We try to study resolution reasoning in the rough logic. Here we will discuss resolution principles and λ-resolution strategies of the logic.

6.1 Resolution Principles

Let C_1 and C_2 be two clauses in rough logic defined in a given information system $IS = (U, A)$. If L_1 and L_2 are two literals[23–25] in C_1 and C_2 respectively, and $T_{I_u}(L_1) = T$ and $T_{I_u}(L_2) = \perp$ or $T_{I_u}(L_1) = \perp$ and $T_{I_u}(L_2) = T$, then L_1 and L_2 are called as complementary literals[19,23–25,23].

Theorem 1. *For $\forall F \in RL_{IS}$, F could be transformed equivalently into clause form*

$$C_1 \wedge \cdots \wedge C_n$$

where each C_i is the disjunction of form attribute a_v or negation of its, $a \in A$ is an attribute on A, $v \in V_a$ is an attribute value in attribute set V_a.

Definition 4. *Let $C_1 : C_1' \vee a_v$ and $C_2 : C_2' \vee b_u$ be two clauses used in the resolution. a_v and b_u are complementary literals in C_1 and C_2 respectively, then they are resolved to produce an empty clause, denoted by ∇. Therefore, resolution principles of C_1 and C_2 are defined as follows*[9,15,16,23,24]*:*

$$\frac{\begin{array}{c} C_1 : C_1' \vee a_v \\ C_2 : C_2' \vee b_u \end{array}}{C : C_1' \vee C_2'} \tag{1}$$

where a_v and b_u are the literal to be resolved upon. The new clause C is called as a rough resolvent $GR(C_1, C_2) : C = C_1' \vee C_2'$.

Example 1. We extract a rough logical formula from following information table 2

$$F(a_5, b_2, b_4, c_3, c_0) = (a_5 \vee c_3) \wedge (b_2 \vee c_0) \wedge b_4 \tag{2}$$

Formula (2) is a rough logical formula defined in the given information system. The ground instances of the formula are as follows:

$$F(a_5, b_2, b_4, c_3, c_0) = (a_5^{\{1,2,3,4,5,6\}} \vee c_3^{\{\}}) \wedge (b_2^{\{4,5,6\}} \vee c_0^{\{1,2,3,4,5,6\}}) \wedge b_4^{\{1,2,3\}} \tag{3}$$

In formula (3) clauses $C_1 : (a_5^{\{1,2,3,4,5,6\}} \vee c_3^{\{\}})$ and $C_2 : (b_2^{\{4,5,6\}} \vee c_0^{\{1,2,3,4,5,6\}})$ include ground literals $c_3^{\{\}}$ and $c_0^{\{1,2,3,4,5,6\}}$ in them respectively. They are exactly complementary ground literal. Therefore, clause $C_1 : (a_5^{\{1,2,3,4,5,6\}} \vee c_3^{\{\}})$ and clause $C_2 : (b_2^{\{4,5,6\}} \vee c_0^{\{1,2,3,4,5,6\}})$ are resolved as follows:

$$\frac{\begin{array}{c} C_1 : (a_5^{\{1,2,3,4,5,6\}}) \vee (c_3^{\{\}})) \\ C_2 : (b_2^{\{4,5,6\}}) \vee (c_0^{\{1,2,3,4,5,6\}})) \end{array}}{C : a_5^{\{1,2,3,4,5,6\}} \vee b_2^{\{4,5,6\}})} \tag{4}$$

The resolvent $GR(C_1, C_2) = a_5^{\{1,2,3,4,5,6\}} \vee b_2^{\{4,5,6\}}$. Thus formula (3) is reduced by resolution, to have

$$F(a_5, b_2, b_4, c_3, c_0) = (a_5^{\{1,2,3,4,5,6\}} \vee b_2^{\{4,5,6\}}) \wedge b_4^{\{1,2,3\}} \tag{5}$$

Table 2. Information Table

U	a	b	c	d	e
1	5	4	0	1	1
2	5	4	0	2	1
3	5	4	0	2	2
4	5	2	0	1	2
5	5	2	0	2	2
6	5	2	0	1	0

6.2 λ-Resolution Strategies of the Rough Logic

Definition 5. Let L_1 and L_2 be literals in the rough logic, where L_1 is true to degree at least λ, L_2 is true to degree at most $1-\lambda$, if $\lambda \geq 0.5$, $T_{I_u}(L_1) > \lambda$ and $T_{I_u}(L_2) < 1-\lambda$; Or L_1 is true to degree at most λ, L_2 is true to degree at least $1-\lambda$, if $\lambda < 0.5$, $T_{I_u}(L_1) < \lambda$ and $T_{I_u}(L_2) \geq 1-\lambda$, then L_1 and L_2 are called as λ-complement literal pair in rough logic on IS[9,15,16,23,24].

Definition 6. Let C_1 and C_2 be without common variable clauses, and L_1 in C_1 and L_2 in C_2 are λ-complement literals, then λ-resolvent of C_1 and C_2 is defined as follows:

$$GR_\lambda(C_1, C_2) = (C_1 - L_1) \vee (C_2 - L_2) = C_1' \vee C_2'$$

where $C_1' = C_1 - L_1$, $C_2' = C_2 - L_2$.

Example 2. Let $IS = (U, A, V, f)$ be an information system, as shown on information table 1 in the above. We may construct a rough logical formula on IS[8–16,21–24]. We extract a formula $F \in RL_{IS}$ from the IS as follows:

$$F(a_5, b_2, b_4, c_0, \neg e_0) = (a_5 \vee b_4) \wedge b_2 \wedge (c_0 \vee \neg e_0) \tag{6}$$

Formula (6) may be written as the following ground clause form:

$$F(a_5, b_2, b_4, c_0, \neg e_0) = (a_5^{\{1,6\}} \vee b_4^{\{1,2,3\}}) \wedge b_2^{\{4,5,6\}} \wedge (c_0^{\{1,2,3,4\}} \vee \neg e_0^{\{2,3,4,5\}}) \tag{7}$$

where each item is a ground clause. When λ is defined as 0.6, obviously, $a_5^{\{1,6\}}$ and $c_0^{\{1,2,3,4\}}$ is a λ-complement ground literal pair. So, the resolvent $GR_\lambda(C_1, C_2)$ of $a_5^{\{1,6\}} \vee b_4^{\{1,2,3\}}$ in C_1 and $c_0^{\{1,2,3,4\}} \vee \neg e_0^{\{2,3,4,5\}}$ in C_2 is computed as follows:

$$GR_\lambda(C_1, C_2) = (a_5^{\{1,6\}} \vee b_4^{\{1,2,3\}} - a_5^{\{1,6\}}) \vee (c_0^{\{1,2,3,4\}} \vee \neg e_0^{\{2,3,4,5\}} - c_0^{\{1,2,3,4\}}) \tag{8}$$

Hence, the formula (7) could be rewritten as

$$(b_4^{\{1,2,3\}} \vee \neg e_0^{\{1,2,3,4,5\}}) \wedge b_2^{\{4,5,6\}} \tag{9}$$

In face, when $\lambda = 0.6$, $a_5^{\{1,6\}}$ and $\neg e_0^{\{1,2,3,4,5\}}$ is also a λ-complement ground literal pair, hence the resolvent $GR_\lambda(C_1, C_2)$ could be obtained as follows:

$$(b_4^{\{1,2,3\}} \vee c_0^{\{1,2,3,4\}}) \wedge b_2^{\{4,5,6\}} \tag{10}$$

Example 3. Let $IS = (U, A, V, f)$ be an information system, as shown on information table 1 in the above. We may prove the following rough logical formula F is λ-true to degree at least λ in this information table 1, where $\lambda = 0.6$

$$F(a_5, c_1, c_0, e_0, e_1, \neg e_0) = e_0 \wedge (a_5 \vee c_1) \wedge (c_0 \vee \neg e_0) \rightarrow e_1 \qquad (11)$$

Formula (11) may be written as the following clause form by Skolem's method, where decision aim e_1 is used as a negation clause $\neg e_1$ to add.

$$F(a_5, c_1, c_0, e_0, e_1, \neg e_0) = e_0 \wedge (a_5 \vee c_1) \wedge (c_0 \vee \neg e_0) \wedge \neg e_1 \qquad (12)$$

Formula (12) may be written as the following ground clause form according to information table 1.

$$F(a_5, c_1, c_0, e_0, e_1, \neg e_0) = e_0^{\{6\}} \wedge (a_5^{\{1,6\}} \vee c_1^{\{5,6\}}) \wedge (c_0^{\{1,2,3,4\}} \vee \neg e_0^{\{1,2,3,4,5\}}) \wedge \neg e_1^{\{3,4,5,6\}} \qquad (13)$$

Hence, by $\lambda = 0.6$-resolution, literals $a_5^{\{1,6\}}$ and $c_0^{\{1,2,3,4\}}$ in the formula (13) are the literals resolved upon, they are eliminated. So, the formula (13) could be rewritten as

$$e_0^{\{6\}} \wedge (c_1^{\{5,6\}} \vee \neg e_0^{\{1,2,3,4,5\}}) \wedge \neg e_1^{\{3,4,5,6\}} \qquad (14)$$

Again by $\lambda = 0.6$-resolution in the formula (14), literals $c_1^{\{5,6\}}$ and $\neg e_1^{\{3,4,5,6\}}$ in the formula (14) are the literals resolved upon, they are eliminated. So, the formula (14) could be rewritten as

$$e_0^{\{6\}} \wedge \neg e_0^{\{1,2,3,4,5\}} \qquad (15)$$

Literals $e_0^{\{6\}}$ and $\neg e_0^{\{1,2,3,4,5\}}$ in the formula (15) are the literals resolved upon, they are eliminated. So, the formula (15) could be obtained an empty after resolution. In fact, we use a method of resolution refutation in classical logic in this example. Because decision aim is negated and to join in resolution, to obtain an empty in conclusion. So, this rough logical formula F from the information table 1 is λ-true to degree at least $\lambda = 0.6$ in the information table 1. In fact, we could compute to obtain this rough logical formula

$$F(a_5, c_1, c_0, e_0, e_1, \neg e_0) = e_0 \wedge (a_5 \vee c_1) \wedge (c_0 \vee \neg e_0) \rightarrow e_1 \qquad (16)$$

to be true in the information table. Namely,
$T_{I_u}(F(a_5, c_1, c_0, e_0, e_1, \neg e_0)) = card(e_0 \wedge (a_5 \vee c_1) \wedge (c_0 \vee \neg e_0) \rightarrow e_1)/card(U)$

So the resolution reasoning of rough logic proposed in this article is valid in information system .

The λ-resolution proof in the above is based on the semantics of the rough logical formulae, no involving predicate symbol. So, the meaning of two different predicate symbols is satisfied by λ-complement, the λ-complement literal pair may be resolved.

In the following we will prove that the resolution is λ-soundness.

Theorem 2. Let $\Delta = \{C_1), \cdots, C_n\}$ be a clause set of the rough logical formula. If there is a resolution deduction of the clause C from Δ, then the Δ implies C logically, that is,

$$C_1 \wedge \cdots \wedge C_n) \rightarrow_\lambda C$$

is true[23,24].

Proof: It is finished by simple induction on length of the resolution deduction. For the deduction, we need only to show that any given resolution step is λ-soundness.

Suppose that C_1 and C_2 are two clauses in the logic at arbitrary step i,
(1). $C_1 = C_1' \vee a_v$
(2). $C_2 = C_2' \vee b_u$
where C_1' and C_2' are still clauses in the logic. Assuming that C_1 and C_2 are λ-true to degree at least λ at step i, a_v and b_u are λ-complement literal pair at the step i, then a_v and b_u are eliminated to produce a resolvent

$$GR_\lambda(C_1, C_2) : C = C_1' \vee C_2'$$

which is a new clause in the logic.

Now let us prove that C is also λ-true to degree at least λ. Suppose that two clauses joined in resolution are C_1 and C_2. If the literal a_v in C_1 is $(1-\lambda)$-true to degree at least $1-\lambda$ on IS and b_u in C_2 is λ-true to degree at most λ on IS, then C_1' in C_1 is λ-true to degree at least λ, so the new clause $C : C_1' \vee C_2'$ is λ-true to degree at least at λ on IS; If the literal b_u in C_2 is $(1-\lambda)$-true to degree at least $1-\lambda$ on IS and a_v in C_1 is λ-true at most on IS, then C_2' is λ-true to degree at least λ on IS, so the new clause $C : C_1' \vee C_2'$ is also λ-true to degree at least λ on IS.

The extracting of resolution step i could be arbitrary, $\lambda \in [0, 1]$ is a real number. If λ is 0 or 1, then the λ-soundness is called as soundness of the resolution in the rough logic. Otherwise, which is called as λ-soundness of the resolution in the rough logic. So, according to induction, proof of the soundness of resolution deduction in the rough logic is finished. Formally, we have

$$C_1 \wedge \cdots \wedge C_n \rightarrow_\lambda C$$

is true{23,24}.

7 Applications of Rough Logic to Problem Resolving in AI

The problem solving in *AI* is an idea of collapsing a complex problem into simple, solvable sub-problems[23–27]. And then the answers of the local problems could be amalgamated into the answer of the original problem. Hence we need study the knowledge representation, namely representing the practical problem by a rough logical formula. And the original global logical formula is decomposed into sub-formulae, till predicates or propositions. The sub-formulae, predicates or propositions are solved, and then the local answers of sub-formulae are amalgamated into the solution of the global logical formula. The problem solving procedure is listed as follows:

1. Gathering a group of data related with the solving problem in the situation;
2. The group of data is constructed into an information table
3. We could extract a rough logical formula F from this information table;
4. By Skolem's way, the rough logical formula F is equivalently transformed into the following clause form:

$$C_1 \wedge C_2 \wedge \cdots \wedge C_n$$

where each clause $C_i = L_{i_1} \vee L_{i_2} \vee \cdots \vee L_{i_m}$ is a disjunction of literals. Literal L_{i_j} is a predicate, proposition or negative of them.

5. The meaning corresponding to C_i is denoted by

$$m(C_i) = \{x \in U : x \models_{IS} C_i\}$$

The symbol $\models_{IS}$ is a satisfiability symbol for rough logical formula[21–30].

Example 4. Let $IS = (U, A)$ be the system of diagnosis and treating in Chinese Traditional Medicine, where U is a set of the patients, A is a set of symptoms (attributes) for the patients. The system possesses the function to test blood viscosity concentration for patients[26,27]. The data set of the patient gathered in clinic is denoted by $P = \{Name_v, Sex_v, Age_v, Testing - Value_v\}$. Such as, $P = \{Name_{Wang}, Sex_{male}, Age_{65}, Testing - Value_{3.5}\}$, where Wang is a value of attribute Name, Male is a value of attribute Sex, 65 is a value of attribute Age, and 3.5 is blood viscosity concentration of testing the patient via testing instruments in clinic. We construct an information table using this group of data as follows:

Table 3. Information Table

U	Name	Sex	Age	T-V
1	Wang	Male	65	3.5
2	Li	Female	30	4.2
3	Zhang	Male	50	5.2

For convenience, we only use a testing index item of blood viscosity concentration $T - V_{3.5}$ in this table 3. In fact, to need testing 18 index items, other items are similar as this testing of $T - V_{3.5}$, but the methods of their processed are different from each other. We extract the following rough logical formula F from this table 3, this F represents the resolving problem in practice.

$$F = Name_{Wang} \wedge Sex_{Male} \wedge Age_{65} \wedge T - V_{3.5}$$

where $T - V$ is an abbreviation for Testing-Value. Now we collapse this formula into sub-formulae or predicates, and to solve these sub-formulae or predicates according to the algorithm given in the system[27]. And then we amalgamate the answers of the sub-formulae into an answer of the global formula. The steps of the solving procedure are listed as follows.

1. F is collapsed into some sub-formulae, i.e. the predicates $P_1 = name_{Wang}$, $P_2 = Sex_{Male}$, $P_3 = Age_{65}$ and $P_4 = TV_{3.5}$. Because this logical formula F is exactly a clause form of Skolem's, we needn't transform the rough logical formula F into its clause form equivalently.
2. We select to solve the sub-formula P_2, It is interpreted as attribute value Sex_{Male}, it is the meaning of sub-formula P_2. In fact, we find out the interval [4.42, 4.79] of blood viscosity concentration corresponding to reference value of health male[26,27]. And then the interval [4.42, 4.79] is granulated by 0.618. The average value and the standard deviation of the blood viscosity concentration interval of normal people reference value are computed according to following formulae. The average value and the standard deviation in this example are computed, we will obtain $AV_2 = 4.70$ and $SD_2 = 0.28$ respectively.
$$AV = (a + \sum_{j=1}^{n-2}(a + j * 0.618) + b)/n \tag{17}$$
$$SD = sqrt(\sum_{j=0}^{n-1}((a + j * 0.618) - AV)^2/n) \tag{18}$$
where $sqrt$ is the square root function symbol, and n is the total number of all small intervals partitioned on the interval [a, b].
3. The level of index item IV corresponding to testing value $TV = 3.5$ in clinic are computed as follows:
$$IV = (TV - AV)/SD, \quad TV > a \tag{19}$$
$$IV = (-TV + AV)/SD, \quad TV \leq a \tag{20}$$
where TV is the testing blood viscosity concentration in clinic via testing instrument. It is 3.5 in the example. Average value AV and standard deviation SD in the example is $AV_2 = 4.27$ and $SD_2 = 0.28$ respectively. a is the lower bound corresponding to the interval. IV in the example is computed to be $IV_2 = -3.74$.
4. The levels of 9 sub-types (Concentration-$ConL$, Viscosity-VL, Aggregation-AL, Coagulation-$CoaL$, Hematocrit-HL, Erythrocyte Aggregation-EAL, Red Cell Rigidity-$RCRL$, Blood Plasma Viscosity-$BPVL$, Platelet Aggregation-$BPAL$)are computed. The levels of blood viscosity concentration ($BHVS$, $BLVS$, $BHLVS$ and $BLHVS$) are also computed[27].
5. By querying case history, IV is decided to plus or minus a correct value according to the record in case history base for this patient. Thus, the patient is diagnosticated as a result of blood viscosity concentration[27].

8 Perspective of Studying for the Rough Logic

The rough logic will inspire what we construct the rough logical formula to be fitted to applications for some specialities. Thus, we tried to resolve the practice problems in various speciality using it[31–35]. Besides, it will offer new theory and methodology for applications of classical logic and other nonstandard logic. So, we will further study the theoretical significance and practice value of the logic. The studying of semantics of rough logical formulae defined in a given information system will offer a new model for studying of granular computing. Studying based on semantics of the rough logic is also an extension of Rough Logic proposed by Pawlak[1,2].

In further work, we will study resolution refutation of the logic, that is, given a set Δ of clauses in the rough logic. A resolution deduction of C from Δ is a finite sequence $C_1, \cdots, C_n$ of clauses such that each C_i either is a clause in Δ or a resolvent preceding C_j and C_k. Finally, $C_n = C$. If there is a deduction of ∇ from Δ, where ∇ is a λ-empty clause, then we call it λ-resolution refutation. We will also further study the related properties and reasoning of the rough logic.

Acknowledgement

We would like to thank the support of Natural Science Fund of China (NSFC-60173054). Thanks are also to Dr. James Kuodo Huang (member IEEE) for his modifying English in this article.

References

1. Pawlak, Z.: Rough logic. Bull. of Polish Acad. of Sci. 35(5-6), 253–259 (1987)
2. Pawlak, Z.: Rough sets-theoretical aspects of reasoning about data. Kluwer Academic Publishers, Dordrecht (1991)
3. Orlowska, E.: A logic of indicernibility relation. In: Skowron, A. (ed.) SCT 1984. LNCS, vol. 208, pp. 177–186. Springer, Heidelberg (1985)
4. Chakraborty, M.K., Banerjee, M.: Rough logic with rough quantifiers. Warsaw University of Technology, ICS Research Report 49/93 (1993)
5. Nakamura, A.: A rough logic based on incomplete information and its applications. International Journal of Approximate Reasoning 15, 367–378 (1996)
6. Nakamura, A., Matsueda, M.: Rough logic on incomplete knowledge systems. In: Lin, T.Y. (ed.) Proceedings of the third International workshop on Rough Sets and Soft Computing (RSSC 1994), San Jose State University, San Jose, California, USA, pp. 56–64 (1994)
7. Nakamura, A.: Graded modalitiesin rough logic. In: Polkowski, L., Skowron, A. (eds.) Rough Sets Knowledge Discovery1, pp. 192–208. Physica-Verlag, Heidelberg (1998)
8. Liu, Q.: Operator rough logic and its resolution principle. Chinese Journal of Computer 21(5), 435–476 (1998) (In Chinese)
9. Liu, Q.: The OI-Resolution of Operator Rough Logic. In: Polkowski, L., Skowron, A. (eds.) RSCTC 1998. LNCS (LNAI), vol. 1424, pp. 432–435. Springer, Heidelberg (1998)

10. Liu, Q.: The resolution for rough propositional logic with lower(L) and upper(H) approximate operators. In: Zhong, N., Skowron, A., Ohsuga, S. (eds.) RSFDGrC 1999. LNCS (LNAI), vol. 1711, pp. 352–356. Springer, Heidelberg (1999)
11. Liu, Q., Wang, Q.Y.: Granular logic with Closeness Relation and Its Reasoning. In: Ślęzak, D., Wang, G., Szczuka, M., Düntsch, I., Yao, Y. (eds.) RSFDGrC 2005. LNCS (LNAI), vol. 3641, pp. 709–717. Springer, Heidelberg (2005)
12. Lin, T.Y., Liu, Q.: First-order rough logic I: Approximate reasoning via rough sets. Fundamenta Informaticae 27(2-3), 137–154 (1996)
13. Yao, Y., Liu, Q.: A Generalized Decision Logic in Interval-Set-Valued Information table. In: Zhong, N., Skowron, A., Ohsuga, S. (eds.) RSFDGrC 1999. LNCS (LNAI), vol. 1711, pp. 285–293. Springer, Heidelberg (1999)
14. Liu, Q.: λ-Level Rough Quality Relation and the Inference of Rough Paramodulation. In: Ziarko, W., Yao, Y. (eds.) RSCTC 2000. LNCS (LNAI), vol. 2005, pp. 462–469. Springer, Heidelberg (2001)
15. Chang, C.L., Lee, R.C.T.: Symbolic logic and machine theorem proving. Academic Press, London (1993)
16. Liu, X.H.: Fuzzy Logic and Fuzzy Reseaning. Press Of Jilin University (1989)
17. Banerjee, M., Khan, A.: Propositional Logic from Rough Set Theory. In: Peters, J.F., Skowron, A., Düntsch, I., Grzymała-Busse, J.W., Orłowska, E., Polkowski, L. (eds.) Transactions on Rough Sets VI. LNCS, vol. 4374, pp. 1–25. Springer, Heidelberg (2007)
18. Hamilton, A.G.: Logic for Mathematicans. Cambridge University Press, Cambridge (1980)
19. Liu, Q., Liu, S.H., Zheng, F.: Rough Logic and Its Applications in Data Mining. Journal of Software 12(3), 415–419 (2001) (In Chinese)
20. Rasiowa, H., Skowron, A.: Rough concepts logic. In: Skowron, A. (ed.) SCT 1984. LNCS, vol. 208, pp. 288–297. Springer, Heidelberg (1985)
21. Liu, Q.: Rough Sets and Rough Reseaning (Third), p. 8. Science Press, Beijing (2005) (In Chinese)
22. Liu, Q.: Granular Language and Its Reasoning, Data Mining and Knowledge Discovery: Theory, Tools, and Technology V. In: Proceeding of SPIE-The International Society for Optical Engineering, Orlando, Florida, USA, April 21-22, pp. 279–287 (2003)
23. Liu, Q., Sun, H.: Theoretical Study of Granular Computing. In: Wang, G.-Y., Peters, J.F., Skowron, A., Yao, Y. (eds.) RSKT 2006. LNCS (LNAI), vol. 4062, pp. 93–102. Springer, Heidelberg (2006)
24. Liu, Q., Huang, Z.H.: G-Logic and Resolution Reasoning. Chinese Journal of Computer 27(7), 865–873 (2004) (In Chinese)
25. Hobbs, J.R.: Granularity. In: Proceedings of IJCAI, Los Angeles, pp. 432–435 (1985)
26. Liu, Q., Liu, Q.: Granules and Applications of Granular Computing in Logical Reseaning. Research and Development of Computer 41(4), 546–551 (2004) (In Chinese)
27. Liu, Q., Jiang, F., Deng, D.Y.: Design and Implement for the Diagnosis Software of Blood Viscosity Syndrome Based on Hemorheology on GrC. LNCS (LNAI), vol. 2639, pp. 413–420. Springer, Berlin (2003)
28. Kripke, S.: Semantic Analysis of Modal Logic. In: Zeitschrift für Mathematische Logik und Grundlagen der Mathematik, pp. 67–96 (1963)
29. Liu, Q., Wang, J.Y.: Semantic Analysis of Rough Logical Formulas Based on Granular Computing. In: Proceedings of IEEE GrC 2005, pp. 393–396. IEEE, Los Alamitos (2006)

30. Zhang, B., Zhang, L.: Theory and Applications for Problem Solving. Publisher of Tsinghua University (1990) (In Chinese)
31. Polkwski, L.: A Calculus on Granules from Rough inclusions in Information Systems, The Proceedings of International Forum on Theory of GrC from Rough Set Perspective. Journal of Nanchang Institute of Technology 25(2), 22–27 (2006)
32. Lin, T.Y.: From Rough Sets and Neighborhood Systems to Information Granulation and Computing in Words. In: European Congress on Intelligent Techniques and Soft Computing, pp. 1602–1606 (1997)
33. Yao, Y.Y.: Three Perspectives of Granular Computing, The Proceedings of International Forum on Theory of GrC from Rough Set Perspective. Journal of Nanchang Institute of Technology 25(2), 22–27 (2006)
34. Liu, Q., Liu, Q.: Approximate Reasoning Based on Granular Computing in Granular Logic. In: The Proceedings of ICMLS 2002, November 4-6, pp. 1258–1262. IEEE, Los Alamitos (2002)
35. Lin, T.Y.: Granular Computing on Binary Relations II: Rough Set Representations and Belief Functions. In: Skowron, A., Polkowski, L. (eds.) Rough Sets in Knowledge Discovery, pp. 121–140. Physica-Verlag, Berlin (1998)
36. Wang, X.J.: Introduction for mathematical logic. Press. Of Beijing Uni., Beijing (1982)
37. Lin, Y.: Granular Computing on Partitions, Coverings Neighborhood Systems, The Proceedings of International Forum on Theory of GrC from Rough Set Perspective. Journal of Nanchang Institute of Technology 25(2), 22–27 (2006)
38. Skowron, A.: Rough-Granular Computing, The Proceedings of International Forum on Theory of GrC from Rough Set Perspective. Journal of Nanchang Institute of Technology 25(2), 22–27 (2006)
39. Yao, Y.Y.: Information granulation and rough set approximation. International Journal of Intelligence Systems 16, 87–104 (2001)
40. Dai, J.H.: Axis Problem of Rough 3-Valued Algebras, The Proceedings of International Forum on Theory of GrC from Rough Set Perspective (IFTGrCRSP 2006), Nanchang, China. Journal of Nanchang Institute of Technology 25(2), 48–51 (2006)
41. Liu, G.L.: The Topological Structure of Rough Sets over Fuzzy Lattices. In: Proceedings of IEEE International Conference on Granular Computing, Beijing, China, July 25-27, vol. I, pp. 535–538 (2005)
42. Pei, D.W.: A Generalized Model of Fuzzy Rough Sets. Int. J. General Systems 34(5), 603–613 (2005)
43. Yao, J.T., Yao, Y.Y.: Induction of classification rules by granular computing. In: Proceedings of the International Conference on Rough Sets and Current Trends in Computing, pp. 331–338. Springer, Berlin (2002)
44. Miao, D.Q.: Rough Group, Rough Subgroup and their Properties. In: Ślęzak, D., Wang, G., Szczuka, M., Düntsch, I., Yao, Y. (eds.) RSFDGrC 2005. LNCS (LNAI), vol. 3641, Part I, pp. 104–113. Springer, Heidelberg (2005)
45. Wu, W.Z.: Rough Set Approximations VS. Measurable Space. In: 2005 Proceedings of IEEE International Conference on Granular Computing, Atlanta, Georgia, USA, pp. 329–332 (2006)
46. Liu, Q., Sun, H.: Studying direction of granular computimg from rough set perspective of development. Journal of Nanchang Institute Technology 25(5), 1–10 (2006) (In Chinese)

On Reduct Construction Algorithms

Yiyu Yao[1], Yan Zhao[1], and Jue Wang[2]

[1] Department of Computer Science, University of Regina
Regina, Saskatchewan, Canada S4S 0A2
{yyao, yanzhao}@cs.uregina.ca

[2] Laboratory of Complex Systems and Intelligence Science, Institute of Automation
Chinese Academy of Sciences, Beijing, China 100080
jue.wang@mail.ia.ac.cn

Abstract. This paper critically analyzes reduct construction methods at two levels. At a high level, one can abstract commonalities from the existing algorithms, and classify them into three basic groups based on the underlying control structures. At a low level, by adopting different heuristics or fitness functions for attribute selection, one is able to derive most of the existing algorithms. The analysis brings new insights into the problem of reduct construction, and provides guidelines for the design of new algorithms.

Keywords: Reduct construction algorithms, deletion strategy, addition-deletion strategy, addition strategy, attribute selection heuristics.

1 Introduction

The theory of rough sets has been applied to data analysis, data mining and knowledge discovery. A fundamental notion supporting such applications is the concept of reducts, which has been studied extensively by many authors [14, 17, 21, 22, 25, 29, 30]. A reduct is a subset of attributes that is jointly sufficient and individually necessary for preserving the same information as provided by the entire set of attributes. It has been proved that finding a reduct with the minimal number of attributes is NP-hard [26]. Research efforts on reduct construction algorithms therefore mainly focus on designing search strategies and heuristics for finding a satisfactory reduct efficiently.

A review of the existing reduct construction algorithms shows that most of them tie together search strategies (i.e., control structures) and attribute selection heuristics. This leads to difficulties in analyzing, comparing, and classifying those algorithms, as well as the trend of introducing new algorithms constantly. With ample research results on this topic, it is perhaps the time for us to pause and to analyze critically those results, in order to gain more insights.

With a clear separation of control structures and attribute selection heuristics, we can critically analyze reduct construction algorithms with respect to the high level control strategies, and the low level attribute selection heuristics,

M.L. Gavrilova et al. (Eds.): Trans. on Comput. Sci. II, LNCS 5150, pp. 100–117, 2008.

respectively. This allows us to conclude that the differences between the existing algorithms lie more on the attribute selection heuristics than on the control strategies.

The rest of the paper is organized as follows. First of all, we discuss the connections between feature selection and reduct construction in Section 2. After that, basic concepts and notations of rough set theory are reviewed in Section 3. Three basic control structures are then presented in Section 4-6 by reformulating the existing algorithms, from which many variations can be generated easily. After these, Section 7 is the conclusion.

2 Feature Selection and Reduct Construction

Reduct computation is related to many disciplines. The same objective of simplifying the attribute domain has been studied in machine learning, pattern recognition, and feature selection in specific [3, 12, 13, 23].

Feature selection is a fundamental task in a number of different disciplines, such as pattern recognition, machine learning, concept learning and data mining. Feature selection is necessary for both description and prediction purposes. In the description process, it can be computationally complex to construct rules by using all available features; in the prediction process, the constructed high dimensional rules can be hard to test and evaluate for new coming instances. From a conceptual perspective, selection of relevant features, and elimination of irrelevant ones, are the main tasks of feature selection. From a theoretical perspective, it can be shown that an optimal feature selection requires an exhaustive search of all possible subsets of the entire feature set. If the cardinality of the entire feature set is large, this exhaustive method is impractical. For practical feature selection applications, the search is normally for a satisfactory set of features instead of an optimal set.

In the domain of feature selection, two methods, *forward selection* and *backward elimination*, have been extensively studied [3, 12, 13]. The forward selection strategy starts with the empty set and consecutively adds one attribute at a time until we obtain a satisfactory set of features. This strategy also can be called as an *addition* strategy for simplicity. On the contrary, the backward elimination strategy starts with the full set and consecutively deletes one attribute at a time until we obtain a satisfactory set of features. In this paper, this strategy also is called a *deletion* strategy. The forward strategy can be extended from the one-by-one sequential-add style to the "plus l - take away r" style. This kind of methods first enlarge the feature subset by l features, then delete r features as long as the remaining attribute set exhibits an improvement compared to the previous feature set. They avoid the nesting problem of feature subsets that are encountered in the sequential style, but need to set the values of l and r [4, 18]. The same idea can be applied to backward strategy variations.

In a consecutive forward selection or a backward elimination process, one can adopt different heuristics for feature selection. A heuristic decides and then adds the best feature, or deletes the worst feature, at each round. As a consequence, variations of the same algorithm can be derived.

The difference between reduct computation and feature selection is their halting strategies. For the purpose of feature selection, one might stop adding or deleting features when the information preservation is satisfied, the classification accuracy is not degrading, or the computation cost is affordable. For reduct construction, the algorithm does not stop until the *minimum* set of features that possesses some particular property is obtained. Reduct construction thus is a special case of feature selection. In fact, many feature selection algorithms can be viewed as performing a biased form of reduct computation. The results are not necessarily being reducts. Obviously, the extensive studies of feature selection, including the identification of relevant, irrelevant and redundant features, and the design, implementation and renovation of the filter and wrapper methods, affect the study of reduct computation.

By considering the properties of reducts, the deletion strategy always results in a reduct [7, 30]. On the other hand, algorithms based on a straightforward application of the addition strategy only produce a superset of a reduct [8, 10, 15, 16, 20]. In order to resolve this problem, many authors have considered a combined strategy by re-applying the deletion strategy on the superset of the reduct produced by the straightforward addition strategy [25]. An interesting question is whether there exists an addition-only strategy that can produce a reduct. A positive answer has been given by Zhao and Wang with an addition algorithm without further deletion [29].

According to the above discussion, we have three control strategies used by reduct construction algorithms. They are the deletion strategy, the addition-deletion strategy, and the addition strategy. We can classify reduct construction algorithms into the corresponding three groups.

3 Basic Concepts and Notations

The basic concepts, notations, and results related to the problem of reduct construction are briefly reviewed in this section.

3.1 Information Table and Attribute Lattice

Suppose data are represented by an information table, where a finite set of objects are described by a finite set of attributes [17].

Definition 1. *An information table S is the tuple:*

$$S = (U, At, \{V_a \mid a \in At\}, \{I_a \mid a \in At\}),$$

where U is a finite nonempty set of objects, At is a finite nonempty set of attributes, V_a is a nonempty set of values for an attribute $a \in At$, and $I_a : U \to V_a$ is an information function. For an object $x \in U$, an attribute $a \in At$, and a value $v \in V_a$, $I_a(x) = v$ means that the object x has the value v on attribute a.

Table 1. An information table

	a	b	c	d	e
o_1	0	0	0	1	1
o_2	0	1	2	0	0
o_3	0	1	1	1	0
o_4	1	2	0	0	1
o_5	0	2	2	1	0
o_6	0	3	1	0	2
o_7	0	3	1	1	1

Example 1. An information table is illustrated in Table 1, which has five attributes and seven objects.

The family of all attribute sets form an attribute lattice under the refinement order. Let $|At|$ denote the cardinality of At. An attribute lattice has $|At| + 1$ levels. The only node on the top level indicates the empty set $\emptyset$. The only node on the bottom level indicates the biggest attribute set At. Nodes on the second level stand for singleton attribute sets. There are $|At|$ nodes in the second level. For the nth level and $n \geq 2$, there are

$$\frac{|At|(|At|-1)\ldots(|At|-n+2)}{(n-1)!}$$

nodes. There are $2^{|At|}$ attribute sets in the entire attribute lattice. An edge connecting a pair of nodes implies the refinement relationship between an attribute set and a subset or superset of the attribute set.

Example 2. Figure 1 illustrates the attribute lattice of the previous information Table 1. It is obvious that totally $2^5 = 32$ attribute sets can be defined for the universe.

3.2 Equivalence Relations

Definition 2. *Given an information table S, for any subset $A \subseteq At$ there is an associated equivalence relation $E_A \subseteq U \times U$, i.e.,*

$$E_A = \{(x, y) \in U \times U \mid \forall a \in A, I_a(x) = I_a(y)\},$$

which partitions U into disjoint subsets, called equivalence classes. An equivalence class containing any object $x \in U$ is defined as: $[x]_A = \{y \in U \mid \forall a \in A, I_a(x) = I_a(y)\}$. Such a partition of the universe is denoted by U/E_A, or U/A for simplicity.

A partition U/E_A is a refinement of another partition U/E_B, or equivalently, U/E_B is a coarsening of U/E_A, denoted by $U/E_A \preceq U/E_B$, if every equivalence class of U/E_A is contained in some equivalence class of U/E_B. The refinement relation is a partial order, i.e, it is reflexive, anti-symmetric and transitive.

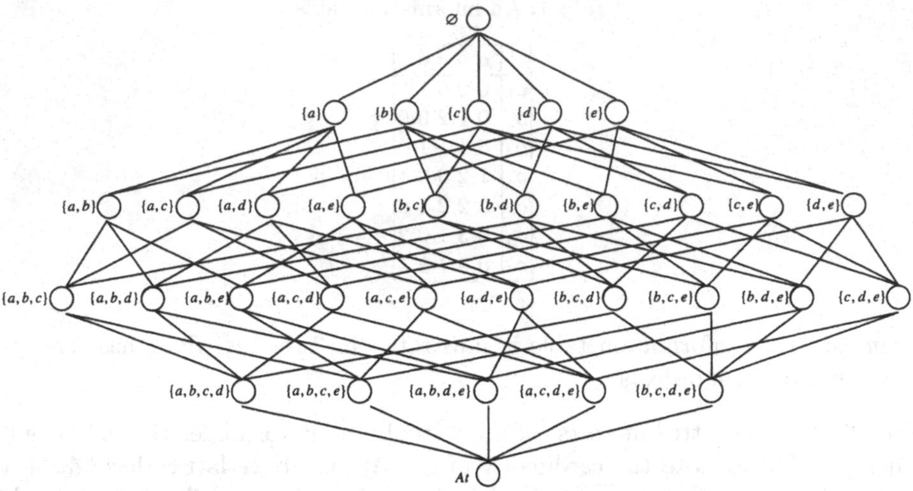

Fig. 1. The attribute lattice of the information Table 1

Given two partitions U/E_A and U/E_B, the meet of their equivalence classes, $U/E_A \wedge U/E_B$, is all nonempty intersections of an equivalence class from U/E_A and an equivalence class from U/E_B. The join of their equivalence classes, $U/E_A \vee U/E_B$, is all unions of an equivalence class from U/E_A and an equivalence class from U/E_B. The meet is the largest refinement partition of both U/E_A and U/E_B; the join is the smallest coarsening partition of both U/E_A and U/E_B. Clearly, $U/E_\emptyset$ is the coarsest partition, and U/E_{At} is the finest partition. For any $A \subseteq At$, we have $U/E_{At} \preceq U/E_A \preceq U/E_\emptyset$. The family of all partitions form a partition lattice under the refinement order.

3.3 Discernibility Matrices

Definition 3. *Given an information table S, for any two objects $(x, y) \in U \times U$, there is an associated discernibility relation $m_{x,y} \subseteq At$, i.e.,*

$$m_{x,y} = \{a \in At \mid I_a(x) \neq I_a(y)\}.$$

The physical meaning of $m_{x,y}$ is that objects x and y can be distinguished by any attribute in $m_{x,y}$.

The family of all discernibility relations can be conveniently stored in a $|U| \times |U|$ matrix, called a discernibility matrix M [21]. A discernibility matrix M is symmetric, i.e., $m_{x,y} = m_{y,x}$, and $m_{x,x} = \emptyset$. The family of all discernibility relations also can be expressed as a set M, collecting only the distinct nonempty elements, i.e., $M = \{m_{x,y} \mid m_{x,y} \neq \emptyset\}$.

Example 3. The discernibility matrix of information Table 1 is illustrated in Table 2. Since the discernibility matrix is symmetric, we only list its lower left half.

Table 2. The discernibility matrix of information Table 1

	o_1	o_2	o_3	o_4	o_5	o_6	o_7
o_1	-	-	-	-	-	-	-
o_2	$\{b,c,d,e\}$	-	-	-	-	-	-
o_3	$\{b,c,e\}$	$\{c,d\}$	-	-	-	-	-
o_4	$\{a,b,d\}$	$\{a,b,c,e\}$	At	-	-	-	-
o_5	$\{b,c,e\}$	$\{b,d\}$	$\{b,c\}$	$\{a,c,d,e\}$	-	-	-
o_6	$\{b,c,d,e\}$	$\{b,c,e\}$	$\{b,d,e\}$	$\{a,b,c,e\}$	$\{b,c,d,e\}$	-	-
o_7	$\{b,c\}$	$\{b,c,d,e\}$	$\{b,e\}$	$\{a,b,c,d\}$	$\{b,c,e\}$	$\{d,e\}$	-

The matrix also can be transformed to a set by collecting distinct elements and eliminating empty elements, such that:

$$\begin{aligned} M = \{&\{b,c\},\{b,d\},\{b,e\},\{c,d\},\{d,e\},\{a,b,d\},\\ &\{b,c,e\},\{b,d,e\},\{a,b,c,d\},\{a,b,c,e\},\\ &\{a,c,d,e\},\{b,c,d,e\},At\}. \end{aligned}$$

The difference of equivalence relations and discernibility relations is obvious. The equivalence relation E_A is based on an attribute set A, indicating all the object pairs that are indiscernible regarding A. The discernibility relation $m_{x,y}$ is based on an object pair (x,y), indicating all the attributes that any of them can distinguish x and y. The relationships between these two relations can be expressed as follows:

$$\begin{aligned} &(x,y) \notin E_{m_{x,y}};\\ &(x,y) \in E_A \Leftrightarrow A \cap m_{x,y} = \emptyset \text{ and } A \cup m_{x,y} = At. \end{aligned}$$

3.4 Reducts

Definition 4. *Given an information table S, a subset $R \subseteq At$ is called a ρ-reduct of At for the property ρ, if R satisfies the two conditions:*

(i). *R and At possess the same property ρ;*
(ii). *for any $a \in R$, $R - \{a\}$ cannot remain the property ρ.*

The first condition indicates the joint sufficiency of the attribute set R, and the second condition indicates that each attribute in R is individually necessary.

The property ρ can be interpreted in different ways. For example, considering the equivalence relations, the property ρ can be expressed as U/E_P, $[x]_P$ or the joint entropy of P, for any $P \subseteq At$ [17]. Also, regarding the family M of discernibility relations, the property ρ can be expressed as $\forall m \in M$, $m \cap P \neq \emptyset$.

According to different interpretations, condition (i) of the reduct definition can be written as:

- $U/E_R = U/E_{At}$,
- for all $x \in U$, $[x]_R = [x]_{At}$,

- $H(R) = H(At)$, where $H(.)$ denotes the joint entropy of the set, or
- $\forall m \in M, m \cap At \neq \emptyset$ and $m \cap R \neq \emptyset$.

It means that the equivalence relations of R and At define the same partition of the universe. For each object x in the universe, x has the same equivalence class defined by R and At. R and At provide the same information grain. Object pairs that can be distinguished by At also can be distinguished by R.

Given an information table, there may exist many reducts. The intersection of all reducts is called the Core.

Definition 5. *An attribute set $R' \subseteq At$ is called a super-reduct of a reduct R, if $R' \supseteq R$; an attribute set $R' \subset At$ $R' \neq \emptyset$ is called a partial reduct of a reduct R, if $R' \subset R$.*

Given a reduct, there exist many super-reducts and many partial reducts.

Figure 2 shows a very simple attribute lattice with 8 nodes in total. Suppose two reducts have been identified, and highlighted by stars on their corresponding nodes. If an attribute set is a reduct, then all its supersets are super-reducts. Here we shade their corresponding nodes in the lattice. At the same time, any subset of a reduct is a partial reduct. In the graph, we use circle with solid line to denote their corresponding nodes.

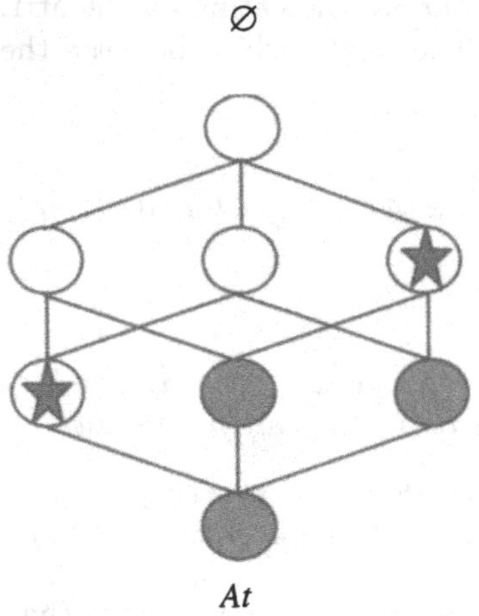

Fig. 2. An illustration of super- and partial reducts in a sample attribute lattice

Reduct computation can be understood as a search in the attribute lattice under the refinement relation. Both the deletion and addition strategies can be practised. The deletion strategy searches from At to $\emptyset$. As long as the condition (i) is met, a reduct or a super-reduct is obtained. If all of its subset are partial reducts, then it is identified as a reduct. A searching heuristic can facilitate the search process by deciding which attribute to be eliminated first, in order to move the search upward.

On the other hand, the addition strategy executes the search from $\emptyset$ to At. When the condition (i) is met, a reduct or a super-reduct is obtained, and the forward selection can be stopped. We need to eliminate the superfluous attributes from a super-reduct, if such is obtained. A searching heuristic decides which attribute to be added first, in order to move the search downward. An enhanced searching heuristic needs to prevent the search from leading to a proper superset of a reduct. By doing so, a backtrack elimination can be saved.

4 Reduct Construction by Deletion

4.1 Control Strategy

By a deletion method, we take At as a super-reduct, which is the largest super-reduct. Deletion methods can be described generally in Algorithm 1.

Algorithm 1. *The deletion method for computing a reduct*
Input: *An information table.*
Output: *A reduct R.*

(1) $R = At, CD = At$.
(2) **While** $CD \neq \emptyset$:
  (2.1) Compute fitness values of all the attributes in CD *regarding the property* ρ *using a fitness function* δ;
  (2.2) Select an attribute a according to its fitness, let $CD = CD - \{a\}$;
  (2.3) **If** $R - \{a\}$ *is jointly sufficient, let* $R = R - \{a\}$.
(3) **Output** R.

Many algorithms are proposed based on this simple deletion control strategy. For example, the algorithms proposed in [5,7,30] are implemented for computing a reduct based on information tables.

A deletion method starts with the trivial super-reduct, i.e., the entire attribute. It has to check all the attributes in At for deletion. It is not efficient in the cases when a reduct is short, and many attributes are eliminated from At after checking.

4.2 Attribute Selection Heuristics

The order of attributes for deletion is essential for reduct construction. Regarding a property ρ, different fitness functions may determine different orders of attributes, that may result in different reducts.

The attribute selection heuristic is given by a fitness function:

$$\delta : At \longrightarrow \Re, \tag{1}$$

where At is the set of attributes in the information table, and $\Re$ is the set of real numbers. The meaning of the function δ is determined by many semantic considerations. For example, it may be interpreted in terms of the cost of testing, the easiness of understanding, the actionability of an attribute, or the information gain an attribute produces.

Example 4. Suppose the fitness function δ is interpreted as an information entropy

$$\delta(a) = H(a) = -\sum_{v \in V_a} p(v) \log p(v). \tag{2}$$

This heuristic can be easily applied to information tables. For the information Table 1, we obtain $H(a) = 0.592$, $H(b) = 1.950$, $H(c) = 1.557$, $H(d) = 0.985$ and $H(e) = 1.449$, which yields an order $b \to c \to e \to d \to a$. According to this entropy-based order, the attribute a that contains least information is most likely to be deleted first, and the attributes d, e, c and b are then considered in turn. As a result, a reduct $\{b, c, e\}$ is computed. The iterative steps are illustrated in Figure 3.

Step 1: check a					
	a	b	c	d	e
o_1		0	0	1	1
o_2		1	2	0	0
o_3		1	1	1	0
o_4		2	0	0	1
o_5		2	2	1	0
o_6		3	1	0	2
o_7		3	1	1	1
$U/E_{\{b,c,d,e\}} = U/E_{At}$, a can be deleted. $R = \{b,c,d,e\}$.					

Step 2: check d				
	b	c	d	e
o_1	0	0		1
o_2	1	2		0
o_3	1	1		0
o_4	2	0		1
o_5	2	2		0
o_6	3	1		2
o_7	3	1		1
$U/E_{\{b,c,e\}} = U/E_{At}$, d can be deleted. $R = \{b,c,e\}$.				

Step 3: check e			
	b	c	e
o_1	0	0	
o_2	1	2	
o_3	1	1	
o_4	2	0	
o_5	2	2	
o_6, o_7	3	1	
$U/E_{\{b,c\}} \neq U/E_{At}$, e cannot be deleted. $R = \{b,c,e\}$.			

Step 4: check c			
	b	e	e
o_1	0		1
o_2, o_3	1		0
o_4	2		1
o_5	2		0
o_6	3		2
o_7	3		1
$U/E_{\{b,e\}} \neq U/E_{At}$, c cannot be deleted. $R = \{b,c,e\}$.			

Step 5: check b			
	b	c	e
o_1, o_4		0	1
o_2, o_5		2	0
o_3		1	0
o_6		1	2
o_7		1	1
$U/E_{\{c,e\}} \neq U/E_{At}$, b cannot be deleted. $R = \{b,c,e\}$.			

Fig. 3. An illustration of using a deletion strategy for the information Table 1

Suppose the fitness function δ is interpreted as the frequency that an attribute appears in any element of the discernibility matrix M, i.e.,

$$\delta(a) = |\{m \in M \mid a \in m\}|. \tag{3}$$

We attempt to first delete an attribute that differentiates a small number of objects. We can obtain a set of quantitative values for our sample discernibility matrix in Table 1, such that $\delta(a) = 6$, $\delta(b) = 18$, $\delta(c) = 16$, $\delta(d) = 12$, and $\delta(e) = 15$. The yielded order is consistent with the order yielded by the information gain. Consequently, the same reduct is computed.

Many algorithms use entropy-based heuristics, such as information gain, conditional entropy, and mutual information [2, 15, 16, 24, 27]. Some algorithms use frequency-based heuristics with respect to the discernibility matrix, such as [6, 17, 22, 25].

Besides a quantitative evaluation, the fitness function δ can be interpreted as a qualitative evaluation. A qualitative method relies on pairwise comparisons of attributes. For any two attributes $a, b \in At$, we assume that a user is able to state whether one is more important than, or is more preferred to, the other. Based on the user preference, usually the preferred attributes are intended to be kept, and the unfavourable attributes are intended to be deleted.

5 Reduct Construction by Addition-Deletion

5.1 Control Strategy

By an addition-deletion strategy, we start the construction from an empty set or the Core, and consequently add attributes until a super-reduct is obtained. The constructed super-reduct contains a reduct, but itself is not necessarily a reduct unless all the attributes in it are individually necessary. We need to delete the superfluous attributes in the super-reduct till a reduct is found [29,30]. The addition-deletion methods can be described generally in Algorithm 2.

Algorithm 2. *The addition-deletion method for computing a reduct*
Input: *An information table.*
Output: *A reduct R.*

Addition:
(1) $R = \emptyset, CA = At$.
(2) **While** *R is not jointly sufficient and $CA \neq \emptyset$:*
  (2.1) Compute fitness values of all the attributes in CA regarding the property ρ using a fitness function σ;
  (2.2) Select an attribute a according to its fitness, let $CA = CA - \{a\}$;
  (2.3) Let $R = R \cup \{a\}$.

Deletion:
(3) $CD = R$.
(4) **While** $CD \neq \emptyset$:
  (4.1) Compute fitness values of all the attributes in CD regarding the property ρ using a fitness function δ;
  (4.2) Select an attribute a according to its fitness, let $CD = CD - \{a\}$;
  (4.3) **If** *$R - \{a\}$ is jointly sufficient, let $R = R - \{a\}$.*
(5) **Output** R.

The addition-deletion strategy has been proposed and studied since the deletion strategy is not efficient, and the straightforward addition process can only find a super-reduct, but not a reduct. A lack of consideration of the latter problem has produced many incomplete reduct construction algorithms, such as the ones reported in [8, 10, 16, 20]. An addition-deletion algorithm based on the discernibility matrix has been proposed by Wang and Wang [25], which can construct the subset of attributes from At, and then reduce it to a reduct efficiently.

Due to the fact that an addition-deletion method computes a relatively precise super-reduct first, the deletion checking process is expected to be more efficient than a straightforward deletion-only method. This is true when regarding some orders, a super-reduct is constructed pretty fast. However, the process of computing a super-reduct itself is also time consuming, as well as the process of deleting the superfluous attributes from the constructed super-reduct.

5.2 Attribute Selection Heuristics

For the addition-deletion strategies, the orders of attributes for addition and deletion are both essential for the result reduct. Regarding a property ρ, by using the fitness function σ, we add the fit attributes to the empty set or the Core to form a super-reduct; by using the fitness function δ, we delete the superfluous attributes from the super-reduct in order to form a reduct. σ and δ can be two different heuristics, or the same heuristic. If one can order the attributes according to a fitness function δ from the most fit attribute to the least fit attribute, then this order can be used for adding them one by one until the sufficient condition is met, and the reversed order can be used for deleting the superfluous attributes. By this means, one heuristic determines two orders, and a reduct composed of more fit attributes is obtained.

Example 5. For the information table in Table 1, suppose the fitness function σ is interpreted as the frequency or information gain as we have defined for the fitness function δ in the previous section. A set of quantitative values are computed according to the chosen heuristic. The attribute b is mostly intended to be added, followed by attributes c, e, d and a. In this case, a super-reduct $\{b, c, e\}$ is computed. After using the reverse order to check the necessity, this super-reduct is identified as a reduct. The iterative steps are illustrated in Figure 4.

Step 1: add *b*

	b
o_1	0
o_2, o_3	1
o_4, o_5	2
o_6, o_7	3

$U/E_{\{b\}} \neq U/E_{At}$, $R = \{b\}$.

Step 2: add *c*

	b	c
o_1	0	0
o_2	1	2
o_3	1	1
o_4	2	0
o_5	2	2
o_6, o_7	3	1

$U/E_{\{b,c\}} \neq U/E_{At}$, $R = \{b,c\}$.

Step 3: add *e*

	b	c	e
o_1	0	0	1
o_2	1	2	0
o_3	1	1	0
o_4	2	0	1
o_5	2	2	0
o_6	3	1	2
o_7	3	1	1

$U/E_{\{b,c,e\}} = U/E_{At}$, $R = \{b,c,e\}$.

Step 4: check *e*

	b	c	e
o_1	0	0	
o_2	1	2	
o_3	1	1	
o_4	2	0	
o_5	2	2	
o_6, o_7	3	1	

$U/E_{\{b,c\}} \neq U/E_{At}$, e cannot be deleted. $R = \{b,c,e\}$.

Step 5: check *c*

	b	c	e
o_1	0		1
o_2, o_3	1		0
o_4	2		1
o_5	2		0
o_6	3		2
o_7	3		1

$U/E_{\{b,e\}} \neq U/E_{At}$, c cannot be deleted. $R = \{b,c,e\}$.

Step 6: check *b*

	b	c	e
o_1, o_4		0	1
o_2, o_5		2	0
o_3		1	0
o_6		1	2
o_7		1	1

$U/E_{\{c,e\}} \neq U/E_{At}$, b cannot be deleted. $R = \{b,c,e\}$.

Fig. 4. An illustration of using an addition-deletion strategy for the information Table 1

6 Reduct Construction by Addition

6.1 Control Strategy

By an addition method, we start the reduct construction process from an empty set or the Core, and consequently add attributes to it until it becomes a reduct. The essential difference between the addition method and the addition-deletion method is that, the addition method takes in one attribute if the constructed set is a partial reduct, while the addition-deletion method continuously adds attributes until a super-reduct is produced. In this case, superfluous attributes can be added by an addition-deletion method, and the deletion process is required to eliminate them. The addition methods can be described generally in Algorithm 3.

Algorithm 3. *The addition method for computing a reduct*
Input: *An information table S.*
Output: *A reduct R.*

(1) $R = \emptyset$, $CA = At$;
(2) **While** $CA \neq \emptyset$:
  (2.1) Compute fitness values of all the attributes in CA regarding the property ρ using a fitness function σ;
  (2.2) Select an attribute a according to its fitness;
  (2.3) **If** *a is a core attribute, then let $R = R \cup \{a\}$ and $CA = CA - \{a\}$*
  else
    (2.3.1) Compute fitness values of all the elements in $Group(a) = \{m \in M \mid a \in m\}$ regarding the property ρ using a fitness function δ';
    (2.3.2) **If** *$Group(a) = \emptyset$, let $CA = CA - \{a\}$ and go to Step (2),* **else**, *select an element $m = \{a\} \cup A$ according to its fitness;*
    (2.3.3) **If** *$CA - A$ is jointly sufficient, let $R = R \cup \{a\}$ and $CA = CA - m$,* **else**, *go to Step (2.3.2).*
    If *a cannot be made necessary regarding all $m \in Group(a)$, let $CA = CA - \{a\}$.*
(3) **Output** R.

For a selected attribute $a \in CA$, $Group(a) = \{m \in M \mid a \in m\}$ is the set of matrix elements that each indicates an object pair that can be distinguished by a. If a is a core attribute then it is individually necessary for constructing a reduct. If a is a non-core attribute, one can make a necessary by eliminating its associated attributes in an element $m \in Group(a)$ from further consideration. Suppose $Group(a) = \{m_1, m_2, \ldots, m_d\}$, $(m_i = A \cup \{a\}) \in Group(a)$ and $A \neq \emptyset$. It means that all the attributes in m can distinguish the object pair associated with m, and the attribute a is not individually necessary for such a task. We can make a necessary by eliminating all the attributes in A. If A is a superset of another element $m' \in M$, then A is necessary for distinguishing the object pair associated with m', which means that A cannot be eliminated. In other words, the attribute a cannot be made necessary regarding m. If a cannot be made necessary regarding all $m_i \in Group(a)$, then a cannot be added to the partial reduct.

Example 6. For our running example, suppose the non-core attribute a is selected. It is easy to obtain from the matrix in Table 2 that $Group(a) = \{m_{o_1,o_4}, m_{o_2,o_4}, m_{o_3,o_4}, m_{o_4,o_5}, m_{o_4,o_6}, m_{o_4,o_7}\}$. Suppose to make a necessary, the matrix element $m_{o_1,o_4} = \{a, b, d\}$ is selected, and thus the attributes in the set $m_{o_1,o_4} - \{a\} = \{b, d\}$ need to be eliminated. However, the elimination will cause the element $m_{o_2,o_5} = \{b, d\}$ becomes empty. In other words, the object pair (o_2, o_5) can no longer be distinguished. Therefore, attribute set $\{b, d\}$ cannot be eliminated, which means that attribute a cannot be added to the partial reduct regarding the matrix element m_{o_1,o_4}. We can easily verify that a cannot be added regarding any matrix element in $Group(a)$, thus a does not belong to the partial reduct.

6.2 Attribute Selection Heuristics

The addition algorithm requires the attributes added to the reduct are individually necessary. To ensure it, the associated attributes are eliminated for consideration. At the same time, the elimination should not change the joint sufficiency of the remaining attributes. Therefore, the general addition algorithm explicitly checks both the sufficiency condition and the necessity condition in Step (2.3.3). Its time complexity is higher than the general deletion algorithm.

Zhao and Wang suggested using a matrix absorption operation to simplify the checking process [29]. The matrix absorption operation is a sequence of all possible element absorption operations on pairs of elements whenever the following condition holds:

$$\emptyset \neq M(x', y') \subset M(x, y).$$

That is, the value of $M(x, y)$ is replaced by the value of $M(x', y')$ in the matrix. We also say $M(x, y)$ is absorbed by $M(x', y')$. The physical meaning of the absorption can be explained as follows. Suppose $M(x', y') \neq \emptyset$ and $M(x', y') \subset M(x, y)$. The set of attributes discerning both pairs (x', y') and (x, y) is given by $M(x, y) \cap M(x', y') = M(x', y')$. After absorption, $M(x, y)$ becomes $M(x', y')$. Attributes in $M(x', y')$ are sufficient to discern both object pairs (x', y') and (x, y). When an attribute from $M(x', y')$ is in a reduct, the same attribute can be used to discern (x, y). Thus, it is not necessary to consider attributes in $M(x, y) - M(x', y')$. After matrix absorption, no element in the matrix is a proper subset of another element.

By using the matrix absorption operation, the general addition algorithm can be much simplified. Let attribute a be selected in Steps (2.1) and (2.2), and $Group(a)$ store the elements contains a from the absorbed matrix. When an element $(m = \{a\} \cup A) \in Group(a)$ is selected in Steps (2.3.1) and (2.3.2), A can be eliminated immediately. Since A is not a proper subset of another element, thus is not necessary for distinguishing any object pair. Since $CA - A$ is ensured jointly sufficient, therefore, m can be eliminated after attribute a being made necessary and added to the partial reduct. We can set $CA = \{a \in At \mid Group(a) \neq \emptyset\}$, and apply the matrix absorption operation every time after the CA is updated.

Example 7. For our running Example 3, we can observe that the distinct matrix element $\{b, c\}$ can distinguish the object pairs (o_1, o_7) and (o_3, o_5) whose corresponding matrix elements equal to $\{b, c\}$, and also the object pairs (o_1, o_2), (o_1, o_3), (o_1, o_5), (o_1, o_6), (o_2, o_4), (o_2, o_6), (o_2, o_7), (o_3, o_4), (o_4, o_6), (o_4, o_7), (o_5, o_6) and (o_5, o_7) whose corresponding matrix elements contain $\{b, c\}$. By applying the matrix absorption operation we can obtain:

$\{d, e\}$ absorbs $\{b, d, e\}, \{a, c, d, e\}, \{b, c, d, e\}, At$;
$\{b, e\}$ absorbs $\{b, c, e\}, \{b, d, e\}, \{a, b, c, e\}, \{b, c, d, e\}, At$;
$\{b, c\}$ absorbs $\{b, c, e\}, \{a, b, c, d\}, \{a, b, c, e\}, \{b, c, d, e\}, At$;
$\{b, d\}$ absorbs $\{a, b, d\}, \{b, d, e\}, \{a, b, c, d\}, \{b, c, d, e\}, At$;
$\{c, d\}$ absorbs $\{a, b, c, d\}, \{a, c, d, e\}, \{b, c, d, e\}, At\}$.

As a result, the absorbed discernibility matrix contains the following distinct elements $\{b, c\}$, $\{b, d\}$, $\{b, e\}$, $\{c, d\}$, $\{d, e\}$. We can use $\widehat{M}$ denote the absorbed matrix in a set representation.

For the absorbed discernibility matrix, if we group the matrix elements, we obtain five overlapped sets:

$$\begin{aligned}
Group(a) &= \emptyset, \\
Group(b) &= \{\{b, c\}, \{b, d\}, \{b, e\}\}, \\
Group(c) &= \{\{b, c\}, \{c, d\}\}, \\
Group(d) &= \{\{b, d\}, \{c, d\}, \{d, e\}\}, \\
Group(e) &= \{\{b, e\}, \{d, e\}\}.
\end{aligned}$$

Attribute set $CA = \{b, c, d, e\}$.

The fitness function σ can be the one that we discussed in Sections 4 and 5. We need to discuss more about the fitness function δ'. We should note that the fitness function δ' of the proposed addition algorithm is different from the fitness function δ of the general deletion algorithm. That is because δ evaluates the fitness of one single attribute at a time, and δ' evaluates the fitness of a matrix element m, which is a set of attributes. Typically, δ' is the summation or the average fitness of all the included attributes.

Quantitatively, the selection of a matrix element for deletion can be described by a mapping:

$$\delta' : \{m_i \in Group(a)\} \longrightarrow \Re. \tag{4}$$

The meaning of the function δ' is determined by many semantic considerations as well.

Example 8. For example, a frequency-based heuristic can be defined as follows. For $m_i = \{a\} \cup A$,

$$\delta'(m_i) = |\{m \in \widehat{M} \mid m \cap A \neq \emptyset\}|. \tag{5}$$

For the running example, if the reduct attribute b is selected according to the information gain measure, we thus focus on $Group(b) = \{\{b, c\}, \{b, d\}, \{b, e\}\}$. Using the former heuristic, we obtain that $\delta'(\{b, c\}) = 2, \delta'(\{b, d\}) = 3$, and $\delta'(\{b, e\}) = 2$ in $\widehat{M}$. Suppose we therefore pick the element $\{b, d\}$. Consequently, a reduct $\{b, c, e\}$ can be computed. The iterative steps are illustrated in Figure 4.

We can also define the fitness function δ' as the information entropy, i.e., the joint entropy of all the attributes in the attribute set $m_i - \{a\}$. For example, if $m_i - \{a\} = \{b, c\}$, then

$$\begin{aligned}
\delta'(m_i) &= H(m_i - \{a\}) \\
&= H(\{b, c\}) \\
&= - \sum_{x \in V_b} \sum_{y \in V_c} p(b, c) \log p(b, c).
\end{aligned} \tag{6}$$

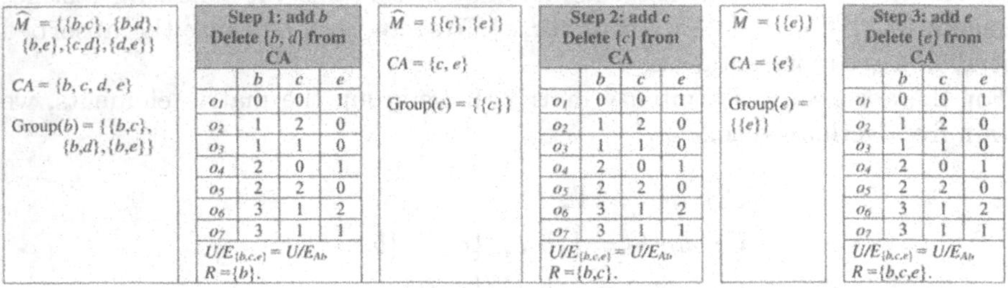
$\widehat{M}$ = {{b,c}, {b,d}, {b,e},{c,d},{d,e}}
CA = {b, c, d, e}
Group(b) = {{b,c}, {b,d},{b,e}}
Step 1: add b Delete {b, d} from CA
b c e
o1 0 0 1
o2 1 2 0
o3 1 1 0
o4 2 0 1
o5 2 2 0
o6 3 1 2
o7 3 1 1
U/E{b,c,e} = U/EAt, R = {b}.
$\widehat{M}$ = {{c}, {e}}
CA = {c, e}
Group(c) = {{c}}
Step 2: add c Delete {c} from CA
b c e
o1 0 0 1
o2 1 2 0
o3 1 1 0
o4 2 0 1
o5 2 2 0
o6 3 1 2
o7 3 1 1
U/E{b,c,e} = U/EAt, R = {b,c}.
$\widehat{M}$ = {{e}}
CA = {e}
Group(e) = {{e}}
Step 3: add e Delete {e} from CA
b c e
o1 0 0 1
o2 1 2 0
o3 1 1 0
o4 2 0 1
o5 2 2 0
o6 3 1 2
o7 3 1 1
U/E{b,c,e} = U/EAt, R = {b,c,e}.

Fig. 5. An illustration of using an addition strategy for the information Table 1

By applying this heuristic to the sample information Table 1, we obtain that $\delta'(\{b,c\}) = H(\{c\}) = 1.557$, $\delta'(\{b,e\}) = H(\{e\}) = 1.449$ and $\delta'(\{b,d\}) = H(\{d\}) = 0.985$. The reduct $\{b,c,e\}$ can be computed if the element $\{b,d\}$ is selected.

Similarly, qualitative evaluation can also be applied here for selecting a matrix element for deletion. This can be based on the user preference on the attribute set, that we have discussed in the previous sub-section. We usually select the most unfavourable matrix element for deletion.

Example 9. For our running example in Table 1, we only find two reducts according to the introduced heuristics. By applying different heuristics, we may be able to find the rest of reducts, like $\{c,d,e\}$ and $\{b,d\}$. The attribute lattice shown in Figure 6 highlights all the reducts by stars, and super-reducts by shadings, partial reducts by circles with solid lines.

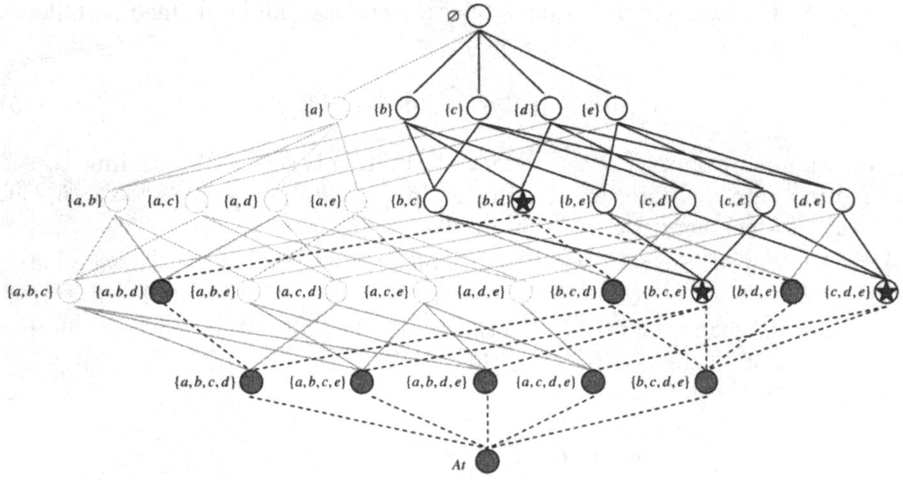

Fig. 6. An illustration of super- and partial reducts in the attribute lattice of Table 1

7 Time Complexity Analysis

Suppose the partition of the information table is chosen for the time complexity analysis. For an attribute $a \in At$, the execution of $U/E_{\{a\}}$ needs to compare each object pair regarding attribute a. It thus requires $\frac{|U|(|U|+1)}{2}$ comparisons, where $|U|$ is the cardinality of U. For an attribute set $A \subset At$, the execution of U/E_A needs to compare each object pair regarding all the attributes in A. It thus requires $\frac{|U|(|U|+1)}{2}|A|$ comparisons.

The attribute deletion operation of the deletion strategy is to check if the remaining attribute set is still jointly sufficient for each iteration. To check the necessity of attribute a_1, one needs to verify if $At - \{a_1\}$ produces the same partition as At does, and thus $\frac{|U|(|U|+1)}{2}(|At| - 1)$ comparisons are required. If a_1 is deleted after the checking, then to verify if $At - \{a_1\} - \{a_2\}$ produces the same partition as At does, one needs $\frac{|U|(|U|+1)}{2}(|At| - 2)$ comparisons. If a_1 is not deleted after the checking, then one still needs $\frac{|U|(|U|+1)}{2}(|At| - 1)$ comparisons, to check the necessity of a_2. Totally, $O(|U|^2|At|^2)$ comparisons are required to check the jointly sufficiency condition and the individual necessity condition for all attributes.

The addition-deletion strategy checks the joint sufficiency condition for a constructed super-reduct, and checks the individual necessity condition for all the attributes in the constructed super-reduct. To verify if $U/E_{\{a_1\}} = U/E_{At}$ one needs $\frac{|U|(|U|+1)}{2}$ comparisons. To verify if $U/E_{\{a_1,a_2\}} = U/E_{At}$ one needs $\frac{2|U|(|U|+1)}{2}$ comparisons, and so on. Totally, $O(|U|^2|At|^2)$ comparisons are required to construct a super-reduct. And same number of comparisons are required to check the necessity of all attributes.

The addition strategy picks an attribute to make it individually necessary by eliminating its associated attributes, at the same time, it ensures the elimination does not change the joint sufficiency of the remaining attributes. To ensure the necessity of attribute a_1, one needs to verify if $At - A_1$ produces the same partition as At does, where $m_{a_1} = \{a_1\} \cup A_1$. This requires $\frac{|U|(|U|+1)}{2}(|At| - |A_1|)$ comparisons. If a_1 is added after the checking, then one needs to verify if $At - m_{a_1} - A_2$ produces the same partition as At does, where $m_{a_2} = \{a_2\} \cup A_2$. This requires $\frac{|U|(|U|+1)}{2}(|At| - |m| - |A_2|)$ comparisons. If a_1 is not added after the checking, then one still needs $\frac{|U|(|U|+1)}{2}(|At| - |A_2| - 1)$ comparisons, to ensure the necessity of a_2. Totally, $O(|U|^2|At|^2)$ comparisons are required to check the jointly sufficiency condition and the individual necessity condition for all attributes.

This analysis is very rough. It should be noted that the addition-deletion algorithm normally does not need to add all attributes in At for a super-reduct. The addition algorithm relies on the absorption operation to simplify the matrix and generate the groups for all attributes. Normally, the addition algorithm is the most inefficient one comparing to the other two.

8 Conclusion

This paper provides a critical study of the existing reduct construction algorithms based on a two-level view: a high level view of control strategy and a low level view of attribute selection heuristics. Three groups of algorithms are discussed based on the deletion strategy, the addition-deletion strategy and the addition strategy.

We define the concepts of super-reducts and partial reducts besides the concept of reduct. A deletion strategy and an addition-deletion strategy strike to find a reduct from a super-reduct. An addition strategy strikes to find a reduct from a partial reduct.

This paper may be considered as an attempt to synthesize the results from existing studies into a general and easy to understand form, with an objective towards a more abstract theory. Any success in such a research will not only produce valuable insights into the problem, but also provide guidelines for the design of new reduct construction algorithms.

References

1. Bazan, J.G., Nguyen, H.S., Nguyen, S.H., Synak, P., Wroblewski, J.: Rough set algorithms in classification problem. In: Polkowski, L., Tsumoto, S., Lin, T.Y. (eds.) Rough Set Methods and Applications, pp. 49–88 (2000)
2. Beaubouef, T., Petry, F.E., Arora, G.: Information-theoretic measures of uncertainty for rough sets and rough relational databases. Information Sciences 109, 185–195 (1998)
3. Blum, A.L., Langley, P.: Selection of relevant features and examples in machine learning. Artificial Intelligence, 245–271 (1997)
4. Devijver, P.A., Kittler, J.: Pattern Recognition: A Statistical Approach. Prentice-Hall, New York (1982)
5. Grzymala-Busse, J.W.: LERS - A system for learning from examples based on rough sets. In: Slowinski, R. (ed.) Intelligent Decision Support, pp. 3–18. Kluwer Academic Publishers, Boston (1992)
6. Hoa, N.S., Son, N.H.: Some efficient algorithms for rough set methods. In: Proceedings of the Conference of Information Processing and Management of Uncertainty in Knowledge-based Systems, pp. 1451–1456 (1996)
7. Hu, X.: Using rough sets theory and database operations to construct a good ensemble of classifiers for data mining applications. In: Proceedings of ICDM, pp. 233–240 (2001)
8. Hu, X., Cercone, N.: Learning in relational databases: a rough set approach. International Journal of Computation Intelligence 11, 323–338 (1995)
9. Jain, A.K., Zongker, D.: Feature selection: evaluation, application and small sample performance. IEEE Transactions on Pattern Analysis and Machine Intelligence 19, 153–158 (1997)
10. Jenson, R., Shen, Q.: A rough set-aided system for sorting WWW bookmarks. In: Zhong, N., et al. (eds.) Web Intelligence: Research and Development, pp. 95–105 (2001)
11. John, G.H., Kohavi, R., Pfleger, K.: Irrelevant features and the subset selection problem. In: Proceedings of the Eleventh International Conference on Machine Learning, pp. 121–129 (1994)

12. Kohavi, R., John, G.H.: Wrappers for feature subset selection. Artificial Intelligence, 273–324 (1997)
13. Koller, D., Sahami, M.: Toward optimal feature selection. In: Proceedings of the Thirteenth International Conference of Machine Learning, pp. 284–292 (1996)
14. Mi, J.S., Wu, W.Z., Zhang, W.X.: Approaches to knowledge reduction based on variable precision rough set model. Information Sciences 159, 255–272 (2004)
15. Miao, D., Hou, L.: A comparison of rough set methods and representative inductive learning algorithms. Fundamenta Informaticae 59, 203–219 (2004)
16. Miao, D., Wang, J.: An information representation of the concepts and operations in rough set theory. Journal of Software 10, 113–116 (1999)
17. Pawlak, Z.: Rough Sets: Theoretical Aspects of Reasoning About Data. Kluwer, Boston (1991)
18. Portinale, L., Saitta, L.: Feature selection, Technical report, D14.1, University of Dortmund (2002)
19. Rauszer, C.: Reducts in information systems. Foundamenta Informaticae 15, 1–12 (1991)
20. Shen, Q., Chouchoulas, A.: A modular approach to generating fuzzy rules with reduced attributes for the monitoring of complex systems. Engineering Applications of Artificial Intelligence 13, 263–278 (2000)
21. Skowron, A., Rauszer, C.: The discernibility matrices and functions in information systems. In: Slowiński, R. (ed.) Intelligent Decision Support, Handbook of Applications and Advances of the Rough Sets Theory. Kluwer, Dordrecht (1992)
22. Slezak, D.: Various approaches to reasoning with frequency based decision reducts: a survey. In: Polkowski, L., Tsumoto, S., Lin, T.Y. (eds.) Rough set methods and applications, pp. 235–285 (2000)
23. Swiniarski, R.W.: Rough sets methods in feature reduction and classification. International Journal of Applied Mathematics and Computer Science 11, 565–582 (2001)
24. Wang, G., Yu, H., Yang, D.: Decision table reduction based on conditional information entropy. Chinese Journal of Computers 25, 759–766 (2002)
25. Wang, J., Wang, J.: Reduction algorithms based on discernibility matrix: the ordered attributes method. Journal of Computer Science and Technology 16, 489–504 (2001)
26. Wong, S., Ziarko, W.: On optimal decision rules in decision tables. Bulletin of the Polish Academy of Sciences and Mathematics, 693–696 (1985)
27. Yu, H., Yang, D., Wu, Z., Li, H.: Rough set based attribute reduction algorithm. Computer Engineering and Applications 17, 22–47 (2001)
28. Zhao, M.: Data Description Based on Reduct Theory, Ph.D. Thesis, Institute of Automation, Chinese Academy of Sciences (2004)
29. Zhao, K., Wang, J.: A reduction algorithm meeting users' requirements. Journal of Computer Science and Technology 17, 578–593 (2002)
30. Ziarko, W.: Rough set approaches for discovering rules and attribute dependencies. In: Klösgen, W., Żytkow, J.M. (eds.) Handbook of Data Mining and Knowledge Discovery, pp. 328–339 (2002)

Attribute Set Dependence in Reduct Computation*

Pawel Terlecki and Krzysztof Walczak

Institute of Computer Science, Warsaw University of Technology,
Nowowiejska 15/19, 00-665 Warsaw, Poland
P.Terlecki, K.Walczak@ii.pw.edu.pl

Abstract. In the paper we propose a novel approach to finding rough set reducts in information systems. Our method combines an apriori-like scheme of space traversing with an efficient pruning condition based on attribute set dependence. Moreover, we discuss theoretical and implementational aspects of our pruning procedure, including adopting a bst and a trie tree for storing set collections. Operation number and execution time tests have been performed in order to demonstrate the efficiency of our approach.

Keywords: rough sets, reduct, apriori, set dependence, trie.

1 Introduction

Cognitive informatics (CI) is a new discipline that has recently emerged from cognitive science (Wang 2002). It combines various existing approaches under the common goal of understanding the gist of intelligent thinking. In fact, CI struggles to construct a set of computational models that could explain particular states and processes in human's brain, like reasoning, abstracting, learning, emotions, etc. Among many other approaches within knowledge discovery the rough set theory seems to be one of the most convenient tools for solving many CI problems (Semeniuk-Polkowska & Polkowski 2003). It provides simple, elegant and powerful way for describing inexact, uncertain and vague information. Moreover, it proposes the concept of a reduct that is widely used in data reduction, feature selection, rule induction and object classification (Bazan, Nguyen, Nguyen, Synak & Wroblewski 2000, Swiniarski 2001). Since the time of introduction (Pawlak 1982), it has gained numerous advocates in various fields of application being combined with statistical methods, neural networks (Swiniarski & Skowron 2003), fuzzy sets and other valuable approaches. The past decade has also brought some insights on relations between data-mining and rough set problems has appeared (Lin 1996, Terlecki & Walczak 2006).

The paper refers to the reduct set problem, which is one of the most fundamental rough set issues, defined as finding all the reducts of an information

* The research has been partially supported by grant No 3 T11C 002 29 received from Polish Ministry of Education and Science.

system. In order to apply the already known solutions, the problem is frequently transformed into the problem of finding the prime implicants of a monotonous boolean function. The classic methods employ the notions of discernibility matrix and discernibility function (Skowron & Rauszer 1992). On the other hand, there are some algorithms that efficiently traverse an attribute set space by means of pruning conditions employing concise representations (Kryszkiewicz 1994, Kryszkiewicz & Cichon 2004). In practical problems, it is often enough to compute only a subset of all the existing reducts. Most basic approaches focus on finding only the best reduct according to some criteria (Hu, Lin & Han 2003) or multiple reduct (Wu, Bell & McGinnity 2005). Moreover, some heuristic, evolutionary ideas have been proposed (Wroblewski 1995).

The algorithms presented in the paper follow the Apriori scheme of set generation (Agrawal & Srikant 1994). We propose a novel pruning condition based on the notion of set dependence. The convexity of complement subspaces of dependent and independent sets has been demonstrated. Our method traverses the subspace of independent sets. We also show how to construct an algorithm in order to test the condition efficiently and to avoid maintaining additional structures. It is worth noticing that several methods in the rough set theory examine the space similarly (Bazan, Nguyen, Nguyen, Synak & Wroblewski 2000, Nguyen 2002).

One of the major challenges is to efficiently employ rough set methods in large databases. In the case of reduct computation the large number of objects increases strongly the cost of discernibility calculation for a given attribute set. Therefore, we performed several tests to prove the usefulness of our pruning strategy in reducing the number of these operations.

Section 2 provides selected elements of the rough set theory and border representations. In Section 3 we consider the notions of discernibility and dependence, and give theoretical background for a proposed pruning approach. The algorithm is described in Section 4 and followed by a brief analysis and comments on their implementation provided in Section 5. A discussion on data structures for storing set collections is presented in Section 6. Section 7 contains results obtained for several popular data sets. Tests focus on the efficiency of our pruning condition measured by the number of dominant operations. Time results for different data structures are covered in Section 8. The paper is summarized in Section 9.

2 Preliminaries

Let an information system be a pair $(\mathcal{U}, \mathcal{A})$, where $\mathcal{U} = \{u_1, .., u_{|\mathcal{U}|}\}$(universum) is a non-empty, finite set of objects and $\mathcal{A}$ is a non-empty finite set of attributes. The domain of an attribute $a \in \mathcal{A}$ is denoted by V_a and its value for an object $u \in \mathcal{U}$ is denoted by $a(u)$.

Consider $B \subseteq \mathcal{A}$. An indiscernibility relation $IND(B)$ is defined as follows: $IND(B) = \{(u, v) \in \mathcal{U} \times \mathcal{U} : \forall_{a \in B}\ a(u) = a(v)\}$. An attribute $a \in B$ is dispensable in B, iff $IND_{B-\{a\}} = IND_B$, otherwise a is indispensable. We call B independent, iff all its members are indispensable, otherwise it is dependent.

An attribute set $B \subseteq \mathcal{A}$ is a super reduct, iff $IND(B) = IND(\mathcal{A})$. An independent super reduct is called a reduct. Finding all the reducts of an information system is called the reduct set problem. For the sake of convenience, we introduce the following collections.

Definition 1. *Independent set collection* $ISC = \{B \subseteq \mathcal{A} : B \text{ is independent}\}$. *Dependent set collection* $DSC = \{B \subseteq \mathcal{A} : B \text{ is dependent}\}$. *Super reduct collection* $URED = \{B \subseteq \mathcal{A} : IND(B) = IND(\mathcal{A})\}$. *Reducts collection* $RED = ISC \cap URED$.

The property of set dependence generates a binary partition $\{ISC, DCS\}$ in $P(\mathcal{A}) = 2^{\mathcal{A}}$. Moreover, it can be easily demonstrated that every subset of an independent set is independent and every superset of a dependent set is dependent. These facts are expressed formally below.

Lemma 1. *Let* $B, S \subseteq \mathcal{A}$*, we have:* $S \subseteq B \wedge B \in ISC \Longrightarrow S \in ISC$.

Lemma 2. *Let* $B, S \subseteq \mathcal{A}$*, we have:* $B \subseteq S \wedge B \in DSC \Longrightarrow S \in DSC$.

A discernibility matrix C is a matrix $|\mathcal{U}| \times |\mathcal{U}|$ with elements $C_{ij} = \{a \in \mathcal{A} : a(u_i) \neq a(u_j)\}$ for $i, j = 1..|\mathcal{U}|$. This matrix can be used to check whether a given attribute set differentiates objects as well as $\mathcal{A}$ does. Let EC be a set of all elements of a matrix C. The following measure allows to make inferences about discernibility avoiding direct usage of comparison of relations.

Definition 2. *Let* $B \subseteq \mathcal{A}$*. We define as:*

$$covcount(B) = |\{X \in EC : X \cap B \neq \emptyset\}|.$$

Lemma 3. *Let* $B, S \subseteq \mathcal{A}$ *such that* $S \subset B$*, we have:* $IND(S) = IND(B) \Longleftrightarrow covcount(S) = covcount(B)$.

In the paper, we decided to use concise set representations to describe regions of the search space $P(\mathcal{A})$. It requires the following notions.

Consider a set S. A border is an ordered pair $<\mathcal{L}, \mathcal{R}>$ such that $\mathcal{L}, \mathcal{R} \subseteq P(S)$ are antichains and $\forall_{X \in \mathcal{L}} \exists_{Z \in \mathcal{R}} X \subseteq Z$. $\mathcal{L}$ and $\mathcal{R}$ are called a left and a right bound, respectively. A border $<\mathcal{L}, \mathcal{R}>$ represents a set interval $[\mathcal{L}, \mathcal{R}] = \{Y \in P(S) : \exists_{X \in \mathcal{L}} \exists_{Z \in \mathcal{R}} X \subseteq Y \subseteq Z\}$. The left and right bounds consist, respectively, of minimal elements and maximal elements of a set, assuming inclusion relation.

The collection $F \subseteq P(S)$ is a convex space (or is interval-closed) if we have: $\forall_{X,Z \in F} \forall_{Y \in P(S)} X \subseteq Y \subseteq Z \Rightarrow Y \in F$. Definitions of a border and a convex space lead to a conclusion that every convex space has a unique border and every collection that has a border is convex.

For brevity, we use the following notation: an expression *k-set* denotes a k-element set. Moreover, for a given set collection F we introduce a convenient notation $F_k = \{B \in F : |B| = k\}$, i.e ISC_k, RED_k, $P_k(\mathcal{A})$, etc. For a given set S, we call its subset (superset) *direct* when it has the cardinality smaller (greater) by 1 than the cardinality of S.

3 Discernibility and Dependency

In the reduct set problem we deal with an exponentially-large search space $P(\mathcal{A})$. Therefore, the algorithms that solve the problem by traversing the space have to use such strategies that avoid examining all possible attribute sets.

These methods are constructed around to main issues. The first one is to give efficient pruning conditions. The basic idea is to visit only those regions about which we cannot infer from the already examined subspace. The second issue is strongly influenced by the pruning strategy and concerns the way of traversing the search space. It has two objectives: to make the pruning stage as efficient as possible and not to generate exponentially-large set collections.

We begin our consideration with a discussion of pruning conditions and then combine them with the appropriate ways of space traversing.

Basic criteria originate from works related to monotonous boolean functions. In particular, the following two conditions are extensively discussed in (Kryszkiewicz 1994).

Theorem 1. *(Kryszkiewicz 1994) Let* $B \subseteq \mathcal{A}$*, we have:* $S \subset B \wedge B \notin URED \Longrightarrow S \notin RED$.

Theorem 2. *(Kryszkiewicz 1994) Let* $B, S \subseteq \mathcal{A}$*, we have:* $B \subset S \wedge B \in URED \Longrightarrow S \notin RED$.

The former uses the notion of discernibility and states that we do not need to examine actual subsets of a non-super reduct B, since they cannot differentiate more object pairs than B does. The latter tells us that actual supersets of a reduct cannot be minimal, so they can be also excluded from examination.

In the text we propose a strategy that is based solely on set dependence. The following theorem refers to convexity and the next one generalizes Theorem 2.

Theorem 3. *Collections* ISC *and* DSC *are convex. There exist subcollections* $MISC, mDSC \subseteq P(\mathcal{A})$ *such that* ISC *has a border* $< \emptyset, MISC >$ *and* DSC *has a border* $< mDSC, \{\mathcal{A}\} >$*, where the symbols* $MISC$ *and* $mDSC$ *stand for maximal independent set collection and minimal dependent set collection, respectively.*

Proof. It is sufficient to show that both collections have specified borders.

Let us focus on ISC first. Consider $ISC \subseteq [\{\emptyset\}, MISC]$. Let $B \in ISC$. Obviously, $B \supseteq \emptyset$. Notice that inclusion partially orders elements in ISC, so also $\exists_{S \in MISC} B \subseteq S$. Conversely, $ISC \supseteq [\{\emptyset\}, MISC]$. Let $B \in [\{\emptyset\}, MISC]$. From the definition of a border we have $\exists_{S \in MISC} \emptyset \subseteq B \subseteq S$. According to Lemma 1 B is independent, so $B \in ISC$. Summing up, we have found that ISC has a border $< \{\emptyset\}, MISC >$ and, consequently, is convex.

A proof for DSC is analogical and employs Lemma 2.

Theorem 4. *Let* $B, S \subseteq \mathcal{A}$*, we have:* $B \subseteq S \wedge B \in DSC \Longrightarrow S \notin RED$.

Proof. Consider $B \in DSC$ and $S \subseteq \mathcal{A}$ such that $B \subseteq S$. From Lemma 2 we have $S \in DSC$. Thus, $S \notin ISC$ and S cannot be a reduct.

According to the definition, it is possible to test set dependence by examining all direct subsets of a given set. In practice, it is convenient to use *covcount* to verify set dependence.

Theorem 5. *Let* $B \subseteq \mathcal{A}$*, we have:* $\exists_{a \in B} covcount(B) = covcount(B - \{a\}) \Longleftrightarrow B \in DSC$.

Proof. From the definition attribute $a \in B$ is dispensable in B iff $IND(B) = IND(B - \{a\})$. From Lemma 3, where $S = B - \{a\}$, we have: $a \in B$ is dispensable iff $covcount(B) = covcount(B - \{a\})$.

However, every *covcount* computation can be costly when very large databases are concerned. Therefore, first we perform pruning using information on dependent sets and reducts visited so far. We check whether all direct subsets of a tested set are independent and are not reducts. Otherwise, the set is dependent basing on Lemma 2 or Theorem 2.

Theorem 6. *Let* $B \subseteq \mathcal{A}$*, we have:* $\exists_{a \in B}(B - \{a\}) \notin (ISC_{|B|-1} - RED_{|B|-1}) \Longrightarrow B \in DSC$.

Proof. Let $B \subseteq \mathcal{A}$ and $a \in B$ such that $(B - \{a\}) \notin (ISC_{|B|-1} - RED_{|B|-1})$. Since $|B - \{a\}| = |B| - 1$, so $(B - \{a\}) \in P_{|B|-1}(\mathcal{A}) - (ISC_{|B|-1} - RED_{|B|-1}) = DSC_{|B|-1} \cup RED_{|B|-1}$. Therefore, $(B - \{a\}) \in DSC_{|B|-1}$ or $(B - \{a\}) \in RED_{|B|-1}$. Let us consider both cases separately.

Table 1. The information system $\mathcal{IS} = (\{u_1, u_2, u_3, u_4, u_5\}, \{a, b, c, d, e\})$

	a	b	c	d	e
u_1	0	0	1	0	0
u_2	1	1	1	1	0
u_3	1	1	1	2	0
u_4	0	2	0	1	0
u_5	2	3	1	1	1

$MISC = \{\{a, c\}, \{a, d\}, \{b, d\}, \{c, d, e\}\}$
$mDSC = \{\{a, b\}, \{a, e\}, \{b, c\}, \{b, e\}\}$
$RED = \{\{a, d\}, \{b, d\} \{c, d, e\}\}$

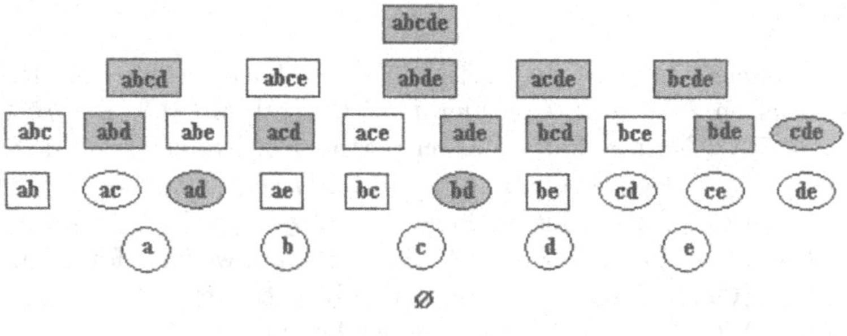

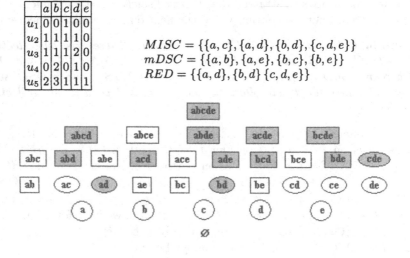

Fig. 1. The search space $P(\{a, b, c, d, e\})$ of the reduct set problem for the information system $\mathcal{IS}$. Independent sets - ovals, dependent sets - rectangles, super reducts - gray background.

Let $(B-\{a\}) \in DSC_{|B|-1} \subseteq DSC$. In accordance with Lemma 2 we have $(B-\{a\}) \subseteq B \wedge (B-\{a\}) \in DSC \Longrightarrow B \in DSC$. Let, now, $(B-\{a\}) \in RED_{|B|-1}$. It means that $IND(B-\{a\}) = IND(\mathcal{A}) = IND(B)$, so a is dispensable in B and $B \in DSC$.

Let us move to a brief example. We classify attribute sets according to two binary characteristics: dependence and discernibility. The information system $\mathcal{IS}$ and its search space are depicted in Table 1 and Fig. 1, respectively.

4 Algorithm Overview

In the paper we present a novel approach to finding all the reducts of an information system. Our method is a combination of apriori-like set generation and an efficient pruning technique based on Theorems 5 and 6.

The general scheme of our algorithm follows the classic apriori structure (Agrawal & Srikant 1994). In every step we generate a family of k-sets and use pruning techniques to remove reducts and dependent sets. The final family of every step L_k contains only independent sets that are not reducts.

The collections RED and ISC are created incrementally. In k-th step all their k-element members are computed. When the algorithm stops we obtain collections: $RED = \bigcup_{k=1..|\mathcal{A}|} RED_k$.

$RED_1 = \{\text{all 1-reducts}\}$
$ISC_1 = \{\text{all 1-sets}\}$
$L_1 = \{\text{all 1-sets}\} - RED_1$
for $(k = 2; L_{k-1} \neq \emptyset; k++)$ **do**
    $C_k = \textit{apriori-gen-join}(L_{k-1})$
    $D_k = \textit{prune-with-subsets}(C_k, L_{k-1})$
    $ISC_k = D_k - \textit{find-dependent}(D_k)$
    $RED_k = \textit{find-RED}(ISC_k)$
    $L_k = ISC_k - RED_k$
end for

4.1 Candidate Set Generation

The function *apriori-gen-join* is responsible for candidate set generation. A new collection of sets C_k is generated according to the join step of *apriori-gen* function described in (Agrawal & Srikant 1994). The generation of C_k is based on a collection of independent sets L_{k-1} obtained in the previous iteration. As a result we obtain a collection of all possible k-element sums of two elements chosen from L_{k-1}.

4.2 Pruning with Subsets

The function *prune-with-subsets* removes from family C_k all members B that are supersets of any dependent attribute set or reduct. Direct pruning by maximal independent sets found so far would be a costly operation. However, in accordance to Theorem 6, it is enough to test whether $\{B - \{a\} \subseteq P_{k-1}(\mathcal{A}) : a \in \mathcal{A}\} \subseteq ISC_{k-1} - RED_{k-1} = L_{k-1}$. It needs at most $|B|$ membership tests in a collection L_{k-1} computed in a previous step.

4.3 Finding Dependent Sets

Even if all actual subsets of a given B are independent, B can be dependent. When we cannot prove dependency basing on Theorem 6, we have to check it by means of Theorem 5. Otherwise, $B \in ISC_k$. This operation requires computing $covcount(B)$. We compare this value with $covcount(S)$, for all S such that S is a direct subset of B. Notice that, each S is an independent non reduct as B passed through a phase of dependent superset pruning. Moreover, the value of $covcount(S)$ will have already been computed to prove the independence of S.

4.4 Finding Reducts

Notice that, $RED_k \subseteq ISC_k$. Thus, in every iteration we have to find these $B \in ISC_k$ for which $covcount(B) = covcount(\mathcal{A})$. Notice that, $covcount$ is already computed for elements of ISC_k, so this step requires simply traversing ISC_k.

4.5 Execution Example

This section provides a short example of an algorithm execution. For brevity, we use a convenient notation $ab-14$ meaning a set $\{a, b\}$ and $covcount(\{a, b\}) = 14$.

We consider the information system $\mathcal{IS} = (\mathcal{U}, \{a, b, c, d, e\})$ (Table 1). We have the following facts:

- $EC = \{\{a, b, e\}, \{a, b, c, e\}, \{a, b, d, e\}, \{a, b, c\}, \{a, b, d\}, \{d\}, \{b, c, d\}, \{a, b, c, d\}\}$
- $covcount(\{a, b, c, d, e\}) = 8$

Our algorithm performs the following steps (Table 2) while solving the reduct set problem for $\mathcal{IS}$.

Table 2. Algorithm execution for $\mathcal{IS}$

k	L_k	RED_k
1	$\{a-6, b-7, c-4, d-5, e-3\}$	$\emptyset$
2	$\{ac-7, cd-7, ce-6, de-7\}$	$\{ad-8, bd-8\}$
3	$\emptyset$	$\{cde-8\}$

5 Algorithm Analysis

5.1 Implementation Details

The algorithm scheme, presented in Section 4, gives a brief overview of our method. We decided on the notation of set sequences to emphasize the connection between the presented theorems and particular algorithmic steps. However, it is easy to notice that steps 6, 7, 8, 9 can be performed during and *apriori-gen-join* function in order to avoid additional computations. Consider k-th iteration and $B \in C_k$ generated from $E, F \in L_{k-1}$. Firstly, we have to examine a collection $DS = \{B - \{a\} : a \in B\}$ that contains direct subsets of B. Obviously, E, F can be omitted, since they are independent, not super reducts. Now, for each direct subset $S \in DS - \{E, F\}$ we check $S \in L_{k-1}$. Finding any S not holding this condition causes a rejection of B and repeating the whole procedure for the next candidate. Otherwise, $covcount(B)$ is calculated and the condition $covcount(B) = max_{S \in DS}(covcount(S))$ is checked. If it holds, we reject B. Otherwise, $B \in ISC$. If, additionally, $covcount(B) = covcount(\mathcal{A})$, B is accepted as an independent, not super reduct. Notice that this maximum can be easily calculated while elements of DS are being examined. Summing up, for a given B we check a membership of $S \in DS$ in collection L_{k-1} exactly once.

Another observation refers to temporary collections stored in memory. Basically, we maintain and successively update the resulting collection RED. Moreover, in every iteration the only historical collection needed is L_{k-1}. It is used both: for candidate generation and for efficient pruning. Notice that we do not have to remember the collection ISC_{k-1}, since pruning by dependent subsets and reducts is performed in the same algorithmic step (Theorem 6) and employs only the whole collection L_{k-1}.

Testing the membership of a set in a collection is also a significant operation, which can be efficiently implemented using a tree structure or hashing methods. One of the propositions is covered further, in Section 6.

5.2 Algorithm Execution Example

This example demonstrates how we examine each candidate set in order to decide whether it is a reduct or a member of L_k collection.

Consider the information system $(\mathcal{U}, \{a, b, c, d, e, f\})$, for which we have computed $covcount(\{a, b, c, d, e, f\}) = 50$.

We focus on a hypothetical iteration for $k = 5$. From the previous iteration we have $L_4 = \{abcd - 26, abce - 35, abcf - 23, abde - 31, acde - 40, bcde - 12\}$. We select two sets $E = \{a, b, c, d\}$, $F = \{a, b, c, e\}$ and generate a candidate $B = E \cup F = \{a, b, c, d, e\}$. We have a collection of direct subsets of B equal to $DS = \{\{b, c, d, e\}, \{a, c, d, e\}, \{a, b, d, e\}, \{a, b, c, e\}, \{a, b, c, d\}\}$.

Firstly, we have to check whether $S \in L_4$ for each $S \in DS - \{E, F\}$. During this examination we also compute $max_{S \in DS}(covcount(S)) = 40$. Because $DS \subset L_4$, so we cannot use Theorem 6. After computation, we obtain $covcount(B) = 40$. Because $covcount(B) = max_{S \in DS}(covcount(S)) = covcount(\{a, c, d, e\})$, according to Theorem 5, we have found that B is dependent and we do not add it to L_5.

5.3 Complexity Analysis

Our algorithm traverses the search region ISC and, additionally, examines not pruned, direct supersets of $MISC - RED$ in order to prove their independence.

Although the approach uses the concept of concise (border) representation we avoided costly checking whether a collection contains a subset/superset of a given set. Pruning is performed by membership tests only.

Thorough emphasis should also be placed on *covcount* computation, which is a basic operation in our algorithm. According to Lemma 3 and the definition of a super reduct, we can infer about dependence and discernibility only by means of *covcount* measure. The operation appears to be costly, when large databases are concerned, so it should be optimized and performed as rarely as possible. For sure, the *covcount* has to be computed at least for the elements of ISC and for $\mathcal{A}$. Notice that we compute *covcount* for each examined set only once.

Moreover, it can be easily demonstrated that for computing *covcount* we do not need to examine all sets from EC but only the minimal elements within a collection of all non-empty elements of EC. Formally, we define this collection as $RC = \{B \in EC : B \neq \emptyset \wedge \forall_{S \in EC}(S = \emptyset \vee S \not\subset B)\}$. Most often, this simple optimization reduces strongly the size of EC, and thus, the operation cost.

However, for very large databases it may be infeasible to construct and reduce the indiscernibility matrix, since these operation have time and space cost of $O(n^2)$. In such a situation, for a given $B \in \mathcal{A}$ the value of $covcount(B)$ can be computed directly from an information system after computing the sizes of blocks of the partition generated by B. This operation involves sorting $\mathcal{U}$ on attribute set B with time cost $O(nlog(n))$, in situ.

6 Data Structures for Storing Attribute Set Collections

As it was described before we consider two dominant operations in our algorithm: *covcount* computation and membership testing. Due to the possibly large databases and large indiscernibility matrices, we have decided to avoid the first one and, according to Theorems 5 and 6, replace it by multiple membership tests. Such a strategy has enforced us to study different data structures for storing an attribute set collection L_{k-1}.

In fact, we perform two basic operations on a collection L_{k-1}. First, we enumerate its members to generate new candidates and test whether a given attribute set belongs to the collection. It should be kept in mind that in apriori-like candidate generation many pairs of attribute sets cannot be joined. Thus, the collection should be traversed in a specific ordering in order to avoid additional examinations.

Apart from optimizing the operations, the second important issue is an economical management of operational memory. In fact, real life datasets usually have hundreds of attributes, which results in large search spaces. In consequence, temporary collections become memory-consuming, especially when each attribute set is stored separately. Thus, concise data structures are preferable.

The most natural approach is to store attribute sets in a vector. The main advantage is that we can easily ensure ordering needed for candidate generation making this stage very efficient. However, a linear time is needed for each membership test.

In a more sophisticated solution each attribute set is treated as a node of a balanced binary search tree. The number of nodes is equal to the cardinality of the collection. This tree structure also stores each attribute set separately and needs to allocate more additional memory than a vector. When we use a prefix-wise node ordering, an in-order method of node traversing allows us to visit attribute sets in order appropriate for candidate generation. Due to dynamical nature, this step is slower than in case of vector. On the other hand, a balanced tree assures a logarithmic time for finding a particular set. In fact, the access time can be worse, when C is large and we cannot treat the cost of attribute set comparison as constant. Notice that when we use prefix-wise node ordering and go down the tree in order to find a member, the time of each comparison increase with a node level. Such a phenomenon is due to longer common prefixes of a given set and a set associated with a node. This property can significantly affect efficiency of tree building stage and later member finding.

The last approach refers to a classic trie structure that was originally designed for storing a collection of strings over a given alphabet S. A trie is a multi-way ordered tree. Each edge is labeled with a letter of the alphabet. There is one node for every common prefix, which can be read by scanning edge labels on the path from the tree root to this node. All leaves and some internal nodes are marked as terminating, since they are associated with prefixes that belong to S. There are two important advantages of a trie structure. The search time is logarithmic with a single letter comparison, not a string comparison, as a basic operation. The structure is more efficient than a list of strings or a balanced tree when large collections are considered. Actually, the difference in search time between a trie and a BST tree decrease with an increase of average number of children of a trie node. Although references used to build nodes involve additional memory, trie remains a very compact structure.

In our case, we assume some linear ordering in C, order elements of each attribute set and treat it as a string of attributes. Now, the attribute set collection L_{k-1} can be stored in a trie tree. Notice that all terminating nodes are leaves, since all attribute sets have the same cardinality $k-1$.

The original structure has been adopted to meet the requirements of the algorithm. First of all, it is unnecessary to store attribute sets explicitly, while they are once inserted to the tree. Information on the *covcount* value of an attribute set can be saved in an appropriate terminating node. Secondly, we have decided to use a balanced tree for storing the references from each node to its children. It allows us to conveniently traverse subsets in candidate generation stage. In fact, each two sets that can be joined share a parent node at the last but one level. Thus, it is possible to perform this step in the same way like using a vector. It is also advisable to use this approach for data sets with a large number of possible attributes $|C|$, e.g. gene expression data (1k-10k attributes). Another option is to store children in a hash table. This solution also ensures an efficient memory usage and outperforms dynamic structures for reasonably small $|C|$. However, it is less efficient when candidate generation is concerned.

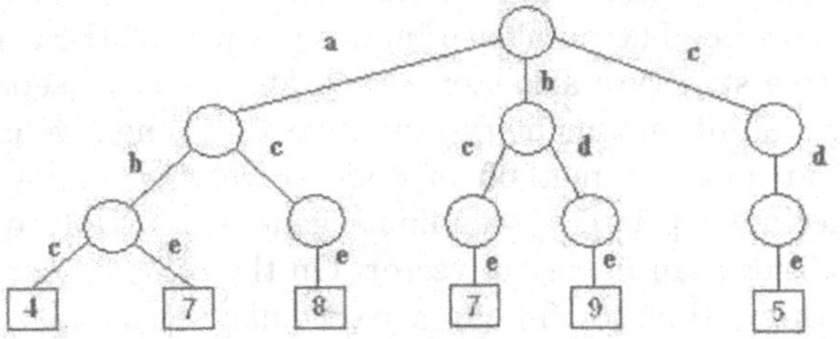

Fig. 2. A modified trie tree for the exemplary collection L_{k-1}. Each path from the root to a leaf refers to an attribute set from this collection. Ovals denote internal nodes and rectangles - terminal nodes. The values of *covcount* of respective attribute sets are given in terminal nodes.

Let us consider the set of possible attributes $\mathcal{C} = \{a, b, c, d, e\}$, the collection $L_{k-1} = \{\{a, b, c\}, \{a, b, e\}, \{a, c, e\}, \{b, c, e\}, \{b, d, e\}, \{c, d, e\}\}$ and the *covcount* values for these attribute sets: 4, 7, 8, 7, 9, 5, respectively. A modified trie tree for this collection is presented in Fig. 2.

Let us consider a set $\{b, c, e\}$. Its *covcount* value can be found in the tree by going three levels down. Since all attribute sets have the same size, the average cost of accessing a factual member of a collection L_{k-1} is similar. On the other hand, when a given set do not belong to a collection, the computation is shorter, e.g. finding a set $\{a, d, e\}$ requires visiting only tow levels.

To sum up, we have used a trie structure in order to get a more economical representation of an attribute set collection and to assure a logarithmic search time. It is worth mentioning that a trie structure remains a basis for many contemporary structures proposed for transaction databases, mainly in association rules and pattern mining. For example, support calculations in Apriori algorithm are often optimized by means of itemset trees (Agrawal & Srikant 1994). On the other hand, a FP-growth algorithm (Han, Pei & Yin 2000) is completely based on another trie-like structure, a frequent pattern tree (FP-Tree). Interesting solutions, involving various modifications of a pattern tree (P-Tree), have been proposed for mining jumping emerging patterns (Bailey, Manoukian & Ramamohanarao 2002), strong emerging patterns (Fan & Ramamohanarao 2006) and chi emerging patterns (Fan & Ramamohanarao 2003).

7 Pruning Efficiency Testing

When we deal with a NP-hard problem, the time cost of algorithms depends strongly on the structure of input data. Thus, we resigned from a comparison with other classic methods and focused mainly on proving the efficiency of our pruning approach in reducing the search space.

Input information systems (Table 3), originating from (D.J. Newman & Merz 1998), are provided with a preliminary hardness assessment. The size of the search space indicates how many sets have to be examined by an exhaustive approach. On the other hand, the minimal reduct length shows when an apriori-like

Table 3. Dataset characteristics

Name	Number of objects	Number of attributes	Size of the search space	Minimal reduct length	$\|RED\|$	$\|RC\|$
austra	690	15	3.0e+04	4	13	7
diab	768	9	5.0e+02	3	27	18
dna	500	21	2.0e+06	9	577	50
geo	402	11	2.0e+03	7	1	7
heart	270	14	1.0e+04	3	55	26
lymn	148	19	5.0e+05	9	132	45
mushroom	8124	23	8.0e+06	15	1	15
vehicle	846	19	5.0e+05	3	1714	240
zoo	101	17	1.0e+05	11	7	12

algorithm starts to find minimal reducts. The time cost of *covcount* computation for a given attribute set is determined by the size of RC and the number of attributes.

Table 4 contains experimental results. Our algorithm (Algorithm 1) is compared with a similar apriori-like algorithm (Algorithm 2), which prunes the candidate collection only with reducts found so far. In other words, we use as a reference an algorithm based on Theorem 2, that is weaker than Theorem 4.

The results advocate the efficiency of our pruning approach. First of all, the sets generated by Algorithm 1 constitute only a small region of the respective search space. More precisely, in the considered cases ISC contains less sets by 1-2 orders of magnitude. Secondly, a comparison with Algorithm 2 shows that Theorem 4 has significantly better pruning properties than Theorem 2. Last but not least, Algorithm 2 is more prone to data set characteristics such as the minimal reduct length and the number of reducts related to the size of the search space. These parameters determine the frequency of pruning. Conversely, the performance of our algorithm depends more on the characteristics of more numerous and diversified collection $MISC$.

Table 4. Experimental results for pruning efficiency testing

Dataset	Algorithm 1					Algorithm 2	
	Generat. sets	Pruned by Theorem 5	Pruned by Theorem 6	*covcount* comput.	Membership tests	Generat. sets	*covcount* comput.
austra	476	64	179	297	500	28855	28825
diab	205	22	45	160	209	288	264
dna	157220	868	59585	97635	526204	2060148	2057656
geo	165	34	0	165	201	2035	2033
heart	1259	130	473	786	1713	7615	7446
lymn	38840	203	1908	36932	175599	517955	515191
mushroom	32923	148	0	32923	180241	8388479	8388353
vehicle	11795	2112	3518	8277	24982	91916	84796
zoo	7910	40	189	7721	30686	130991	130903

The time cost is described by two dominant operations: *covcount* computation and testing the membership of a set in a collection. As a result of stronger space reduction the number of *covcount* computations performed by Algorithm 1 is much lower in comparison to Algorithm 2, often by 1-2 orders of magnitude. Moreover, we do not compute *covcount* for these generated sets, which are pruned by a condition based on Theorem 6. In presented data sets this condition holds more often than the one based on Theorem 5.

8 Execution Time Testing

Section 7 gives insights on efficiency of pruning based on attribute set dependency in terms of operation numbers. However, when algorithms use quite different basic operations or the operations are differently implemented it is hard to use such an approach to compare their overall performance. Therefore, we have decided to perform additional tests on execution time.

We study time efficiency of our algorithm implemented with three different structures for storing a collection L_{k-1}: a vector, a balanced tree and a trie tree.

All the algorithms are implemented with common data structures derived from basic types of Java 1.5. Tests have been performed on Intel Pentium M 1.5GHz with 512MB of RAM, switched to a stable maximum performance, using the Windows XP operating system and Sun JRE 1.5.0.06. We have repeated executions for each data set and each method to obtain a reliable average execution time.

The results are given in Table 5. As we can see, the approach based on vector tends to be slower when the algorithm performs many membership tests related to the number of generated sets (dna, lymn, mushroom, vehicle, zoo). On the other hand, both solutions that employ tree structures behave similarly. However, in case of a trie tree, common prefixes are represented in a compact way, which makes it a more economical structure in terms of memory.

Table 5. Experimental results for execution time testing [ms]

Dataset	Algorithm 1 with an array	Algorithm 1 with a balanced BST	Algorithm 1 with a trie
austra	2179	2183	2180
diab	277	263	267
dna	753534	525816	523693
geo	163	153	153
heart	624	664	607
lymn	29876	7354	7624
mushroom	129092	113323	114611
vehicle	4122	2897	2941
zoo	1131	270	313

9 Summary

In the paper we have proposed an apriori-like algorithm for the reduct set problem. It employs a novel pruning method based on the notion of attribute set dependence. We have demonstrated that supersets of the independent set collection (ISC) cannot be reducts. Moreover, it has been explained how to efficiently perform a pruning test and avoid maintaining ISC.

According to operation-wise tests, introduction of a new pruning approach reduces greatly the search space and the number of discernibility computations for attribute sets, important aspects when large databases are concerned. The execution time tests indicate that the algorithm performs relatively faster when an attribute set collection is implemented by means of dynamic structures: a balanced BST tree or a trie tree, rather than by means of a vector. Since a trie structure takes advantage of common prefixes, it tends to be more compact and temporary collections are less probable to exceed memory limits.

An apriori-like scheme used for candidate set generation allows efficiently to infer about set dependence and prune large regions of the search space. However, such an approach precludes the use of discernibility pruning conditions. In fact, a different space traversing employing method combined of both pruning approaches may perform better. On the other hand, the presented ideas can be adopted in a natural way for finding other types of reducts, i.e reducts related to a decision or approximate reducts. In future work we plan to address both these issues.

References

Agrawal, R., Srikant, R.: Fast algorithms for mining association rules in large databases. In: VLDB 1994, pp. 487–499 (1994)

Bailey, J., Manoukian, T., Ramamohanarao, K.: Fast algorithms for mining emerging patterns. In: Elomaa, T., Mannila, H., Toivonen, H. (eds.) PKDD 2002. LNCS (LNAI), vol. 2431, pp. 39–50. Springer, Heidelberg (2002)

Bazan, J., Nguyen, H.S., Nguyen, S.H., Synak, P., Wroblewski, J.: Rough set algorithms in classification problem. Rough set methods and applications: new develop. in knowl. disc. in inf. syst., 49–88 (2000)

Newman, D.J., Hettich, S., Merz, C.: UCI repository of machine learning databases (1998)

Fan, H., Ramamohanarao, K.: Efficiently mining interesting emerging patterns. In: Dong, G., Tang, C.-j., Wang, W. (eds.) WAIM 2003. LNCS, vol. 2762, pp. 189–201. Springer, Heidelberg (2003)

Fan, H., Ramamohanarao, K.: Fast discovery and the generalization of strong jumping emerging patterns for building compact and accurate classifiers. IEEE Transactions on Knowledge and Data Engineering 18(6), 721–737 (2006)

Han, J., Pei, J., Yin, Y.: Mining frequent patterns without candidate generation. In: SIGMOD 2000: Proceedings of the 2000 ACM SIGMOD international conference on Management of data, pp. 1–12. ACM Press, New York (2000)

Hu, X., Lin, T.Y., Han, J.: A new rough sets model based on database systems. In: Wang, G., Liu, Q., Yao, Y., Skowron, A. (eds.) RSFDGrC 2003. LNCS (LNAI), vol. 2639, pp. 114–121. Springer, Heidelberg (2003)

Kryszkiewicz, M.: The Algorithms of Knowledge Reduction in Information Systems. PhD thesis, Warsaw University of Technology (1994)

Kryszkiewicz, M., Cichon, K.: Towards scalable algorithms for discovering rough set reducts. In: Peters, J.F., Skowron, A., Grzymała-Busse, J.W., Kostek, B.z., Świniarski, R.W., Szczuka, M.S. (eds.) Transactions on Rough Sets I. LNCS, vol. 3100, pp. 120–143. Springer, Heidelberg (2004)

Lin, T.: Rough set theory in very large databases. In: Proc. of CESA IMACS 1996, Lille, France, vol. 2, pp. 936–941 (1996)

Nguyen, H.S.: Scalable Classification Method Based on Rough Sets (2002)

Pawlak, Z.: Rough sets. International Journal of Computer and Information Sciences 11, 341–356 (1982)

Semeniuk-Polkowska, M., Polkowski, L.: Conjugate information systems: Learning cognitive concepts in rough set theory. In: Wang, G., Liu, Q., Yao, Y., Skowron, A. (eds.) RSFDGrC 2003. LNCS (LNAI), vol. 2639, pp. 255–258. Springer, Heidelberg (2003)

Skowron, A., Rauszer, C.: The discernibility matrices and functions in information systems. In: Slowinski, R. (ed.) Intelligent Decision Support, pp. 331–362. Kluwer, Dordrecht (1992)

Swiniarski, R.W.: Rough set methods in feature reduction and classification. International Journal of Applied Mathematics and Computer Science 11(3), 565–582 (2001)

Swiniarski, R.W., Skowron, A.: Rough set methods in feature selection and recognition. Pattern Recognition Letters 24(6), 833–849 (2003)

Terlecki, P., Walczak, K.: On the relation between rough set reducts and jumping emerging patterns. Information Sciences (2006) (to be published)

Wang, Y.: On cognitive informatics. icci 00, 34 (2002)

Wroblewski, J.: Finding minimal reducts using genetic algorithm. In: Proc. of the 2nd Annual Join Conference on Information Sciences, pp. 186–189 (1995)

Wu, Q., Bell, D.A., McGinnity, T.M.: Multiknowledge for decision making. Knowl. Inf. Syst. 7(2), 246–266 (2005)

A General Model for Transforming Vague Sets into Fuzzy Sets

Yong Liu[1], Guoyin Wang[1,2], and Lin Feng[2,3]

[1] Institute of Computer Science and Technology, Chongqing University of Posts and Telecommunications, Chongqing, 400065, P. R. China
{liuyong,wanggy}@cqupt.edu.cn

[2] School of Information Science and Technology, Southwest Jiaotong University, Chengdu, 610031, P.R. China

[3] Department of Engineering and Technology, Sichuan Normal University, Chengdu, 610072, P.R. China
mgyfl@tom.com

Abstract. The relationship between vague sets and fuzzy sets are analyzed in this paper. A general model for transforming vague sets into fuzzy sets is proposed. The transformation of a vague set into a fuzzy set is proved to be a many-to-one mapping relation. The validity of this transformation model is also discussed. It establishes a mathematical relationship between the vague set theory and fuzzy set theory.

Keywords: Fuzzy sets, Vague sets, Membership function, Transformation-model.

1 Introduction

Since the theory of fuzzy sets [1] was proposed in 1965, it has been applied in many uncertain information processing problems successfully. A fuzzy set F of an universe of discourse $U=\{u_1,u_2,\cdots,u_n\}$ is a set of ordered pairs $\{(u_1,\mu_F(u_1)),(u_2,\mu_F(u_2)),\cdots,(u_n,\mu_F(u_n))\}$, where μ_F is the membership function of the fuzzy set F, $\mu_F:U\rightarrow[0,1]$, and $\mu_F(u_i)$ is the membership of u_i belonging to F. It is obvious that $\forall_{u_i\in U}(0\le\mu_F(u_i)\le 1)$.

In the type 2 fuzzy sets theory [2], there is a membership function that maps a set X to an interval [0,1]. A type 2 fuzzy set is a ϕ-fuzzy set or interval-valued fuzzy set. Such an interval-valued fuzzy set [3] is characterized by a membership function $\mu_F(x)$, $x\in X$, which assigns to each object a grade of membership which is a continuous interval of real numbers in [0,1], rather than a single value. In [4], Gau and Buehrer also assigned to each object a grade of membership which is a subinterval of [0,1].

Quinlan [5] introduced two values, $t(P)$ and $f(P)$, characterizing a proposition P. $t(P)$ is the greatest lower bound of the probability of P derived from the evidence for

M.L. Gavrilova et al. (Eds.): Trans. on Comput. Sci. II, LNCS 5150, pp. 133–144, 2008.

P and $f(P)$ is the greatest lower bound of the probability of $\sim P$ derived from the evidence against P.

In [4], Gau and Buehrer pointed out that a single value $\mu_F(u_i)$ combined the evidence for $u_i \in U$ and the evidence against $u_i \in U$ ($u_i \notin U$), without indicating how much there was of each, could not tell us anything about its accuracy. Therefore, they proposed the concept of vague sets and compared it with the other interval-valued fuzzy sets. Grattan-Guiness [6] defined an interval containment in a natural way, that is $[a,b] \le [c,d]$ iff $a \ge c$ and $b \le d$. Gau and Buehrer also defined a vague set containment in another way [4]. A membership of a vague set is a subinterval of [0,1]. It includes three kinds of information about an element $(x \in U)$, that is, support degree, against degree and unsure degree.

Vague sets are more accurate for describing some vague information than fuzzy sets [7, 8, 9, 10, 11]. Many researchers are interested in the vague sets theory in recent years, and have some good results [12, 13, 14, 15, 16]. Similarity measures between vague sets and between elements are studied in [12, 14] and Multicriteria fuzzy decision-making problems based on vague set theory is proposed in [16]. Several methods have been developed for transforming vague sets into fuzzy sets in order to study the properties of vague sets and the relationship between vague sets and fuzzy sets [17, 18, 19].

In this paper, the relationship between vague sets and fuzzy sets is analyzed and the problem of transforming vague sets into fuzzy sets is also studied. It is found to be a many-to-one mapping relation to transform a vague set into a fuzzy set. A new general model for transforming vague sets into fuzzy sets is proposed. The two transformation methods developed by F Li in [8] are proved to be two special cases of this general transformation model. The validity of this transformation model is also discussed. This general transformation model could be used in uncertain information processing systems, especially in vague information systems. It could transform vague information systems into fuzzy information systems. Fuzzy knowledge could then be further extracted.

The rest of this paper is organized as follows. In section 2, we introduce the definition of vague sets. In section 3, we discuss about some existed methods for transforming vague sets into fuzzy sets. In section 4, we propose a new general model for transforming vague sets into fuzzy sets, and prove its properties. In section 5, the validity of the general transformation model is explained by examples. In section 6, some conclusions are drawn.

2 Vague Set

Definition.1 [4] (vague set) Let U be a space of points(objects), with a generic element of U denoted by u. A vague set A in U is characterized by a truth-membership function $t_A(u)$ and a false-membership function $f_A(u)$, $t_A(u)$ is a lower bound on the grade of membership of u derived from the evidence for u, and $f_A(u)$ is a lower bound on the negation of u derived from the evidence against u. Both $t_A(u)$ and $f_A(u)$ associate a real number in the interval [0,1] with each point in U, where $t_A(u) + f_A(u) \le 1$.

That is,

$$t_A : U \rightarrow [0,1],$$
$$f_A : U \rightarrow [0,1].$$

When U is continuous , a vague set A can be written as

$$A = \int_U [t_A(u), 1 - f_A(u)] / u du .$$

When U is discrete, a vague set A can be written as

$$A = \sum_{i=1}^{n} [t_A(u_i), 1 - f_A(u_i)] / u_i .$$

Here, $[t_A(u), 1 - f_A(u)]$ is the vague value of an element u .

The vague set theory can be interpreted using a voting model. Suppose that A is a vague set in U , $u \in U$, and the vague value is [0.6,0.8], that is, $t_A(u) = 0.6$, $f_A(u) = 1 - 0.8 = 0.2$. Then, we can say that the degree of $u \in A$ is 0.6, and the degree of $u \notin A$ is 0.2. The vague value [0.6,0.8] can be interpreted as "the vote for a resolution is 6 in favor, 2 against, and 2 abstentions(The number of total voting people is assumed to be 10)".

3 Related Methods for Transforming Vague Sets into Fuzzy Sets

F Li proposed two methods for transforming vague sets into fuzzy sets[17]. For the convenience of illustration, we call them Method 1 and Method 2 respectively.

Method 1[17]: $\forall A \in V(U)$ ($V(U)$ is all vague sets of the universe of discourse U), let $u \in U$, and its vague value is $[t_A(u), 1 - f_A(u)]$, then the fuzzy membership function of u to A^F (A^F is the fuzzy set corresponding to the vague set A) is defined as:

$$\mu_{A^F} = t_A(u) + [1 - t_A(u) - f_A(u)]/2 = \frac{1 + t_A(u) - f_A(u)}{2} . \quad (1)$$

The Method 1 can be interpreted using the following voting model: value "1" means the vote for a resolution of favor, while "0" for against, and "0.5" for abstention. For example, a vague value [0.3, 0.7] means that the vote for a resolution is 3 in favor, 3 against, and 4 abstentions. It corresponds to a fuzzy membership value $(3 \times 1 + 4 \times 0.5 + 3 \times 0)/10 = 0.5$.

The Method 1 is simple in problem solving. However, some information might be lost when it is used to describe some special cases. In the Method 1, the fuzzy membership value for abstention is assigned to be "0.5". The influence of other voting values (favor or against) are not considered. In the above example, it is reasonable to assign the value "0.5" to the 4 abstention persons' voting attitude since the voting for against is equal to that for favor. Now, let's look at another example. If the vote for a resolution is 8 in favor, 1 against, and 1 abstention. Usually, the attitude of a person voting for abstention might not be absolutely neutral. His/Her attitude might be influenced by the other voting people. It is more likely that he/she might tend to vote in favor instead of against in this case, since there are more affirmative votes than negative votes. It is unreasonable to assign "0.5" to the abstentions in this case.

F Li proposed another method(Method 2) for solving this problem.

Method 2[17]: $\forall A \in V(U)$ ($V(U)$ is all vague sets of the universe of discourse U), let $u \in U$, and its vague value is $[t_A(u), 1-f_A(u)]$, then the fuzzy membership function of u to A^F (A^F is the fuzzy set corresponding to the vague set A) is defined as:

$$\mu_{A^F} = t_A(u) + [1 - t_A(u) - f_A(u)] \cdot t_A(u) / [t_A(u) + f_A(u)] = \frac{t_A(u)}{t_A(u) + f_A(u)} \tag{2}$$

There will be some unreasonable problems for some special cases when this method is used to transform vague sets into fuzzy sets. For example, a vague value [0,0.2], in this voting model, there are 0 vote in favor, 8 against. The abstention persons' voting attitude might tend to vote against instead of in favor since there are much more negative votes than affirmative votes. However, the abstention persons' favorite voting attitude in this model is 0. It means that abstention persons' voting attitude is absolutely against. Obviously, it is unreasonable. For this reason, Z G Lin proposed a new transformation method in [18]. We call it Method 3 in this paper.

Method 3 [18]: $\forall A \in V(U)$ ($V(U)$ is all vague sets in the universe of discourse U), let $u \in U$, and its vague value is $[t_A(u), 1-f_A(u)]$, then the fuzzy membership function of u to A^F (A^F is the fuzzy set corresponding to the vague set A) is defined as :

$$\mu_{A^F} = \begin{cases} t_A(u) + [1 - t_A(u) - f_A(u)] \bullet \dfrac{1 - f_A(u)}{t_A(u) + f_A(u)} & t_A(u) = 0 \\ t_A(u) + [1 - t_A(u) - f_A(u)] \bullet \dfrac{t_A(u)}{t_A(u) + f_A(u)} & 0 < t_A(u) \le 0.5 \\ t_A(u) + [1 - t_A(u) - f_A(u)] \bullet (0.5 + \dfrac{t_A(u) - 0.5}{t_A(u) + f_A(u)}) & 0.5 < t_A(u) \le 1 \end{cases} \tag{3}$$

Let's consider the following cases using this model.

Case 1 : $t_A(u) = 0$, there is 0 vote in favor, the abstention persons' favorite voting attitude will be $[1 - f_A(u)] \cdot \dfrac{1 - f_A(u)}{f_A(u)}$.

Case 2: $0 < t_A(u) \le 0.5$, the Method 3 is the same as the Method 2.

Case 3: $0.5 < t_A(u) \le 1$, the abstention persons' favorite voting attitude will be $[1 - t_A(u) - f_A(u)] \cdot (0.5 + \dfrac{t_A(u) - 0.5}{t_A(u) + f_A(u)})$. In this case, the abstention persons' voting attitude will tend to vote in favor instead of against since there are more affirmative votes than negative ones.

The Method 3 is an improvement of the Method 1 and Method 2. Lin illuminated the validity of this method. However, we find that there are still some unreasonable cases in this model. Let's look at the following example.

Example 1: $\forall A \in V(U)$, if u = [0, 0.9], the fuzzy membership values of u to A^F using the above three transformation methods are shown in Table 1.

Table 1. Fuzzy values generated by the Methods 1, 2 and 3

Vague Value	Fuzzy Value		
	Method 1	Method 2	Method 3
[0,0.9]	0.45	0	8.1

The domain of μ_{A^F} is between 0 and 1. So, we know $\mu_{A^F} = 8.1$ resulted from the Method 3 is incorrect.

There will be some problems when the Method 3 is used to calculate the fuzzy membership when $t_A(u)=0$. If $t_A(u)=0$, we know that μ_{A^F} should satisfy the following conditions according to formula (3):

$$\begin{cases} 0 \le \mu_{A^F} \le 1 - f_A(u) \\ \mu_{A^F} = t_A(u) + [1 - t_A(u) - f_A(u)] \bullet \dfrac{1 - f_A(u)}{t_A(u) + f_A(u)} = \dfrac{[1 - f_A(u)]^2}{f_A(u)} \end{cases}$$

Then, we could have $\dfrac{(1 - f_A(u))^2}{f_A(u)} \le 1 - f_A(u)$,

Thus, $f_A(u) \ge \dfrac{1}{2}$.

Since $f_A(u) \in [0,1]$, then, $\dfrac{1}{2} \le f_A(u) \le 1$.

So, if $t_A(u)=0$ and $\dfrac{1}{2} \le f_A(u) \le 1$, the Method 3 will be reasonable.

If $t_A(u)=0$ and $0 \le f_A(u) < \dfrac{1}{2}$, it will be unreasonable.

4 A General Model for Transforming Vague Sets into Fuzzy Sets

In this section, we will analyze the mapping relation between the elements of vague sets and the points on a plane, and propose a general model for transforming vague sets into fuzzy sets.

$\forall A \in V(U)$, u is an element in a universe of discourse U. Its vague value is $[t_A(u), 1 - f_A(u)]$. We take $t_A(u)$ and $f_A(u)$ as the axes of ordinate and abscissa respectively on a plane.

Here, $0 \le t_A(u) \le 1$, $0 \le f_A(u) \le 1$, and $t_A(u) + f_A(u) \le 1$.

So, each element of a vague set A can be mapped to a point on the plane. All points are in the area of the triangle OAB of Fig.1.

All elements of a vague set A can be shown in the area of the isosceles right-angle triangle AOB of Fig.2. It is obvious that |OA|=|OB|=1. All points on the border line segment AB correspond to fuzzy values. These points' coordinates satisfy $t_A(u) + f_A(u) = 1$. Each point on the line segment OA corresponds to a vague value

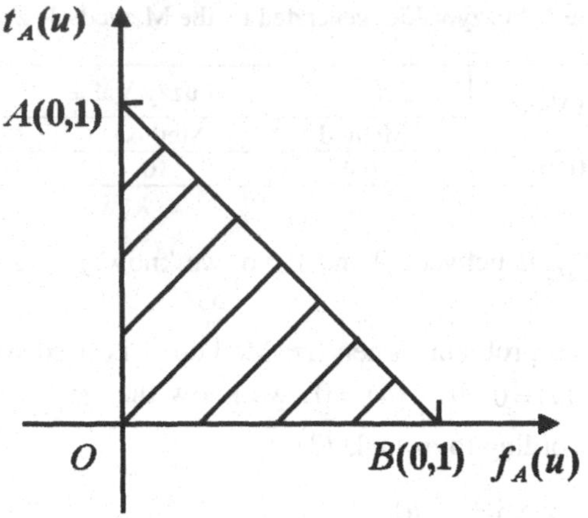

Fig. 1. The mapping between vague values and points on a plane

with a false membership of zero, that is, $f_A(u) = 0$. Each point on the line segment OB corresponds to a vague value with a truth membership of zero, that is, $f_A(u) = 0$.

In Fig.2, the radial OC is the bisector of the first quadrant. The point C is the intersection point of the radial OC and the line segment AB. It is obvious that $t_A(u) = f_A(u)$ for all points on the line segment OC. The point C corresponds to a vague value[0.5, 0.5]. By intuitive understanding, we can assign all points on the line OC the same fuzzy membership value 0.5 as the point C. In a voting model, the attitude of a person voting for abstention might not always be absolutely neutral. His/Her attitude might be influenced by the others. It is more likely that he/she might tend to vote in favor instead of against when there are more affirmative votes than negative ones, while vote against instead of favor when there are more negative votes than affirmative ones.

In Fig.2, in order to map the same fuzzy membership value 0.5 to all points on the line segment OC, we extend the line segment CO to D(it will be discussed later about how to choose the point D). We scan the triangle AOB with a radial l. D is the end point of l. The line segment FG is the intersection of the radial l and the triangle AOB. We assign all points on the line segment FG the same fuzzy membership value as the point G. Assume that the length of the line segment OD is λ, that is, $|OD| = \lambda$, $\lambda \geq 0$. When the point G moves from point A to B along the line segment AB, the radial l will exactly scan the whole triangle AOB. We can transform all vague values into fuzzy values in this way. In this method, it is more possible that a person voting for abstention might tend to vote in favor instead of against when there are more affirmative votes than negative ones, while vote against instead of favor when there are more negative votes than affirmative ones.

According to the above discussion, we could develop a general model for transforming vague sets into fuzzy sets(Method 4). $\forall A \in V(U)$, where $V(U)$ is all vague sets in a universe of discourse U. $\forall u \in U$, and the vague value is $[t_A(u), 1 - f_A(u)]$. Let λ be the

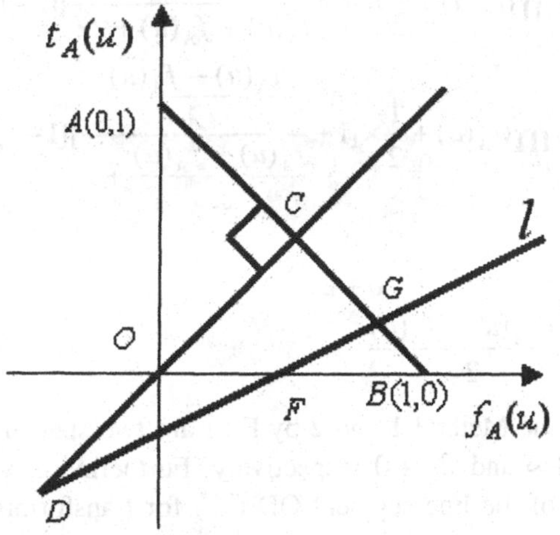

Fig. 2. A general model for transforming vague values into fuzzy values

length of the line segment OD in Fig.2, and $\lambda > 0$. The fuzzy membership function of u to A^F (A^F is the fuzzy set corresponding to the vague set A) is defined as :

$$\mu_{A^F} = t_A(u) + \frac{1}{2}[1 + \frac{t_A(u) - f_A(u)}{t_A(u) + f_A(u) + 2\lambda}][1 - t_A(u) - f_A(u)] \tag{4}$$

In Fig.2, when we transform vague sets into fuzzy sets, we assign the vague values of all points on the line segment FG to the same fuzzy membership value of the point G. The above formula (4) is thus derived.

From formula (4), we can find that it is more possible that a person voting for abstention might tend to vote in favor instead of against when there are more affirmative votes than negative ones, while vote against instead of favor when there are more negative votes than affirmative ones.

According to the formula (4), we can calculate the limits of μ_{A^F} when $\lambda \to 0$ and $\lambda \to +\infty$ respectively, that is,

$$\lim_{\lambda\to 0}\mu_{A^F} = \lim_{\lambda\to 0}\{t_A(u) + \frac{1}{2}\times[1 + \frac{t_A(u) - f_A(u)}{t_A(u) + f_A(u) + 2\lambda}][1 - t_A(u) - f_A(u)]\}$$

$$= t_A(u) + \frac{1}{2}\times[1 + \frac{t_A(u) - f_A(u)}{t_A(u) + f_A(u) + 2\times 0}][1 - t_A(u) - f_A(u)]$$

$$= \frac{t_A(u)}{t_A(u) + f_A(u)}$$

$$\lim_{\lambda\to+\infty}\mu_{A^F}=\lim_{\lambda\to+\infty}\{t_A(u)+\frac{1}{2}\times[1+\frac{t_A(u)-f_A(u)}{t_A(u)+f_A(u)+2\lambda}][1-t_A(u)-f_A(u)]\}$$

$$=\lim_{\lambda\to+\infty}\{t_A(u)+\frac{1}{2}\times[1+\frac{\frac{t_A(u)-f_A(u)}{\lambda}}{\frac{t_A(u)+f_A(u)}{\lambda}+2}][1-t_A(u)-f_A(u)]\}$$

$$=t_A(u)+\frac{1}{2}\times[1+\frac{0}{0+2}][1-t_A(u)-f_A(u)]$$

$$=\frac{1+t_A(u)-f_A(u)}{2}$$

It is obvious that the Method 1 and 2 by F Li are two special cases of our general model when $\lambda\to+\infty$ and $\lambda\to 0$ respectively. Furthermore, we could analyze the effect of the length of the line segment OD (λ) for transforming vague values into fuzzy values.

Theorem 3.1. In Method 4, if $t_A(u)>f_A(u)$, then $\mu_{A^F}\in(t_A(u)+\frac{1-t_A(u)-f_A(u)}{2},\frac{t_A(u)}{t_A(u)+f_A(u)})$.

Proof: $\frac{d\mu_{A^F}}{d\lambda}=-\frac{[t_A(u)-f_A(u)]\cdot[1-t_A(u)-f_A(u)]}{[t_A(u)+f_A(u)+2\lambda]^2}$.

If $t_A(u)>f_A(u)$, then $\frac{d\mu_{A^F}}{d\lambda}\le 0$.

Thus, μ_{A^F} descends monotonically with λ.

Since $\lim_{\lambda\to 0}\mu_{A^F}=\frac{t_A(u)}{t_A(u)+f_A(u)}$, and $\lim_{\lambda\to+\infty}\mu_{A^F}=t_A(u)+\frac{1-t_A(u)-f_A(u)}{2}$,

then, $\mu_{A^F}\in(t_A(u)+\frac{1-t_A(u)-f_A(u)}{2},\frac{t_A(u)}{t_A(u)+f_A(u)})$.

The greater the value of λ is, the smaller the value of μ_{A^F} will be. A person voting for abstention might tend to vote less favorably if λ is greater. It is obvious that the voting tendency of a person voting for abstention tends to vote in favor if $t_A(u)>f_A(u)$ according Theorem 3.1.

Theorem 3.2. In Method 4, if $t_A(u)<f_A(u)$, then $\mu_{A^F}\in(\frac{t_A(u)}{t_A(u)+f_A(u)},t_A(u)+\frac{1-t_A(u)-f_A(u)}{2})$.

Proof: $\frac{d\mu_{A^F}}{d\lambda}=-\frac{[t_A(u)-f_A(u)]\cdot[1-t_A(u)-f_A(u)]}{[t_A(u)+f_A(u)+2\lambda]^2}$.

If $t_A(u)<f_A(u)$, then $\frac{d\mu_{A^F}}{d\lambda}\ge 0$.

Thus, μ_{A^F} increases monotonically with λ.

Since $\lim_{\lambda\to 0}\mu_{A^F}=\frac{t_A(u)}{t_A(u)+f_A(u)}$, and

$$\lim_{\lambda\to+\infty}\mu_{A^F}=t_A(u)+\frac{1-t_A(u)-f_A(u)}{2},$$

then $\mu_{A^F}\in(\frac{t_A(u)}{t_A(u)+f_A(u)},t_A(u)+\frac{1-t_A(u)-f_A(u)}{2})$.

The greater the value of λ is, the greater the value of μ_{A^F} will be. A person voting for abstention might tend more to vote in favor if λ is greater. It is obvious that the voting tendency of a person voting for abstention is likely to vote against if $t_A(u)<f_A(u)$ according to Theorem 3.2.

Theorem 3.3. In Method 4, $\forall A,B\in V(U)$ ($V(U)$ is all vague sets in the universe of discourse U), if $A\subseteq B$ then $A^F\subset B^F$ (A^F, B^F are the fuzzy sets corresponding to the vague sets A and B).

Proof: Since $A\subseteq B$,

then $\forall_{u\in U}\ (t_A(u)\le t_B(u)\wedge 1-f_A(u)\le 1-f_B(u))$.

So, $t_A(u)-t_B(u)\le 0$, $f_B(u)-f_A(u)\le 0$, $t_A(u)f_B(u)\le t_B(u)f_A(u)$,

$$\mu_{A^F}=t_A(u)+\frac{1}{2}[1+\frac{t_A(u)-f_A(u)}{t_A(u)+f_A(u)+2\lambda}][1-t_A(u)-f_A(u)],$$

$$\mu_{B^F}=t_B(u)+\frac{1}{2}[1+\frac{t_B(u)-f_B(u)}{t_B(u)+f_B(u)+2\lambda}][1-t_B(u)-f_B(u)].$$

Then,

$$\mu_{A^F}-\mu_{B^F}=t_A(u)+\frac{1}{2}[1+\frac{t_A(u)-f_A(u)}{t_A(u)+f_A(u)+2\lambda}][1-t_A(u)-f_A(u)]-\{t_B(u)+\frac{1}{2}[1+\frac{t_B(u)-f_B(u)}{t_B(u)+f_B(u)+2\lambda}][1-t_B(u)-f_B(u)]\}$$

$$=\frac{\lambda[2t_A(u)f_B(u)-2f_A(u)t_B(u)+t_A(u)-t_B(u)+f_B(u)-f_A(u)]}{[t_A(u)+f_A(u)+2\lambda][t_B(u)+f_B(u)+2\lambda]}+\frac{2\lambda^2[t_A(u)-t_B(u)+f_B(u)-f_A(u)]}{[t_A(u)+f_A(u)+2\lambda][t_B(u)+f_B(u)+2\lambda]}+\frac{t_A(u)f_B(u)-t_B(u)f_A(u)}{[t_A(u)+f_A(u)+2\lambda][t_B(u)+f_B(u)+2\lambda]}\le 0.$$

That is, $A^F\subset B^F$.

In the process of transforming vague values into fuzzy values, the value of λ is the length of the line segment OD. It adjusts the influence degree of the voting tendency of persons voting for abstention affected by the others.

In this general transformation model, if there are more affirmative votes than negative ones, and the value of λ is much greater, the voting tendency of a person voting for abstention to favor will be less. If there are more negative votes than affirmative ones, and the value of λ is much greater, the voting tendency of a person voting for abstention to against will be less. The result of our general transformation model is always reasonable in any case.

5 Case Study for the General Transformation Model

λ is assigned to be 1 in the general transformation model in order to compare it with the other transformation methods. Thus, we can have the Method 4 from our general transformation model.

Method 4: $\forall A \in V(U)$ ($V(U)$ is all vague sets in the universe of discourse U), let $u \in U$, and its vague value is $[t_A(u), 1-f_A(u)]$, then the fuzzy membership function of u to A^F (A^F is the fuzzy set corresponding to the vague set A) is defined as:

$$\mu_{A^F} = t_A(u) + \frac{1}{2} \times [1 + \frac{t_A(u) - f_A(u)}{t_A(u) + f_A(u) + 2}][1 - t_A(u) - f_A(u)] \quad (5)$$

In order to compare it with the other methods, the 3 examples used by Cai [8] are considered here.

Example 2: $\forall A \in V(U)$, let u be an element in the universe of discourse U, it's vague value be [0, 0.9], the fuzzy membership values of u to A^F generated by the four transformation methods are shown in the 1st line of Table 2.

Table 2. Comparative results of Methods 1, 2, 3, and 4

Vague Value	Fuzzy Value			
	Method 1	Method 2	Method 3	Method 4
[0,0.9]	0.45	0	8.1	0.429
[0,0.3]	0.15	0	0.129	0.111
[0.9,1]	0.95	1	0.994	0.966

Obviously, the result of the Method 4 is reasonable. It is similar with the result of the Method 1. The results of the Method 2 and Method 3 are unreasonable.

Example 3: $\forall A \in V(U)$, let u be an element in the universe of discourse U, it's vague value be [0, 0.3], the fuzzy membership values of u to A^F generated by the four transformation methods are shown in the 2nd line of Table 2.

In the Method 2, the voting tendency of a person voting for abstention is absolutely against. It is unreasonable. The result of the Method 4 is similar with the results of the Method 1 and 3. It is rather reasonable.

Example 4: $\forall A \in V(U)$, let u be an element in the universe of discourse U, it's vague value be [0.9, 1], the fuzzy membership values of u to A^F generated by the four transformation methods are shown in the 3rd line of Table 2.

It is unreasonable that the voting tendency of a person voting for abstention is taken as absolutely in favor in the Method 2. The results of the Method 4, Method 1 and

Method 3 are reasonable. The results of the Method 4 and Method 3 are much better than that of Method 1, since the voting tendency of a person voting for abstention has no relation with the other people in the Method 1.

Now, let's analyze the effect of the parameter λ on the fuzzy membership of u to A^F.

Example 5: $\forall A \in V(U)$, let u be an element in the universe of discourse U, and it's vague value be [0.2, 0.7], the fuzzy membership values of u to A^F generated by the Method 4 with different values λs are shown in Table 3.

Table 3. Fuzzy values generated by the Method 4 with different λ

λ	0.5	0.8	10
Fuzzy Value	0.433	0.438	0.449

The vague value [0.2, 0.7] can be interpreted as "the vote for a resolution is 2 in favor, 3 against, and 5 abstentions". There are more negative votes than affirmative ones. The greater the value of the parameter λ is, the higher the fuzzy membership of u to A^F will be. The voting tendency of favor of a person voting for abstention is increasing with λ monotonically. In real applications, one can choose a suitable value for the parameter λ according to the characteristics of the problem to be processed.

6 Conclusion

The relationship between vague sets and fuzzy sets is studied in this paper. The many-to-one mapping relation for transforming a vague set into a fuzzy set is discovered. A general model for transforming vague sets into fuzzy sets is also proposed. The validity of this transformation model is analyzed. The two transformation methods proposed by F Li in [17] are proved to be its two special cases. The transformation method in [18] is found to be unreasonable for some special cases. The relationship among vague sets, rough sets, fuzzy sets and other non-classical sets could also be further studied in a similar way.

Acknowledgement. This paper is supported by National Natural Science Foundation of P.R.China under grants No.60573068 and No.60773113, Program for New Century Excellent Talents in University (NCET), Natural Science Foundation of Chongqing of China under grant No.2008BA2017.

References

1. Zadeh, L.A.: Fuzzy sets. Information and Control 3(8), 338–353 (1965)
2. Dubois, D., Prade, H.: Fuzzy Sets and Systems: Theory and Applications. Academic, New York (1980)
3. Klir, G.J., Folger, T.A.: Fuzzy Sets, Uncertainty, and information. Prentice-Hall, Englewood Cliffs (1988)

4. Gau, W.L., Buehrer, D.J.: Vague sets. IEEE Transactions on Systems, Man and Cybemetics 2(23), 610–614 (1993)
5. Quinlan, J.R.: Inferno: A cautious approach to uncertain inference. Computer J. 3(26), 255–268 (1983)
6. Grattan-Guiness, I.: Fuzzy membership mapped onto intervals and many-valued quantities. Zeitschr. Math. Logik. and Grundlagen d. Math., Bd. 22, 149–160 (1975)
7. Xu, J.C., An, Q.S., Wang, G.Y., Shen, J.Y.: Disposal of information with uncertain borderline-fuzzy sets and vague sets. Computer Engineering and Applications 16(38), 24–26 (2002) (in Chinese)
8. Cai, L.J., Lv, Z.H., Li, F.: A three-dimension expression of vague set and similarity measure. Computer Science 5(30), 76–77 (2003) (in Chinese)
9. Li, F., Lu, A., Yu, Z.: A construction method with entropy of vague sets based on fuzzy sets. Journal of Huazhong University of Science and Technology (Nature Science Edition) 9(29), 1–2 (2001) (in Chinese)
10. Li, F., Xu, Z.Y.: Measure of similarity between vague sets. Journal of Software 6(12), 922–926 (2001) (in Chinese)
11. Ma, Z.F., Xing, H.C.: Strategies of ambiguous rule acquisition from vague decision table. Chinese Journal of Computers 4(24), 382–389 (2001) (in Chinese)
12. Chen, S.M.: Measures of similarity between vague sets. Fuzzy Sets and Systems 2(74), 217–223 (1995)
13. Bustince, H., Burillo, P.: Vague sets are intuitionistic fuzzy sets. Fuzzy Sets and Systems 3(79), 403–405 (1996)
14. Chen, S.M.: Similarity Measures Between Vague Sets and Between Elements. IEEE Transactions on Systems, Man, and Cybernetics, Part B: Cybernetics 1(27), 153–158 (1997)
15. Hong, D.H., Chul, K.: A note on similarity measures between vague sets and between elements. Information Sciences 115(1-4), 83–96 (1999)
16. Hong, D.H., Choi, C.H.: Multicriteria fuzzy decision-making problems based on vague set theory. Fuzzy Sets and Systems 114(1), 103–113 (2000)
17. Li, F., Lu, Z.H., Cai, L.J.: The entropy of vague sets based on fuzzy sets. Journal of Huazhong University of Science and Technology (Nature Science Edition) 1(31), 1–3 (2003) (in Chinese)
18. Lin, Z.G., Liu, Y.P., Xu, L.Z., Shen, Z.Y.: A method for transforming vague sets into fuzzy sets in fuzzy information processing. Computer Engineering and Applications 9(40), 24–25 (2004) (in Chinese)
19. Li, F., Lu, A., Cai, L.J.: Fuzzy entropy of vague sets and its construction method. Computer Applications and Software 2(19), 10–12 (2002) (in Chinese)

Quantifying Knowledge Base Inconsistency Via Fixpoint Semantics

Du Zhang

Department of Computer Science
California State University
Sacramento, CA 95819-6021
USA
zhangd@ecs.csus.edu

Abstract. Inconsistency and its handling are very important in the real world and in the fields of computer science and artificial intelligence. When dealing with inconsistency in a knowledge base (KB), there is a whole host of deeper issues we need to contend with in order to develop rational and robust intelligent systems. In this paper, we focus our attention on one of the issues in coping with KB inconsistency: how to measure the information content and the significance of inconsistency in a KB. Our approach is based on a fixpoint semantics for KB. The approach reflects each inconsistent set of rules in the least fixpoint of a KB and then measures the inconsistency in the context of the least fixpoint for the KB. Compared with the existing results, our approach has some unique benefits.

Keywords: inconsistency, fixpoint semantics, KB coherence, significance of inconsistency.

1 Introduction

Inconsistency in knowledge and information is ubiquitous in the real world and in the fields of computer science and artificial intelligence. Occurrences of inconsistency often serve as an important indicator and trigger a whole host of possible cognitive activities. In the context of an intelligent agent system, a crucial component is its knowledge base (KB) that contains knowledge about a problem domain [2], [4], [8], [21]. Inconsistency in a KB can affect the correctness and performance of an intelligent system that deploys the KB.

When dealing with inconsistency in a KB, merely labeling the KB as consistent or inconsistent (a binary delineation), and removing conflicting rules in the event that the KB is inconsistent, is no longer adequate and may be counterproductive [15]. There are circumstances where some inconsistent cases are considered more significant than others. There are also circumstances where inconsistency is even deemed useful and desirable [7]. Hence, we need to have a better understanding of the issues that underpins inconsistency before making the appropriate decisions with inconsistency in an agent's knowledge. There are the following dimensions for inconsistency:

M.L. Gavrilova et al. (Eds.): Trans. on Comput. Sci. II, LNCS 5150, pp. 145–160, 2008.

- What are the causes for inconsistency?
- What does inconsistency tell us with regard to each different circumstance?
- Where can inconsistency occur?
- How do we define and represent inconsistency?
- What types of inconsistency do we have to contend with?
- How do we identify inconsistency in knowledge?
- How do we measure and differentiate the significance of inconsistency?
- What can we do with inconsistency once it is identified and when to do something about it?
- What are the circumstances under which inconsistency is considered useful and desirable?

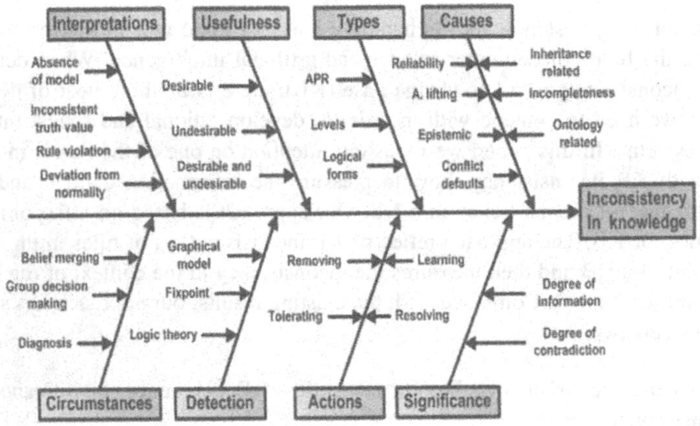

Fig. 1. Dimensions of KB inconsistency

It is important to understand what the aforementioned issues entail and what the state-of-the-practice is in this research area. We will take a brief look at each of the issues.

Causes for inconsistency. There are a number of causes for inconsistency. (1) Ontology related reasons include: either the lack of explicit constraints in the ontology specification (e.g., an ontology does not specify that animals and vegetables are mutually exclusive and jointly exhaustive in living things), or the discrepancy in terminology and its usage (e.g., complementary, mutual exclusive, or incompatible literals are explicitly sanctioned in the KB, polysemy, or antonymy). (2) Epistemic conflicts refer to the fact that different sources have different beliefs [16]. (3) Conflicting defaults in default reasoning [32]. (4) Lack of complete information, reliability of information source, or obsolescence of information [3]. (5) Assertion lifting. The need for lifting or importing assertions from one context to another [20]. (6) Redundancy-induced circumstances where some redundant rule is modified, but others do not [31]. (7) Defeasible inheritance induced [32].

What inconsistency tells us. The presence of inconsistency indicates the existence of errors, disagreement, misunderstanding, miscommunication, or lack of information.

Depending on the nature of inconsistency, resolution or tolerance actions will be in order. It is important to bear in mind that the type of cognitive activities in reasoning with inconsistent information needs to be commensurate with the nature and type of inconsistency.

Where inconsistency occurs. Inconsistency can occur in many situations. Following are some sample circumstances cited in [19]. (1) Diagnosis and testing. (2) Preference elicitation. (3) Belief merging and revision (e.g., merging and revising KB from heterogeneous sources). (4) Group decision making (e.g., negotiation in multi-agent systems, common sense reasoning in autonomic systems). The issue of inconsistency has found its way into database systems, knowledge-based systems, agent and autonomic systems, software engineering requirement specifications, and digital libraries and search engines, to name a few.

Inconsistency definitions. In the context of knowledge-based systems, various definitions have been given in the literature on inconsistency [9], [27], [30], [31]. In [15], ten different inconsistency definitions were enumerated, covering the grounds from classical logic (inconsistency as explosive reasoning, inconsistency as conflicting inferences, inconsistency as inference of a contradiction formula, inconsistency as trivial reasoning, and inconsistency as a lack of a model), to non-classical logics such as multi-valued logics (inconsistency as an inconsistent truth value, and inconsistency as delineated falsity), and to inconsistency under certain operational semantics (inconsistency as unrealisability, inconsistency as rule violation, and inconsistency as violation of normality). How to capture the essence of inconsistency in a single description that unifies all of the aforementioned definitions remains an open issue.

Types of inconsistency. There are different ways to classify inconsistency. In [32], inconsistency is examined from: different levels of granularity in knowledge, different categories of knowledge, and different logic forms. There are eighteen syntactic cases of inconsistency in rule-based systems in terms of *complementary*, *mutually exclusive* and *incompatible* literals [30]. In a given context space [20], inconsistency can be local within a context or global among several contexts. Temporally, it can be transient with regard to some instance(s) of the working memory in a knowledge-based system or persistent for all of the working memory instances. There are logical consistency and output consistency (whether the same set of inputs produces the same set of outputs or several sets of outputs) [27].

How to identify inconsistency. Early work in rule-base verification and validation treated inconsistency as specific deficiencies, and focused on devising algorithms to detect them. The approaches were based on formalizing KB either in terms of some graphical model [29] or as a quasi logic theory [9] or in some knowledge representation formalism [26]. Additional results and references can be found in [23].

How to handle inconsistency. Depending on the nature and the type of inconsistency, we may just tolerate it, isolate it, or resolve it through uncertainty reasoning or meta-level reasoning, or adopt certain inconsistency management strategy. Inconsistency proves to be useful in directing reasoning, and in helping initiate argumentation, seeking additional information, multi-agent negotiation, knowledge acquisition and refinement, and learning [15].

When to handle inconsistency. The timing for actions to be taken to resolve or tolerate inconsistency is yet another issue to be contending with. Immediate resolution of inconsistency via removal of conflicting rules without proper deliberation may be counterproductive, yielding loss of valuable information [15].

In this paper, our focus is on measuring the information content and the significance of inconsistency in a KB. Our approach is based on a fixpoint semantics for KB. The approach reflects each inconsistent set of rules in the least fixpoint of a KB and then measures the inconsistency in the least fixpoint for the KB.

The rest of the paper is organized as follows. Section 2 summarizes the existing work on measuring inconsistency. A brief overview is provided in Section 3 of the fixpoint semantics for a KB and the inconsistency definition in the context of the fixpoint semantics. Section 4 discusses the information content of a KB in terms of the concept of KB coherence. Section 5 describes our proposed approach to measuring the significance of KB inconsistency. A comparison with related work is given in Section 6. Finally Section 7 concludes the paper with remark on future work.

2 Related Work

Not all inconsistent cases are of the same significance. Some are more important or prominent than others. Here is a salient example. When Pat Tillman gave up his NFL career as a star player to join the army in May 2002 and was killed in Afghanistan on April 22, 2004, there were inconsistent conclusions as to whether Pat was killed by enemy fire or by friendly fire [28]. The inconsistent information about where the tragedy took place, whether near the village of Sperah [24] or Manah [28], definitely pales in comparison in its significance with the conflicting causes of his death.

Current approaches to measuring inconsistency include [15]: (1) Consistency-based analysis that reasons with consistent subsets of a KB. (2) Information theoretic approach that relies on using information theory to gauge the information content of a KB containing inconsistent rules [22]. (3) Probabilistic approach that adopts a probabilistic distribution for rules in a KB [17], [18]. (4) Epistemic approach that gauges the degree of information in a KB based on the number of actions needed to identify the truth value of an atomic proposition, and the degree of contradiction in terms of the number of actions needed to restore consistency in the KB [19]. (5) Model theoretic approach that evaluates a KB using three or four valued models that allows an inconsistent truth value [13], [14]. (6) Possibility theoretic approach that attaches a weight to each rule in a KB and establishes an α-cut of a possibilistic KB so as to describe the degree of inconsistency [3].

There are two important measures: the amount of information or *degree of information*, and the amount of contradiction or *degree of contradiction*. Typically, those measures are numerical values and vary as a result of the representation formalism utilized [19]. The conjoint use of both measures plays a pivotal role in inconsistency resolution or tolerance [15].

There are two approaches toward characterizing the importance of inconsistency. The first approach focuses on the number of formulas in a KB involved in the contradiction, whereas the second approach emphasizes on the number of atoms which exhibit conflicting information [15].

3 Overview of KB Fixpoint Semantics

As described in [32], there are a number of different types of inconsistency that can exist in a knowledge base. At the core of those types of inconsistency are the possible logical expressions involving the inconsistency-causing literals: *complementary*, *mutual exclusive*, *incompatible*, *anti*-subtype, *anti-supertype*, *asymmetric*, *anti-inverse*, *mismatching*, *disagreeing*, *contradictory*, and *iProbVal* (for inconsistent probabilistic truth value). In addition, a violation of some domain constraint may also constitute an occurrence of inconsistency. Hence, for a given KB, if its augmented fixpoint (see Equation 1) contains any of the aforementioned literals, then the KB contains inconsistent knowledge.

Since the main focus of this paper (measuring the information content and the significance of inconsistency in a KB) does not hinge on the specific definition of any of the inconsistency types, we will only summarize the notations we use to represent those eleven types of inconsistency in Table 1 (L, L_i, L_j and L_k in Table 1 are literals, $⩔L_k$ is a disjunction of literals, $⩓L_k$ is a conjunction of literals, and Δ is a set of rules). Explanations on some inconsistency types can be found in examples in Sections 4 and 5 below.

There are a number of fixpoint semantics for a logical theory [5], [6]: classical two-valued, two-valued with stratification, three-valued for handling negation, four-valued for dealing with inconsistency and incompleteness, and the truth value space of [0, 1].

In our previous results [31], we adopted the four-valued logic *FOUR* as defined in [1], [10]. *FOUR* has the truth value set of {*true*, *false*, $\bot$, $\top$} where *true* and *false* have their canonical meanings in the classical two-valued logic, $\bot$ indicates *undefined* or *don't know*, and $\top$ is *overdefined* or *contradiction*.

The four-valued logic *FOUR* is the smallest nontrivial bilattice, a member in a family of similar structures called bilattices [10]. Bilattices offer a general framework for reasoning with multi-valued logics and have many theoretical and practical benefits [10].

According to [1], there are two natural partial orders in *FOUR*: *knowledge* ordering $\leq_k$ (vertical) and *truth* ordering $\leq_t$ (horizontal) such that:

$$\bot \leq_k \mathit{false} \leq_k \top, \quad \bot \leq_k \mathit{true} \leq_k \top \ \text{and}$$
$$\mathit{false} \leq_t \top \leq_t \mathit{true}, \quad \mathit{false} \leq_t \bot \leq_t \mathit{true}.$$

Table 1. Types of inconsistency

Inconsistency Type	**Notation**
Complementary	$L_i \neq L_j$
Mutually exclusive	$L_i \not\asymp L_j$
Incompatible	$L_i \not\equiv L_j$
Anti-subtype	$L_i \not\sqsubseteq L_j$
Anti-supertype	$L \not\Leftrightarrow (\biguplus L_k)$
Asymmetric	$L_i \not\succcurlyeq L_j$
Anti-inverse	$L_i \nsim L_j$
Mismatching	$L \not\equiv (\bigsqcap L_k)$
Disagreeing	$L_i \ngeqq L_j$
Contradictory	$L_i \not\simeq L_j$
iProbVal	$Prob(L) \notin Prob(\Delta)$

Both partial orders offer a complete lattice. The meet and join for $\leq_k$, denoted as $\otimes$ and $\oplus$, respectively, yield: $false \otimes true = \bot$ and $false \oplus true = \top$. The meet and join for $\leq_t$, denoted as $\wedge$ and $\vee$, respectively, result in: $\top \wedge \bot = false$ and $\top \vee \bot = true$.

The knowledge negation reverses the $\leq_k$ ordering while preserving the $\leq_t$ ordering. The truth negation reverses $\leq_t$ ordering while preserving $\leq_k$ ordering.

For a knowledge base Ω, we define a transformation T_Ω, which is a "revision operator" [6] that revises our beliefs based on the rules in RB and established facts in WM. The interpretation of T_Ω can be understood in the following sense. A single step of T_Ω to Ω amounts to generating a set of ground literals, denoted as $\vdash_{WMi}$RB, which is obtained by firing all enabled rules in RB under WM_i. Let WM_0 denote the initial state for WM. We use WM_i (i = 1, 2, 3,…) to represent subsequent states of WM as a result of firing all enabled rules under the state of WM_{i-1}. It can be shown that T_Ω is monotonic and has a least fixpoint $lfp(T_\Omega)$ with regard to $\leq_k$ [5], [6].

Since a monotonic operator also has a greatest fixpoint denoted as *gfp*(), $gfp(T_\Omega)$ exists and can be expressed as follows:

$$gfp(T_\Omega) = \bigcup \{B \mid T_\Omega(B) = B\}.$$

Because of the definition of T_Ω, $lfp(T_\Omega)$ is identical to $gfp(T_\Omega)$ for a given knowledge base Ω. Operationally, the fixpoint of T_Ω for a KB can be obtained as follows. Given a set G of initial facts, WM_0 gets initialized based on G.

$i = 0;$

$\Phi_0 = G;$

$\Phi_1 = \Phi_0 \cup \vdash_{WM0}RB;$

$$
\begin{array}{l}
\text{while } (\Phi_{i+1} \mathrel{!=} \Phi_i) \text{ do } \{ \\
\quad i{+}{+}; \\
\quad \Phi_{i+1} = \Phi_i \cup \vdash_{\mathrm{WM}_i}\mathrm{RB} \\
\}; \\
lfp(\mathrm{T}_\Omega) = gfp(\mathrm{T}_\Omega) = \Phi_i;
\end{array}
$$

$lfp(\mathrm{T}_\Omega)$ ($gfp(\mathrm{T}_\Omega)$) contains all the derivable conclusions from the KB through some inference method. Let Γ_Ω be the set of domain (ontological) constraints for a given KB Ω, We define an augmented fixpoint $lfp^+(\mathrm{T}_\Omega)$ for Ω as follows :

$$lfp^+(\mathrm{T}_\Omega) = lfp(\mathrm{T}_\Omega) \cup \Gamma_\Omega \tag{1}$$

Thus, $lfp^+(\mathrm{T}_\Omega)$ constitutes the semantics for the KB. $lfp^+(\mathrm{T}_\Omega)$ allows us to deal with the following situations: (1) derived facts in $lfp(\mathrm{T}_\Omega)$ are conflicting with each other; and (2) a derived fact is not conflicting with any other facts in $lfp(\mathrm{T}_\Omega)$, but is contradicting with a domain or ontological constraint in Γ_Ω. For instance, if we have a derived fact: $age(\text{john}, 250) \in lfp(\mathrm{T}_\Omega)$, even if it does not contradict with any literal in $lfp(\mathrm{T}_\Omega)$, it still constitutes an inconsistent case because it violates the following domain constraint: $age(x, y) \wedge human(x) \wedge (y \leq 150)$ (a human being's age should be less than or equal to 150). We use the following to denote that a derived fact violates some domain constraint

$$\mathrm{A} \not\rightsquigarrow \Gamma_\Omega \ \text{ where } \mathrm{A} \in lfp(\mathrm{T}_\Omega) \tag{2}$$

The following fixpoint description for inconsistency is given in terms of $lfp^+(\mathrm{T}_\Omega)$.

Let ν be a mapping from ground atomic formulas to *FOUR*. Given a ground atomic formula A,

$$
\begin{array}{l}
\nu(\mathrm{A}) = \mathit{true}, \text{ if} \\
\quad \mathrm{A} \in lfp^+(\mathrm{T}_\Omega) \wedge \neg\mathrm{A} \notin lfp^+(\mathrm{T}_\Omega) \\
\quad \wedge \nexists \neg\mathrm{A}' \in lfp^+(\mathrm{T}_\Omega)\ [\mathrm{A} \cong \mathrm{A}'] \\
\quad \wedge \nexists \mathrm{A}'' \in lfp^+(\mathrm{T}_\Omega)\ [(\mathrm{A} \not\simeq \mathrm{A}'') \vee (\mathrm{A} \not\approx \mathrm{A}'') \\
\qquad\qquad \vee (\mathrm{A}'' \not\sqsubseteq \mathrm{A}) \vee (\mathrm{A} \not\downarrow \mathrm{A}'') \vee (\mathrm{A} \gneqq \mathrm{A}'') \vee (\mathrm{A} \not\asymp \mathrm{A}'') \vee (\mathrm{A} \not\rightsquigarrow \mathrm{A}'')] \\
\quad \wedge \nexists (\Cup \mathrm{L}_i) \subseteq lfp^+(\mathrm{T}_\Omega)\ [\mathrm{A} \not\Leftrightarrow (\Cup \mathrm{L}_i)] \\
\quad \wedge \nexists (\Cap \mathrm{L}_j) \subseteq lfp^+(\mathrm{T}_\Omega)\ [\mathrm{A} \not\equiv (\Cap \mathrm{L}_j)] \\
\quad \wedge \nexists \Delta \subseteq lfp^+(\mathrm{T}_\Omega)\ [\mathrm{Prob}(\mathrm{A}) \notin \mathrm{Prob}(\Delta)].
\end{array}
\tag{3}
$$

$$
\begin{array}{l}
\nu(\mathrm{A}) = \mathit{false}, \text{ if} \\
\quad \neg\mathrm{A} \in lfp^+(\mathrm{T}_\Omega) \wedge \mathrm{A} \notin lfp^+(\mathrm{T}_\Omega) \\
\quad \wedge \nexists \mathrm{A}' \in lfp^+(\mathrm{T}_\Omega)\ [(\mathrm{A} \cong \mathrm{A}') \vee (\mathrm{A} \downarrow \mathrm{A}') \vee (\mathrm{A}' \sqsubseteq \mathrm{A}) \vee (\mathrm{A} \geqq \mathrm{A}') \vee (\mathrm{A} \asymp \mathrm{A}')] \\
\quad \wedge \nexists \neg\mathrm{A}'' \in lfp^+(\mathrm{T}_\Omega)\ [(\mathrm{A} \not\simeq \mathrm{A}'')] \\
\quad \wedge \nexists (\Cap \mathrm{L}_i) \subseteq lfp^+(\mathrm{T}_\Omega)\ [(\mathrm{A} \equiv (\Cap \mathrm{L}_i)] \\
\quad \wedge \nexists (\Cup \mathrm{L}_j) \subseteq lfp^+(\mathrm{T}_\Omega)\ [(\mathrm{A} \Leftrightarrow (\Cup \mathrm{L}_j)].
\end{array}
\tag{4}
$$

$$
\begin{array}{l}
\nu(\mathrm{A}) = \top, \text{ if} \\
\quad \mathrm{A} \in lfp^+(\mathrm{T}_\Omega) \wedge [(\neg\mathrm{A} \in lfp^+(\mathrm{T}_\Omega)) \vee [\exists \mathrm{A}' \in lfp^+(\mathrm{T}_\Omega)\ [(\mathrm{A} \not\simeq \mathrm{A}') \vee (\mathrm{A} \not\equiv \mathrm{A}') \\
\qquad\qquad \vee (\mathrm{A} \not\approx \mathrm{A}') \vee (\mathrm{A} \not\downarrow \mathrm{A}') \vee (\mathrm{A} \gneqq \mathrm{A}') \vee (\mathrm{A} \not\asymp \mathrm{A}') \vee (\mathrm{A} \not\rightsquigarrow \mathrm{A}')]]
\end{array}
\tag{5}
$$

$$
\begin{aligned}
&\vee\ \exists \neg A'' \in lfp^{+}(T_{\Omega})\ [(A \sqsubseteq A'')] \\
&\vee\ \exists (\Cup L_i) \subseteq lfp^{+}(T_{\Omega})\ [(A \nLeftrightarrow (\Cup L_i)] \\
&\vee\ \exists\ (\Cap L_j) \subseteq lfp^{+}(T_{\Omega})\ [(A \ncong (\Cap L_j)] \\
&\vee\ \exists \Delta \subseteq lfp^{+}(T_{\Omega})\ [\mathrm{Prob}(A) \notin \mathrm{Prob}(\Delta)]].
\end{aligned}
$$

$$
\begin{aligned}
&\nu(A) = \bot, \text{ if} \qquad\qquad\qquad\qquad\qquad\qquad\qquad\qquad\qquad\qquad\qquad\qquad\qquad\qquad (6) \\
&A \notin lfp^{+}(T_{\Omega}) \wedge \neg A \notin lfp^{+}(T_{\Omega}) \wedge [(\nexists A' \in lfp^{+}(T_{\Omega}))\ (\nexists \neg A' \in lfp^{+}(T_{\Omega})) \\
&\qquad\qquad\qquad [(A \cong A') \vee (A \downarrow A') \vee (A \geqq A') \vee (A \sqsubseteq A') \vee (A \not\asymp A')]] \\
&\wedge [(\nexists (\Cup L_i) \subseteq lfp^{+}(T_{\Omega}))\ (\nexists \neg (\Cup L_i) \subseteq lfp^{+}(T_{\Omega}))\ (A \Leftrightarrow (\Cup L_i))] \\
&\wedge [(\nexists (\Cap L_j) \subseteq lfp^{+}(T_{\Omega}))\ (\nexists \neg (\Cap L_j) \subseteq lfp^{+}(T_{\Omega}))\ (A \equiv (\Cup L_i))].
\end{aligned}
$$

Definition 1. Given a knowledge base Ω, we obtain its least fixpoint $lfp(T_{\Omega})$. KB is said to contain inconsistent knowledge if for $h_i \in lfp^{+}(T_{\Omega})$, the following holds:

$$\exists\ h_i \in lfp^{+}(T_{\Omega})\ (\nu(h_i) = \top) \tag{7}$$

4 KB Coherence

Given a KB Ω and a set of initial facts G, we use a set $\mathbf{D}$ to denote the facts derived from Ω and G:

$$\mathbf{D}(\Omega) = lfp(T_{\Omega}) - G \tag{8}$$

In general, KB inconsistency may involve several instances of the same predicate. For example, in the following set {*p*(a), ¬*p*(a), *p*(b), ¬*p*(b)}, there are two instances of the *complementary* type (atom and its negation) of inconsistency involving the predicate *p*.

Definition 2. Given a predicate *p*, We use $\vartheta(p)$ to denote the number of inconsistent cases involving *p*.

Definition 3. Given a KB Ω, we define its *conflict set* **Ψ** as follows:

$$
\begin{aligned}
\boldsymbol{\Psi}(\Omega) = &\{A \mid A \in \mathbf{D}(\Omega) \wedge \neg A \in \mathbf{D}(\Omega)\} \qquad\qquad\qquad\qquad\qquad\qquad\qquad\qquad (9) \\
&\cup \{A \mid A, A' \in \mathbf{D}(\Omega) \wedge [(A \not\doteq A') \vee (A \ncong A') \vee (A \not\sqsubseteq A') \vee (A \not\downarrow A') \\
&\qquad\qquad\qquad \vee (A \not\asymp A') \vee (A \gneqq A') \vee (A \Box A')]\} \\
&\cup \{A \mid A \in \mathbf{D}(\Omega) \wedge \Cup A_i \subseteq \mathbf{D}(\Omega) \wedge (A \nLeftrightarrow (\Cup A_i))\} \\
&\cup \{A \mid A \in \mathbf{D}(\Omega) \wedge \Cap A_i \subseteq \mathbf{D}(\Omega) \wedge (A \ncong (\Cap A_i))\} \\
&\cup \{A \mid A \in \mathbf{D}(\Omega) \wedge \Delta \subseteq \mathbf{D}(\Omega) \wedge (Prob(A) \notin Prob(\Delta))\} \\
&\cup \{A \mid A \in \mathbf{D}(\Omega) \wedge A \nsim \Gamma_{\Omega}\}.
\end{aligned}
$$

Ψ(Ω) in a nutshell contains the representatives of all the inconsistent cases found in a given KB Ω.

Let **Θ**(Ω) be a set as specified below:

$$
\begin{aligned}
\boldsymbol{\Theta}(\Omega) = &\{A, \neg A \mid A \in \mathbf{D}(\Omega) \wedge \neg A \in \mathbf{D}(\Omega)\} \qquad\qquad\qquad\qquad\qquad\qquad\qquad (10) \\
&\cup \{A, A' \mid A, A' \in \mathbf{D}(\Omega) \wedge [(A \not\doteq A') \vee (A \ncong A') \vee (A \not\sqsubseteq A') \vee (A \not\downarrow A') \\
&\qquad\qquad\qquad \vee (A \not\asymp A') \vee (A \gneqq A') \vee (A \not\approx A')]\}
\end{aligned}
$$

$$\cup \{A, \Cup A_i \mid A \in \boldsymbol{D}(\Omega) \wedge \Cup A_i \subseteq \boldsymbol{D}(\Omega) \wedge (A \not\Leftrightarrow (\Cup A_i))\}$$
$$\cup \{A, \Cap A_i \mid A \in \boldsymbol{D}(\Omega) \wedge \Cap A_i \subseteq \boldsymbol{D}(\Omega) \wedge (A \not\equiv (\Cap A_i))\}$$
$$\cup \{A, \Delta \mid A \in \boldsymbol{D}(\Omega) \wedge \Delta \subseteq \boldsymbol{D}(\Omega) \wedge (Prob(A) \notin Prob(\Delta))\}$$
$$\cup \{A \mid A \in \boldsymbol{D}(\Omega) \wedge A \nsim \Gamma_\Omega\}.$$

$\boldsymbol{\Theta}(\Omega)$ includes all inconsistent literals that fit the profiles as defined in Table 1 and literals that violate domain constraints.

Definition 4. Given a KB Ω, we define its *base set* $\boldsymbol{\Phi}$ as follows:

$$\boldsymbol{\Phi}(\Omega) = \boldsymbol{\Psi}(\Omega) \cup (\boldsymbol{D}(\Omega) - \boldsymbol{\Theta}(\Omega)). \tag{11}$$

$\boldsymbol{\Phi}(\Omega)$ contains all the representatives of the inconsistent cases and those derived facts that are not part of any inconsistent case.

Example 1. Given the following $\boldsymbol{D}(\Omega)$ for a KB Ω

$\boldsymbol{D}(\Omega)$ = {*p*(b), ¬*p*(b), *p*(d), ¬*p*(d), *q*(e), ¬*q*(f), *r*(a), *Expensive*(a), ¬*HighPriced*(a),
Animal(c), *Vegetable*(c), *Surgeon*(john), ¬*Doctor*(john),
Connected(agent1, agent2), *Connected*(agent2, agent3),
SendMsgTo(agent1, agent2), *ReceiveMsgFrom*(agent2, agent3),
SpaceAvail(agent1, 5GB), *SpaceAvail*(agent1, 3500MB),
Deployed(agent1, 12-1-2007), *InService*(agent1, 10-1-2006)}

there are nine inconsistent cases in $\boldsymbol{D}(\Omega)$:

- *p*(b) and ¬*p*(b), and *p*(d) and ¬*p*(d) are *complementary*;
- *Expensive*(a) and ¬*HighPriced*(a) are *incompatible*;
- *Animal*(c) and *Vegetable*(c) are *mutually exclusive*;
- *Surgeon*(john) and ¬*Doctor*(john) are *anti-subtype*;
- Assuming that *Connected* represents a symmetric relation, then *Connected*(agent1, agent2) and *Connected*(agent2, agent3) are *asymmetric*;
- *SendMsgTo* and *ReceiveMsgFrom* are inverse predicates. Thus, *SendMsgTo*(agent1, agent2) and *ReceiveMsgFrom*(agent2, agent3) are *anti-inverse*;
- *SpaceAvail*(agent1, 5GB) and *SpaceAvail*(agent1, 3500MB) are *disagreeing*; and
- *Deployed*(agent1, 12-1-2007) and *InService*(agent1, 10-1-2006) are *contradictory*.

Therefore for Example 1, we have the following $\boldsymbol{\Psi}(\Omega)$ and $\boldsymbol{\Phi}(\Omega)$, respectively:

$\boldsymbol{\Psi}(\Omega)$ = {*p*(b), *p*(d), *Expensive*(a), *Animal*(c), *Surgeon*(john),
Connected(agent1, agent2), *SendMsgTo*(agent1, agent2),
SpaceAvail(agent1, 5GB), *Deployed*(agent1, 12-1-2007)}

$\boldsymbol{\Phi}(\Omega)$ = {*p*(b), *p*(d), *q*(e), ¬*q*(f), *r*(a), *Expensive*(a), *Animal*(c), *Surgeon*(john),
Connected(agent1, agent2), *SendMsgTo*(agent1, agent2),
SpaceAvail(agent1, 5GB), *Deployed*(agent1, 12-1-2007)}

Definition 5. The *coherence* $\zeta \in [0, 1]$ for a given KB Ω is defined as follows.

$$\zeta(\Omega) = 1 - \frac{|\Psi(\Omega)|}{|\Phi(\Omega)|} \tag{12}$$

If Ω is free of inconsistency, then $\zeta(\Omega) = 1$. On the other hand, if $\zeta(\Omega) = 0$, then Ω is entirely incoherent. When $0 < \zeta(\Omega) < 1$, it indicates that Ω contains inconsistency, but is not completely incoherent[1].

Example 2. The coherence for the KB in Example 1 is 0.25.

Example 3. Following is an example from [25]. Given a medical diagnosis KB consisting of six rules $r_1, \ldots, r_6$ where $r_1, \ldots, r_4$ are from doctor one and $r_5, \ldots, r_6$ from doctor two, d_1 and d_2 indicate two different diagnosis results and s_1 through s_4 represent different symptoms.

r_1: $s_1(x) \wedge s_2(x) \rightarrow d_1(x)$
r_2: $s_1(x) \wedge s_3(x) \rightarrow d_2(x)$
r_3: $d_2(x) \rightarrow \neg d_1(x)$
r_4: $d_1(x) \rightarrow \neg d_2(x)$
r_5: $s_1(x) \wedge s_4(x) \rightarrow d_1(x)$
r_6: $\neg s_1(x) \wedge s_3(x) \rightarrow d_2(x)$

Now a set of lab test results is made available about John and Bill as follows:

$G = \{s_1(john), \neg s_1(bill), \neg s_2(john), \neg s_2(bill), s_3(john), s_3(bill), s_4(john), \neg s_4(bill)\}$.

The least fixpoint for the KB is obtained below:

$lfp(T_\Omega) = \{s_1(john), \neg s_1(bill), \neg s_2(john), \neg s_2(bill), s_3(john), S_3(bill), s_4(john), \neg s_4(bill), d_1(john), d_2(john), d_2(bill), \neg d_1(john), \neg d_2(john), \neg d_1(bill)\}$.

And the conflict set and the base set are:

$\Psi(\Omega) = \{d_1(john), d_2(john)\}$
$\Phi(\Omega) = \{d_1(john), d_2(john), \neg d_1(bill), d_2(bill)\}$

Thus we have $\zeta(\Omega) = 0.5$.

5 Significance Measure of Inconsistency

When a KB contains inconsistency, it's desirable to have a mechanism whereby the relative significance of the inconsistent cases can be established before committing to either reasoning under inconsistency or restoring consistency. There are a number of approaches to measuring the degree or significance of inconsistency as was briefly summarized in Section 2. In this paper, we adopt an approach that establishes the importance of contradictory information based on the number of conflicting atoms in

[1] Note here we exclude the initial given facts in the definition for KB coherence.

the least fixpoint of a given KB. The approach is inspired by the results in Hunter's mass-based significance function [14].

An n-ary predicate represents an n-ary relation that is a subset of the Cartesian product $D_1 \times \ldots \times D_n$, where each D_i is a set of domain elements. We define the cardinality of a predicate p, denoted $\phi(p)$, to be the number of elements in a relation R for which p represents. For instance in Example 3, the diagnosis results d_1 and d_2 are unary predicates that are defined over the set of domain elements {john, bill}, thus

$$\phi(d_1) = \phi(d_2) = 2.$$

For a predicate $p \in \Psi(\Omega)$, we use $\{p \upharpoonleft\!\downharpoonright p'\}$ as a shorthand to indicate either of the following:

- $\{p, \neg p\}$ where p' represents $\neg p$;
- $\{p, dC\}$ where $(p \in lfp(T_\Omega)) \wedge (dC \in \Gamma_\Omega) \wedge (p \nsim dC)$ and p' denotes dC (dC is the domain constraint p violates);
- $\{p, p'\}$ where
  $(p \not\simeq p') \vee (p \not\equiv p') \vee (p \not\sqsubseteq p') \vee (p \not\trianglerighteq p') \vee (p \neq p') \vee (p \ngeqq p') \vee (p \not\approx p') \vee$
  $(p \not\Leftrightarrow (\Cup p')) \vee (p \not\equiv (\Cap p')) \vee (Prob(p) \notin Prob(p'))$[2].

Definition 6. For each element in the conflict set for a KB, a weight $\omega(\{p \upharpoonleft\!\downharpoonright p'\})$ can be defined which amounts to providing a piece of meta-knowledge from the problem domain about the importance of such inconsistent case [14].

The ω function has the following properties:

- Meaningful when $\Psi(\Omega) \neq \varnothing$;
- $\forall p \in \Psi(\Omega)\ \omega(\{p \upharpoonleft\!\downharpoonright p'\}) \in [0, 1]$;
- $\sum_{p \in \Psi(\Omega)} \omega(\{p \upharpoonleft\!\downharpoonright p'\}) = 1$.

In Example 3, if the diagnosis result d_2 is more important than that of d_1, then any inconsistency involving d_2 would be more significant than that of d_1. Based on this, a higher weight can be assigned to d_2.

Definition 7. Two measures are defined for the significance of KB inconsistency: $\hat{s}(\{p \upharpoonleft\!\downharpoonright p'\}) \in [0, 1]$ to denote the *significance* for an inconsistent predicate instance and $\hat{S}(\{p \upharpoonleft\!\downharpoonright p'\}) \in [0, 1]$ to denote the *significance* for an inconsistent predicate.

$$\hat{s}(\{p \upharpoonleft\!\downharpoonright p'\}) = \frac{\omega(\{p \upharpoonleft\!\downharpoonright p'\})}{\phi(p)} \tag{13}$$

---

[2] We use $\{p, \neg p\}$ as a shorthand for a pair of complementary literals, and $\{p, p'\}$ for the remaining ten cases of conflicting literals. For $\{p, p'\}$ in the third case, we assume that it's the case that $\phi(p) = \phi(p')$ and that p and p' are defined over the same set of domain elements. p' in cases of *anti-supertype*, *mismatching* and *iProbVal* represents a set of predicates for the involved literals.

$$\hat{S}(\{p \uparrow\downarrow p'\}) = \frac{\vartheta(p)\,\omega(\{p \uparrow\downarrow p'\})}{\phi(p)} \tag{14}$$

$\hat{s}$ and $\hat{S}$ have the following properties:

- $\hat{s}(\{p \uparrow\downarrow p'\}) = \hat{S}(\{p \uparrow\downarrow p'\})$ when $\vartheta(p) = 1$;
- $\hat{S}$ can be considered as a cumulative measure with regard to a predicate;
- The higher the $\hat{S}$ ($\hat{s}$) value is, the more significant the inconsistency becomes.

In the significance definitions above, we assume that each inconsistent predicate instance is equally important. This of course can be revised to reflect the situation in which some inconsistent predicate instance is more important than the others.

Example 4. Continuing the KB in Example 3, if we have the weight assignments below:

$\omega(\{d_1, \neg d_1\}) = 0.2$
$\omega(\{d_2, \neg d_2\}) = 0.8$

then the following significance measures for the two inconsistent cases can be obtained:

$\hat{s}(\{d_1(\text{john}), \neg d_1(\text{john})\}) = 0.2 \times 0.5 = 0.1$, and
$\hat{s}(\{d_2(\text{john}), \neg d_2(\text{john})\}) = 0.8 \times 0.5 = 0.4$.

Example 5. Given a KB Ω, suppose we have the following:

$\Psi(\Omega) = \{p(\text{a}), q(\text{d}), q(\text{e}), q(\text{g})\}$
$\Phi(\Omega) = \{p(\text{a}), p(\text{b}), p(\text{c}), q(\text{d}), q(\text{e}), q(\text{f}), q(\text{g})\}$
$\omega(\{p, \neg p\}) = 0.6$
$\omega(\{q, \neg q\}) = 0.4$
$\phi(p) = 3$
$\phi(q) = 4$
$\vartheta(p) = 1$
$\vartheta(q) = 3$.

Thus

$\hat{s}(\{p, \neg p\}) = \hat{S}(\{p, \neg p\}) = 0.2$
$\hat{S}(\{q, \neg q\}) = 0.3$.

This example illustrates that even though any inconsistency with regard to p is generally considered more important than that of q. However, when there are more inconsistent cases involving q, its cumulative significance may outweigh that of p.

With the significance measure in place, we can now revise the contradiction case in the valuation function $\nu(A)$ defined in our earlier work for inconsistency. Instead of just concluding that a ground atom A has the truth value of contradiction ($\top$), there is a chain of truth values describing the degrees or the level of significance for the contradiction [14].

Definition 8. We use the notation $\top_\lambda$ to represent a truth value of contradiction with a significance of λ. For a ground atom A, if p is the predicate in A, then

$$\nu(A) = \top_{\lambda(p)} \text{ iff } \nu(A) = \top \wedge \lambda(p) = \S(\{p \uparrow\downarrow p'\}). \tag{15}$$

Example 6. For the KB in Example 4, the truth value for d_1(john) and d_2(john) are as follows:

$$\nu(d_1(\text{john})) = \top_{0.1}$$
$$\nu(d_2(\text{john})) = \top_{0.4}.$$

This extension makes it possible to transform a four-valued semantics to a multi-valued semantics [14]. The significance of inconsistency can be quantified in a similar way as the degree to which a vague predicate is satisfied in fuzzy logic.

6 Discussions and Comparison

Even though our proposed measures are based on the number of literals which exhibit conflicting information (rather than the number of rules in a KB that are involved in the contradiction), we can easily extend the algorithm for establishing the fixpoint semantics for a KB (Section 3) to finding conflicting chains of rules that lead to the derivation of contradictory literals. Though we will not provide the algorithmic details in this paper, but this can be accomplished by keeping track of the rule number and the LHS of the rule for each newly derived ground atom, and checking for a non-empty intersection of the two supporting sets of conditions for the given pair of conflicting atoms. For instance, in Example 4, with regard to the pair of contradictory atoms {d_1(john), $\neg d_1$(john)} in $lfp(T_\Omega)$, we can trace back to the following supporting sets of evidence, respectively:

r_2: {s_1(john), s_3(john)} for $\neg d_1$(john), and
r_5: {s_1(john), s_4(john)} for d_1(john).

Since the intersection of the two sets is {s_1(john)}, the contradictory information in {d_1(john), $\neg d_1$(john)} stems from one of the inconsistent patterns: "rules with shared condition result in complementary conclusions" [30].

Compared with the existing results on measuring inconsistency, our proposed approach to quantifying inconsistency has the following benefits.

- It deals with the language in first order logic.
- It accommodates a larger set of conflict cases by considering not only complementary cases, but also many other cases, including domain constraint violation cases.
- Fixpoint semantics provides a tool that circumvents the issue of selecting a particular paraconsistent logic before properties for contradiction measures can be established.

- It makes it possible to identify rules in a KB that contribute to inconsistency, thus allowing both conflicting rules and contradictory atoms to be utilized in the analysis.

7 Conclusion

When developing a rational and robust intelligent system, how to properly handle inconsistency in its KB becomes an issue of great importance. In this paper, we focus our attention on one of the issues pertaining to KB inconsistency: how to measure the information content and the significance of inconsistency in a KB. Our approach is based on a fixpoint semantics for KB. The approach reflects each inconsistent set of rules in the least fixpoint of a KB and then measures the inconsistency in the context of the least fixpoint for the KB. Compared with the existing results, our approach has some salient features.

Future work can be pursued in the following directions. (1) Properties for both information and contradiction measures (in the context of fixpoint semantics). (2) Possible definitions of new information and contradiction measures that are based on a framework where both the number of conflicting rules in a KB and the number of contradictory atoms are utilized. (3) A conjoint way to make use of degree of information and degree of contradiction in inconsistency tolerance or resolution.

Acknowledgments. The author would like to acknowledge the comments from anonymous reviewers.

References

1. Belnap, N.D.: A Useful Four-Valued Logic. In: Epstein, G., Dunn, J. (eds.) Modern Uses of Multiple-Valued Logic, pp. 8–37. D. Reidel, Dordrecht (1977)
2. Brachman, R.J., Levesque, H.J.: Knowledge Representation and Reasoning. Morgan Kaufmann Publishers, San Francisco (2004)
3. Dubois, D., Lang, J., Prade, H.: Possibilistic Logic. In: Handbook of Logic in Artificial Intelligence and Logic Programming, vol. 3, pp. 439–513. Oxford University Press, Oxford (1994)
4. Fagin, R., Halpern, J.Y., Moses, Y., Vardi, M.Y.: Reasoning about Knowledge. MIT Press, Cambridge (1995)
5. Fitting, M.: Bilattices and the Semantics of Logic Programming. Journal of Logic Programming 11, 91–116 (1991)
6. Fitting, M.: Fixpoint Semantics for Logic Programming: a Survey. Theoretical Computer Science 278(1-2), 25–51 (2002)
7. Gabby, D., Hunter, A.: Making Inconsistency Respectable 2: Meta-Level Handling of Inconsistent Data. In: Moral, S., Kruse, R., Clarke, E. (eds.) ECSQARU 1993. LNCS, vol. 747, pp. 129–136. Springer, Heidelberg (1993)
8. Genesereth, M.R., Nilsson, N.J.: Logical Foundations of Artificial Intelligence. Morgan Kaufmann Publishers, Los Altos (1987)

9. Ginsberg, A., Williamson, K.: Inconsistency and Redundancy Checking for Quasi-First-Order-Logic Knowledge Bases. International Journal of Expert Systems 6(3), 321–340 (1993)
10. Ginsberg, M.L.: Multivalued Logics: a Uniform Approach to Inference in Artificial Intelligence. Computational Intelligence 4(3), 265–316 (1988)
11. Grant, J., Hunter, A.: Measuring Inconsistency in Knowledge Bases. Journal of Intelligent Information Systems 27, 159–184 (2006)
12. Huang, Z., van Harmelen, F., ten Teije, A.: Reasoning with Inconsistent Ontologies. In: The Proceedings of the Nineteenth International Joint Conference on Artificial Intelligence, Edinburgh, Scotland, pp. 454–459 (2005)
13. Hunter, A.: Measuring Inconsistency in Knowledge via Quasi-classical Models. In: The Proceedings of the National Conference on Artificial Intelligence, pp. 68–73 (2002)
14. Hunter, A.: Evaluating Significance of Inconsistencies. In: The Proceedings of the Eighteenth International Joint Conference on Artificial Intelligence, Acapulco, Mexico, pp. 468–473 (2003)
15. Hunter, A., Konieczny, S.: Approaches to Measuring Inconsistent Information. In: Bertossi, L., Hunter, A., Schaub, T. (eds.) Inconsistency Tolerance. LNCS, vol. 3300, pp. 189–234. Springer, Heidelberg (2005)
16. Hunter, A., Summerton, R.: A Knowledge-Based Approach to Merging Information. Knowledge-Based Systems 19, 647–674 (2006)
17. Knight, K.: Measuring Inconsistency. Journal of Philosophical Logic 31, 77–98 (2001)
18. Knight, K.: Two Information Measures for Inconsistent Sets. Journal of Logic, Language and Information 12, 227–248 (2003)
19. Konieczny, S., Lang, J., Marquis, P.: Quantifying Information and Contradiction in Propositional Logic through Test Actions. In: The Proceedings of the Eighteenth International Joint Conference on Artificial Intelligence, Acapulco, Mexico, pp. 106–111 (2003)
20. Lenat, D.: The Dimensions of Context-Space. CYCorp Report (1998)
21. Levesque, H.J., Lakemeyer, G.: The Logic of Knowledge Bases. MIT Press, Cambridge (2000)
22. Lozinskii, E.: Information and Evidence in Logic Systems. Journal of Experimental and Theoretical Artificial Intelligence 6, 163–193 (1994)
23. Menzies, T., Pecheur, C.: Verification and Validation and Artificial Intelligence. In: Zelkowitz, M. (ed.) Advances in computers, vol. 65. Elsevier, Amsterdam (2005)
24. MSNBC: (April 26, 2004), `http://www.msnbc.msn.com/id/4815441` (accessed December 31, 2007)
25. Murata, T., Subrahmanian, V.S., Wakayama, T.: A Petri Net Model for Reasoning in the Presence of Inconsistency. IEEE Transactions on Knowledge and Data Engineering 3(3), 281–292 (1991)
26. Nguyen, T.A., Perkins, W.A., Laffey, T.J., Pecora, D.: Knowledge Base Verification. AI Magazine 8(2), 69–75 (1987)
27. Rushby, J., Whitehurst, R.A.: Formal verification of AI software. NASA Contractor Report 181827 (1989)
28. SFC: (September 25, 2005), `http://www.sfgate.com/cgi-bin/article.cgi?f=/c/a/2005/09/25/MNGD7ETMNM1.DTL` (accessed December 31, 2007)
29. Zhang, D., Nguyen, D.: PREPARE: A Tool for Knowledge Base Verification. IEEE Transactions on Knowledge and Data Engineering 6(6), 983–989 (1994)

30. Zhang, D., Luqi: Approximate Declarative Semantics for Rule Base Anomalies. Knowledge-Based Systems 12(7), 341–353 (1999)
31. Zhang, D.: Fixpoint Semantics for Rule Base Anomalies. In: Proceedings of the Fourth IEEE International Conference on Cognitive Informatics, Irvine, CA, pp. 10–17 (2005)
32. Zhang, D.: On Classifying Inconsistency in Autonomic Agent Systems. Technical Report, December 2007. Department of Computer Science, California State University, Sacramento (submitted, 2007)

Contingency Matrix Theory I: Rank and Statistical Independence in a Contigency Table

Shusaku Tsumoto and Shoji Hirano

Department of Medical Informatics, Faculty of Medicine, Shimane University
89-1 Enya-cho, Izumo
Shimane 693-8501 Japan
tsumoto@computer.org

Abstract. A contingency table summarizes the conditional frequencies of two attributes and shows how these two attributes are dependent on each other with the information on a partition of universe generated by these attributes. This paper discusses statistical independence in a contingency table from the viewpoint of matrix theory. Statistical independence is equivalent to linear dependence of all columns or rows. Also, the equations of statistical independence are equivalent to those on collinearity of projective geometry.

1 Introduction

Statistical independence between two attributes is a very important concept in data mining[1] and statistics[2]. The definition $P(A, B) = P(A)P(B)$ show that the joint probability of A and B is the product of both probabilities. This gives several useful formula, such as $P(A|B) = P(A)$, $P(B|A) = P(B)$. In a data mining context, these formulae show that these two attributes may not be correlated with each other. Thus, when A or B is a classification target, the other attribute may not play an important role in its classification.

Although independence is a very important concept, it has not been fully and formally investigated as a relation between two attributes.

In this paper, a statistical independence in a contingency table is focused on from the viewpoint of granular computing[3,4].

The first important observation is that a contingency table compares two attributes with respect to information granularity. It is shown from the definition that statistical independence in a contingency table is a special form of linear depedence of two attributes. Especially, when the table is viewed as a matrix, the above discussion shows that the rank of the matrix is equal to 1.0. Also, the results also show that partial statistical independence can be observed.

The second important observation is that matrix algebra is a key point of analysis of this table. A contingency table can be viewed as a matrix and several operations and ideas of matrix theory are introduced into the analysis of the contingency table.

The paper is organized as follows: Section 2 discusses the characteristics of contingency tables. Section 3 shows the conditions on statistical independence for a 2×2 table. Section 4 gives those for a $2 \times n$ table. Section 5 extends these

M.L. Gavrilova et al. (Eds.): Trans. on Comput. Sci. II, LNCS 5150, pp. 161–179, 2008.

results into a $m \times n$ contingency table. Section 6 extends the ideas into a multi-way contingency cube. Section 7 defines a contingency matrix and shows the relation between rank and statistical independence. Section 8 and 9 discusses pseudo statistical independence, which is an intermediate dependency between statistical independence and dependence. Finally, Section 10 concludes this paper.

2 Contingency Tables

2.1 Rough Sets Notations

In the subsequent sections, the following notations is adopted, which is introduced in [5]. Let U denote a nonempty, finite set called the universe of the discourse (dataset) and A denote a nonempty, finite set of attributes, i.e., $a : U \rightarrow V_a$ for $a \in A$, where V_a is called the domain of a, respectively. Then, a data table is defined as an information system, $\mathbf{A} = (U, A)$. The atomic formulas over $B \subseteq A$ and V are expressions of the form $[a = v]$, called descriptors over B, where $a \in B$ and $v \in V_a$. The set $F(B, V)$ of formulas over B is the least set containing all atomic formulas over B and closed with respect to disjunction, conjunction and negation. For each $f \in F(B, V)$, f_A denote the meaning of f in A, i.e., the set of all objects in U with property f, defined inductively as follows.

1. If f is of the form $[a = v]$ then, $f_A = \{s \in U | a(s) = v\}$
2. $(f \wedge g)_A = f_A \cap g_A$; $(f \vee g)_A = f_A \vee g_A$; $(\neg f)_A = U - f_a$

2.2 Contingency Table (2 × 2)

From the viewpoint of information systems, a contingency table summarizes the relation between two attributes with respect to frequencies. This viewpoint has already been discussed in [6,7]. However, this study focuses on more statistical interpretation of this table.

Definition 1. *Let R_1 and R_2 denote binary attributes in an attribute space A. A contingency table is table of a set of the meaning of the following formulas:* $|[R_1 = 0]_A|$,$|[R_1 = 1]_A|$, $|[R_2 = 0]_A|$,$|[R_2 = 1]_A|$, $|[R_1 = 0 \wedge R_2 = 0]_A|$,$|[R_1 = 0 \wedge R_2 = 1]_A|$, $|[R_1 = 1 \wedge R_2 = 0]_A|$,$|[R_1 = 1 \wedge R_2 = 1]_A|$, $|[R_1 = 0 \vee R_1 = 1]_A|(= |U|)$. *This table is arranged into the form shown in Table 1, where:* $|[R_1 = 0]_A| = x_{11} + x_{21} = x_{\cdot 1}$, $|[R_1 = 1]_A| = x_{12} + x_{22} = x_{\cdot 2}$, $|[R_2 = 0]_A| = x_{11} + x_{12} = x_{1\cdot}$, $|[R_2 = 1]_A| = x_{21} + x_{22} = x_{2\cdot}$, $|[R_1 = 0 \wedge R_2 = 0]_A| = x_{11}$, $|[R_1 = 0 \wedge R_2 = 1]_A| = x_{21}$, $|[R_1 = 1 \wedge R_2 = 0]_A| = x_{12}$, $|[R_1 = 1 \wedge R_2 = 1]_A| = x_{22}$, $|[R_1 = 0 \vee R_1 = 1]_A| = x_{\cdot 1} + x_{\cdot 2} = x_{\cdot\cdot}(= |U|)$.

From this table, accuracy and coverage[8] for $[R_1 = 0] \rightarrow [R_2 = 0]$ are defined as:

$$\alpha_{[R_1=0]}([R_2 = 0]) = \frac{|[R_1 = 0 \wedge R_2 = 0]_A|}{|[R_1 = 0]_A|} = \frac{x_{11}}{x_{\cdot 1}},$$

and

$$\kappa_{[R_1=0]}([R_2 = 0]) = \frac{|[R_1 = 0 \wedge R_2 = 0]_A|}{|[R_2 = 0]_A|} = \frac{x_{11}}{x_{1\cdot}}.$$

Table 1. Two way Contingency Table

	$R_1 = 0$	$R_1 = 1$	
$R_2 = 0$	x_{11}	x_{12}	$x_{1\cdot}$
$R_2 = 1$	x_{21}	x_{22}	$x_{2\cdot}$
	$x_{\cdot 1}$	$x_{\cdot 2}$	$x_{\cdot\cdot}$ $(= \|U\| = N)$

Table 2. A Small Dataset

a	b	c	d	e
1	0	0	0	1
0	0	1	1	1
0	1	2	2	0
1	1	1	2	1
0	0	2	1	0

Example 1. Let us consider an information table shown in Table 2. The relationship between b and e can be examined by using the corresponding contingency table as follows. First, the frequencies of four elementary relations are counted, called *marginal distributions*: $[b = 0]$, $[b = 1]$, $[e = 0]$, and $[e = 1]$. Then, the frequencies of four kinds of conjunction are counted: $[b = 0] \wedge [e = 0]$, $[b = 0] \wedge [e = 1]$, $[b = 1] \wedge [e = 0]$, and $[b = 1] \wedge [e = 1]$. Then, the following contingency table is obtained (Table 3). From this table, accuracy and coverage for $[b = 0] \rightarrow [e = 0]$ are obtained as $1/(1 + 2) = 1/3$ and $1/(1 + 1) = 1/2$.

One of the important observations from granular computing is that a contingency table shows the relations between two attributes with respect to intersection of their supporting sets. For example, in Table 3, both b and e have two different partitions of the universe and the table gives the relation between b and e with respect to the intersection of supporting sets. It is easy to see that this idea can be extended into $m \times n$ contingency tables, which can be viewed as $n \times n$-matrix.

2.3 Contingency Table ($m \times n$)

Two-way contingency table can be extended into a contingency table for multinominal attributes.

Table 3. Corresponding Contingency Table

	b=0	b=1	
e=0	1	1	2
e=1	2	1	3
	3	2	5

Table 4. Contingency Table ($m \times n$)

	A_1	A_2	$\cdots$	A_n	Sum
B_1	x_{11}	x_{12}	$\cdots$	x_{1n}	$x_{1\cdot}$
B_2	x_{21}	x_{22}	$\cdots$	x_{2n}	$x_{2\cdot}$
$\vdots$	$\vdots$	$\vdots$	$\ddots$	$\vdots$	$\vdots$
B_m	x_{m1}	x_{m2}	$\cdots$	x_{mn}	$x_{m\cdot}$
Sum	$x_{\cdot 1}$	$x_{\cdot 2}$	$\cdots$	$x_{\cdot n}$	$x_{\cdot\cdot} = \|U\| = N$

Definition 2. *Let R_1 and R_2 denote multinominal attributes in an attribute space A which have m and n values. A contingency tables is a table of a set of the meaning of the following formulas: $|[R_1 = A_j]_A|$, $|[R_2 = B_i]_A|$, $|[R_1 = A_j \wedge R_2 = B_i]_A|$, $|[R_1 = A_1 \vee R_1 = A_2 \vee \cdots \vee R_1 = A_m]_A|$, $|[R_2 = B_1 \vee R_2 = A_2 \vee \cdots \vee R_2 = A_n]_A| = |U|$ ($i = 1, 2, 3, \cdots, n$ and $j = 1, 2, 3, \cdots, m$). This table is arranged into the form shown in Table 1, where: $|[R_1 = A_j]_A| = \sum_{i=1}^{m} x_{1i} = x_{\cdot j}$, $|[R_2 = B_i]_A| = \sum_{j=1}^{n} x_{ji} = x_{i\cdot}$, $|[R_1 = A_j \wedge R_2 = B_i]_A| = x_{ij}$, $|U| = N = x_{\cdot\cdot}$ ($i = 1, 2, 3, \cdots, n$ and $j = 1, 2, 3, \cdots, m$).*

3 Statistical Independence in 2 × 2 Contingency Table

Let us consider a contingency table shown in Table 1. Statistical independence between R_1 and R_2 gives:

$$
\begin{aligned}
P([R_1 = 0], [R_2 = 0]) &= P([R_1 = 0]) \times P([R_2 = 0]) \\
P([R_1 = 0], [R_2 = 1]) &= P([R_1 = 0]) \times P([R_2 = 1]) \\
P([R_1 = 1], [R_2 = 0]) &= P([R_1 = 1]) \times P([R_2 = 0]) \\
P([R_1 = 1], [R_2 = 1]) &= P([R_1 = 1]) \times P([R_2 = 1])
\end{aligned}
$$

Since each probability is given as a ratio of each cell to N, the above equations are calculated as:

$$
\begin{aligned}
\frac{x_{11}}{N} &= \frac{x_{11} + x_{12}}{N} \times \frac{x_{11} + x_{21}}{N} \\
\frac{x_{12}}{N} &= \frac{x_{11} + x_{12}}{N} \times \frac{x_{12} + x_{22}}{N} \\
\frac{x_{21}}{N} &= \frac{x_{21} + x_{22}}{N} \times \frac{x_{11} + x_{21}}{N} \\
\frac{x_{22}}{N} &= \frac{x_{21} + x_{22}}{N} \times \frac{x_{12} + x_{22}}{N}
\end{aligned}
$$

Since $N = \sum_{i,j} x_{ij}$, the following formula will be obtained from these four formulae.

$$x_{11}x_{22} = x_{12}x_{21} \text{ or } x_{11}x_{22} - x_{12}x_{21} = 0$$

Thus,

Theorem 1. *If two attributes in a contingency table shown in Table 1 are statistical indepedent, the following equation holds:*

$$x_{11}x_{22} - x_{12}x_{21} = 0 \tag{1}$$

□

It is notable that the above equation corresponds to the fact that the determinant of a matrix corresponding to this table is equal to 0. Also, when these four values are not equal to 0, the equation 1 can be transformed into:

$$\frac{x_{11}}{x_{21}} = \frac{x_{12}}{x_{22}}.$$

Let us assume that the above ratio is equal to $C(constant)$. Then, since $x_{11} = Cx_{21}$ and $x_{12} = Cx_{22}$, the following equation is obtained.

$$\frac{x_{11} + x_{12}}{x_{21} + x_{22}} = \frac{C(x_{21} + x_{22})}{x_{21} + x_{22}} = C = \frac{x_{11}}{x_{21}} = \frac{x_{12}}{x_{22}}. \tag{2}$$

This equation also holds when we extend this discussion into a general case. Before getting into it, let us cosndier a 2 × 3 contingency table.

4 Statistical Independence in 2 × 3 Contingency Table

Let us consider a 2 × 3 contingency table shown in Table 5. Statistical independence between R_1 and R_2 gives:

$$\begin{aligned}
P([R_1 = 0], [R_2 = 0]) &= P([R_1 = 0]) \times P([R_2 = 0])\\
P([R_1 = 0], [R_2 = 1]) &= P([R_1 = 0]) \times P([R_2 = 1])\\
P([R_1 = 0], [R_2 = 2]) &= P([R_1 = 0]) \times P([R_2 = 2])\\
P([R_1 = 1], [R_2 = 0]) &= P([R_1 = 1]) \times P([R_2 = 0])\\
P([R_1 = 1], [R_2 = 1]) &= P([R_1 = 1]) \times P([R_2 = 1])\\
P([R_1 = 1], [R_2 = 2]) &= P([R_1 = 1]) \times P([R_2 = 2])
\end{aligned}$$

Table 5. Contingency Table (2 × 3)

	$R_1 = 0$	$R_1 = 1$	$R_1 = 2$	
$R_2 = 0$	x_{11}	x_{12}	x_{13}	$x_{1\cdot}$
$R_2 = 1$	x_{21}	x_{22}	x_{23}	$x_{2\cdot}$
	$x_{\cdot 1}$	$x_{\cdot 2}$	$x_{\cdot\cdot 3}$	$x_{\cdot\cdot}$
				$(= \lvert U \rvert = N)$

Since each probability is given as a ratio of each cell to N, the above equations are calculated as:

$$\frac{x_{11}}{N} = \frac{x_{11}+x_{12}+x_{13}}{N} \times \frac{x_{11}+x_{21}}{N} \tag{3}$$
$$\frac{x_{12}}{N} = \frac{x_{11}+x_{12}+x_{13}}{N} \times \frac{x_{12}+x_{22}}{N} \tag{4}$$
$$\frac{x_{13}}{N} = \frac{x_{11}+x_{12}+x_{13}}{N} \times \frac{x_{13}+x_{23}}{N} \tag{5}$$
$$\frac{x_{21}}{N} = \frac{x_{21}+x_{22}+x_{23}}{N} \times \frac{x_{11}+x_{21}}{N} \tag{6}$$
$$\frac{x_{22}}{N} = \frac{x_{21}+x_{22}+x_{23}}{N} \times \frac{x_{12}+x_{22}}{N} \tag{7}$$
$$\frac{x_{23}}{N} = \frac{x_{21}+x_{22}+x_{23}}{N} \times \frac{x_{13}+x_{23}}{N} \tag{8}$$

From equation (3) and (6),

$$\frac{x_{11}}{x_{21}} = \frac{x_{11}+x_{12}+x_{13}}{x_{21}+x_{22}+x_{23}}$$

In the same way, the following equation will be obtained:

$$\frac{x_{11}}{x_{21}} = \frac{x_{12}}{x_{22}} = \frac{x_{13}}{x_{23}} = \frac{x_{11}+x_{12}+x_{13}}{x_{21}+x_{22}+x_{23}} \tag{9}$$

Thus, we obtain the following theorem:

Theorem 2. *If two attributes in a contingency table shown in Table 5 are statistical indepedent, the following equations hold:*

$$\begin{aligned} x_{11}x_{22} - x_{12}x_{21} &= x_{12}x_{23} - x_{13}x_{22} \\ &= x_{13}x_{21} - x_{11}x_{23} = 0 \end{aligned} \tag{10}$$

□

It is notable that this discussion can be easily extended into a $2 \times n$ contingency table where $n > 3$. The important equation 9 will be extended into

$$\begin{aligned} \frac{x_{11}}{x_{21}} &= \frac{x_{12}}{x_{22}} = \cdots = \frac{x_{1n}}{x_{2n}} \\ &= \frac{x_{11}+x_{12}+\cdots+x_{1n}}{x_{21}+x_{22}+\cdots+x_{2n}} = \frac{\sum_{k=1}^{n} x_{1k}}{\sum_{k=1}^{n} x_{2k}} \end{aligned} \tag{11}$$

Thus,

Theorem 3. *If two attributes in a $2 \times k$ contingency table ($k = 2, \cdots, n$) are statistical indepedent, the following equations hold:*

$$\begin{aligned} x_{11}x_{22} - x_{12}x_{21} &= x_{12}x_{23} - x_{13}x_{22} = \cdots \\ &= x_{1n}x_{21} - x_{11}x_{n3} = 0 \end{aligned} \tag{12}$$

□

It is also notable that this equation is the same as the equation on collinearity of projective geometry [9].

5 Statistical Independence in $m \times n$ Contingency Table

Let us consider a $m \times n$ contingency table shown in Table 4. Statistical independence of R_1 and R_2 gives the following formulae:

$$P([R_1 = A_i, R_2 = B_j]) = P([R_1 = A_i])P([R_2 = B_j])$$
$$(i = 1, \cdots, m, j = 1, \cdots, n).$$

According to the definition of the table,

$$\frac{x_{ij}}{N} = \frac{\sum_{k=1}^{n} x_{ik}}{N} \times \frac{\sum_{l=1}^{m} x_{lj}}{N}. \tag{13}$$

Thus, we have obtained:

$$x_{ij} = \frac{\sum_{k=1}^{n} x_{ik} \times \sum_{l=1}^{m} x_{lj}}{N}. \tag{14}$$

Thus, for a fixed j,

$$\frac{x_{i_a j}}{x_{i_b j}} = \frac{\sum_{k=1}^{n} x_{i_a k}}{\sum_{k=1}^{n} x_{i_b k}}$$

In the same way, for a fixed i,

$$\frac{x_{i j_a}}{x_{i j_b}} = \frac{\sum_{l=1}^{m} x_{l j_a}}{\sum_{l=1}^{m} x_{l j_b}}$$

Since this relation will hold for any j, the following equation is obtained:

$$\frac{x_{i_a 1}}{x_{i_b 1}} = \frac{x_{i_a 2}}{x_{i_b 2}} \cdots = \frac{x_{i_a n}}{x_{i_b n}} = \frac{\sum_{k=1}^{n} x_{i_a k}}{\sum_{k=1}^{n} x_{i_b k}}. \tag{15}$$

Since the right hand side of the above equation will be constant, thus all the ratios are constant. Thus,

Theorem 4. *If two attributes in a contingency table shown in Table 4 are statistical indepedent, the following equations hold:*

$$\frac{x_{i_a 1}}{x_{i_b 1}} = \frac{x_{i_a 2}}{x_{i_b 2}} \cdots = \frac{x_{i_a n}}{x_{i_b n}} = const. \tag{16}$$

for all rows: i_a and i_b $(i_a, i_b = 1, 2, \cdots, m)$.

□

6 Statistical Independence with m-Way Tables

6.1 Three-Way Table

Let "•" denote as the sum over the row or column of a contingency matrix. That is ,

$$x_{i\bullet} = \sum_{j=1}^{n} x_{ij} \tag{17}$$

$$x_{\bullet j} = \sum_{i=1}^{m} x_{ij}, \tag{18}$$

where (17) and (18) shows marginal column and row sums. Then, it is easy to see that

$$x_{\bullet\bullet} = N,$$

where N denotes the sample size.

Then, Equation (14) is reformulated as:

$$\frac{x_{ij}}{x_{\bullet\bullet}} = \frac{x_{i\bullet}}{x_{\bullet\bullet}} \times \frac{x_{\bullet j}}{x_{\bullet\bullet}} \tag{19}$$

That is,

$$x_{ij} = \frac{x_{i\bullet} \times x_{\bullet j}}{x_{\bullet\bullet}}$$

Or

$$x_{ij}x_{\bullet\bullet} = x_{i\bullet}x_{\bullet j}$$

Thus, statistical independence can be viewed as the specific relations between assignments of i,j and "$\cdot$". By use of the above relation, Equation (16) can be rewritten as:

$$\frac{x_{i_1 j}}{x_{i_2 j}} = \frac{x_{i_1 \bullet}}{x_{i_2 \bullet}},$$

where the right hand side gives the ratio of marginal column sums.

Equation (19) can be extended into multivariate cases. Let us consider a three attribute case.

Statistical independence with three attributes is defined as:

$$\frac{x_{ijk}}{x_{\bullet\bullet\bullet}} = \frac{x_{i\bullet\bullet}}{x_{\bullet\bullet\bullet}} \times \frac{x_{\bullet j\bullet}}{x_{\bullet\bullet\bullet}} \times \frac{x_{\bullet\bullet k}}{x_{\bullet\bullet\bullet}}, \tag{20}$$

Thus,

$$x_{ijk}x_{\bullet\bullet\bullet}^2 = x_{i\bullet\bullet}x_{\bullet j\bullet}x_{\bullet\bullet k}, \tag{21}$$

which corresponds to:

$$P(A = a, B = b, C = c) = P(A = a)P(B = b)P(C = c), \tag{22}$$

where A,B,C correspond to the names of attributes for i,j,k, respectively.

In statistical context, statistical independence requires hiearchical model. That is, statistical independence of three attributes requires that all the two pairs of three attributes should satisfy the equations of statistical independence. Thus, for Equation (22), the following equations should satisfy:

$$\begin{aligned} &P(A = a, B = b) = P(A = a)P(B = b), \\ &P(B = b, C = c) = P(B = b)P(C = c), \textit{ and} \\ &P(A = a, C = c) = P(A = a)P(C = c). \end{aligned}$$

Thus,

$$x_{ij\bullet}x_{\bullet\bullet\bullet} = x_{i\bullet\bullet}x_{\bullet j\bullet} \tag{23}$$

$$x_{i\bullet k}x_{\bullet\bullet\bullet} = x_{i\bullet\bullet}x_{\bullet\bullet k} \tag{24}$$

$$x_{\bullet jk}x_{\bullet\bullet\bullet} = x_{\bullet j\bullet}x_{\bullet\bullet k} \tag{25}$$

From Equation (21) and Equation (23),

$$x_{ijk}x_{\bullet\bullet\bullet} = x_{ij\bullet}x_{\bullet\bullet k},$$

Therefore,

$$\frac{x_{ijk}}{x_{ij\bullet}} = \frac{x_{\bullet\bullet k}}{x_{\bullet\bullet\bullet}} \tag{26}$$

In the same way, the following equations are obtained:

$$\frac{x_{ijk}}{x_{i\bullet k}} = \frac{x_{\bullet j\bullet}}{x_{\bullet\bullet\bullet}} \tag{27}$$

$$\frac{x_{ijk}}{x_{\bullet jk}} = \frac{x_{i\bullet\bullet}}{x_{\bullet\bullet\bullet}} \tag{28}$$

In summary, the following theorem is obtained.

Theorem 5. *If a three-way contingency table satisfy statistical independence, then the following three equations should be satisfied:*

$$\frac{x_{ijk}}{x_{ij\bullet}} = \frac{x_{\bullet\bullet k}}{x_{\bullet\bullet\bullet}}$$

$$\frac{x_{ijk}}{x_{i\bullet k}} = \frac{x_{\bullet j\bullet}}{x_{\bullet\bullet\bullet}}$$

$$\frac{x_{ijk}}{x_{\bullet jk}} = \frac{x_{i\bullet\bullet}}{x_{\bullet\bullet\bullet}}$$

□

Thus, the equations corresponding to Theorem 4 are obtained as follows.

Corollary 1. *If three attributes in a contingency table shown in Table 4 are statistical indepedent, the following equations hold:*

$$\frac{x_{ijk_a}}{x_{ijk_b}} = \frac{x_{\bullet\bullet k_a}}{x_{\bullet\bullet k_b}}$$

$$\frac{x_{ij_ak}}{x_{ij_bk}} = \frac{x_{\bullet j_a\bullet}}{x_{\bullet j_b\bullet}}$$

$$\frac{x_{i_ajk}}{x_{i_bjk}} = \frac{x_{i_a\bullet\bullet}}{x_{i_b\bullet\bullet}}$$

for all i,j, *and* k.

□

6.2 Multi-way Table

The above discussion can be easily extedned into a multi-way contingency table.

Theorem 6. *If a m-way contingency table satisfy statistical independence, then the following equation should be satisfied for any k-th attribute* i_k *and* j_k $(k = 1, 2, \cdots, n)$ *where n is the number of attributes.*

$$\frac{x_{i_1 i_2 \cdots i_k \cdots i_n}}{x_{i_1 i_2 \cdots j_k \cdots i_n}} = \frac{x_{\bullet\bullet \cdots i_k \cdots \bullet}}{x_{\bullet\bullet \cdots j_k \cdots \bullet}}$$

Also, the following equation should be satisfied for any i_k:

$$x_{i_1 i_2 \cdots i_n} \times x^{n-1}_{\bullet\bullet \cdots \bullet} = x_{i_1 \bullet \cdots \bullet} x_{\bullet i_2 \cdots \bullet} \times \cdots \times x_{\bullet\bullet \cdots i_k \cdots \bullet} \times \cdots \times x_{\bullet\bullet \cdots \bullet i_n} \qquad \square$$

7 Contingency Matrix

The meaning of the above discussions will become much clearer when we view a contingency table as a matrix.

Definition 3. *A corresponding matrix $C_{T_{a,b}}$ is defined as a matrix the element of which are equal to the value of the corresponding contingency table $T_{a,b}$ of two attributes a and b, except for marginal values.*

Definition 4. *The rank of a table is defined as the rank of its corresponding matrix. The maximum value of the rank is equal to the size of (square) matrix, denoted by r.*

The contingency matrix of Table $4(T(R_1, R_2))$ is defined as $C_{T_{R_1,R_2}}$ as below:

$$\begin{pmatrix} x_{11} & x_{12} & \cdots & x_{1n} \\ x_{21} & x_{22} & \cdots & x_{2n} \\ \vdots & \vdots & \ddots & \vdots \\ x_{m1} & x_{m2} & \cdots & x_{mn} \end{pmatrix}$$

7.1 Independence of 2 × 2 Contingency Table

The results in Section 3 corresponds to the degree of independence in matrix theory. Let us assume that a contingency table is given as Table 1. Then the corresponding matrix ($C_{T_{R_1,R_2}}$) is given as:

$$\begin{pmatrix} x_{11} & x_{12} \\ x_{21} & x_{22} \end{pmatrix},$$

Then,

Proposition 1. *The determinant of $det(C_{T_{R_1,R_2}})$ is equal to $x_{11}x_{22} - x_{12}x_{21}$,*

Proposition 2. *The rank will be:*

$$rank = \begin{cases} 2, & if\ det(C_{T_{R_1,R_2}}) \neq 0 \\ 1, & if\ det(C_{T_{R_1,R_2}}) = 0 \end{cases}$$

From Theorem 1,

Theorem 7. *If the rank of the corresponding matrix of a 2×2 contingency table is 1, then two attributes in a given contingency table are statistically independent. Thus,*

$$rank = \begin{cases} 2, & dependent \\ 1, & statistical\ independent \end{cases}$$

It is easy to see that this discussion can be extended into $2 \times n$ tables.

7.2 Independence of 3 × 3 Contingency Table

When the number of rows and columns are larger than 3, then the situation is a little changed. It is easy to see that the rank for statistical independence of a $m \times n$ contingency table is equal 1.0 as shown in Theorem 4. Also, when the rank is equal to $\min(m, n)$, two attributes are dependent.

Then, what kind of structure will a contingency matrix have when the rank is larger than 1,0 and smaller than $\min(m, n) - 1$? For illustration, let us consider the following $3times3$ contingecy table.

Example 2. Let us consider the following corresponding matrix:

$$A = \begin{pmatrix} 1\,2\,3 \\ 4\,5\,6 \\ 7\,8\,9 \end{pmatrix}.$$

The determinant of A is:

$$\begin{aligned} det(A) &= 1 \times (-1)^{1+1} det \begin{pmatrix} 5\,6 \\ 8\,9 \end{pmatrix} \\ &\quad +2 \times (-1)^{1+2} det \begin{pmatrix} 4\,6 \\ 7\,9 \end{pmatrix} \\ &\quad +3 \times (-1)^{1+3} det \begin{pmatrix} 4\,5 \\ 7\,8 \end{pmatrix} \\ &= 1 \times (-3) + 2 \times 6 + 3 \times (-3) = 0 \end{aligned}$$

Thus, the rank of A is smaller than 2. On the other hand, since $(123) \neq k(456)$ and $(123) \neq k(789)$, the rank of A is not equal to 1.0 and thus, the rank of A is equal to 2.0. Actually, one of three rows can be represented by the other two rows. For example,

$$(4\,5\,6) = \frac{1}{2}\{(1\,2\,3) + (7\,8\,9)\}.$$

Therefore, in this case, we can say that two of three pairs of one attribute are dependent to the other attribute, but one pair is statistically independent of the other attribute with respect to the linear combination of two pairs. It is easy to see that this case includes the cases when two pairs are statistically independent of the other attribute, but the table becomes statistically dependent with the other attribute.

In other words, the corresponding matrix is a mixure of statistical dependence and independence. We call this case *contextual independent.* From this illustration, the following theorem is obtained:

Theorem 8. *If the rank of the corresponding matrix of a* 3×3 *contigency table is 1, then two attributes in a given contingency table are statistically independent. Thus,*

$$rank = \begin{cases} 3, & dependent \\ 2, & contextual\ independent \\ 1, & statistical\ independent \end{cases}$$

It is easy to see that this discussion can be extended into $3 \times n$ contingency tables.

7.3 Independence of $m \times n$ Contingency Table

Finally, the relation between rank and independence in a multi-way contingency table is obtained from Theorem 4.

Theorem 9. *Let the corresponding matrix of a given contingency table be a* $m \times n$ *matrix. If the rank of the corresponding matrix is 1, then two attributes in a given contingency table are statistically independent. If the rank of the corresponding matrix is* $\min(m, n)$, *then two attributes in a given contingency table are dependent. Otherwise, two attributes are contextual dependent, which means that several conditional probabilities can be represented by a linear combination of conditional probabilities. Thus,*

$$rank = \begin{cases} \min(m,n) & dependent \\ 2, \cdots, & \\ \quad \min(m,n) - 1 & contextual\ independent \\ 1 & statistical\ independent \end{cases}$$

8 Pseudo Statistical Independence: Example

The next step is to investigate the characteristics of linear independence in a contingency matrix. In other words, a $m \times n$ contingency table whose rank is not equal to $\min(m, n)$. Since two-way matrix (2×2) gives a simple equation whose rank is equal to 1 or 2, let us start our discussion from 3×3-matrix, whose rank is equal to 2, first.

8.1 Three-Way Contingency Table (Rank: 2)

Let $M(m, n)$ denote a contingency matrix whose row and column are equal to m and n, respectively. Then, a three-way contingency table is defined as:

$$M(3,3) = \begin{pmatrix} x_{11} & x_{12} & x_{13} \\ x_{21} & x_{22} & x_{23} \\ x_{31} & x_{32} & x_{33} \end{pmatrix}$$

When its rank is equal to 2, it can be assumed that the third row is represented by the first and second row:

$$(x_{31}\; x_{32}\; x_{33}) = p(x_{11}\; x_{12}\; x_{13}) + q(x_{21}\; x_{22}\; x_{23}) \tag{29}$$

Then, we can consider the similar process in Section 5 (13). In other words, we can check the difference defined below.

$$\Delta(i,j) = \frac{x_{ij}}{N} - \frac{\sum_{k=1}^{n} x_{ik}}{N} \times \frac{\sum_{l=1}^{m} x_{lj}}{N}. \tag{30}$$

Then, the following three types of equations are obtained by simple calculation.

$$\Delta(1,j) = (1+q)\left\{x_{1j}\sum_{k=1}^{3} x_{2k} - x_{2j}\sum_{k=1}^{3} x_{1k}\right\}$$

$$\Delta(2,j) = (1+p)\left\{x_{2j}\sum_{k=1}^{3} x_{1k} - x_{1j}\sum_{k=1}^{3} x_{2k}\right\}$$

$$\Delta(3,j) = (p-q)\left\{x_{1j}\sum_{k=1}^{3} x_{2k} - x_{2j}\sum_{k=1}^{3} x_{1k}\right\}$$

According to Theorem 4, if $M(3,3)$ is not statistically independent, the formula: $x_{1j}\sum_{k=1}^{3} x_{2k} - x_{2j}\sum_{k=1}^{3} x_{1k}$ is not equal to 1.0. Thus, the following theorem is obtained.

Theorem 10. *The third row represened by a linear combination of first and second rows will satisfy the condition of statistical independence if and only if* $p = q$.

We call the above property *pseudo statistical independence.* This means that if the third column satisfies the following constraint:

$$(x_{31}\; x_{32}\; x_{33}) = (x_{11}\; x_{12}\; x_{13}) + (x_{21}\; x_{22}\; x_{23}),$$

the third column will satisfy the condition of statistical independence. In other words, when we merge the first and second row and construct a 2×3 contingency table, it will become statistical independent. For example,

$$D = \begin{pmatrix} 1 & 2 & 3 \\ 4 & 5 & 6 \\ 10 & 14 & 18 \end{pmatrix}$$

can be transformed into

$$D' = \begin{pmatrix} 5 & 7 & 9 \\ 10 & 14 & 18 \end{pmatrix},$$

where D' is statistically independent. Conversely, if D' is provided, it can be decomposed into D. It is notable that the decomposition cannot be uniquely determined. It is also notable that the above discussion does not use the information about the columns of a contingency table. Thus, this discussion can be extended into a $3 \times n$ contingency matrix.

8.2 Four-Way Contingency Table (Rank: 3)

From four-way tables, the situation becomes more complicated. In the similar way to Subsection 8.1, a four-way contingency table is defined as:

$$M(4,4) = \begin{pmatrix} x_{11} & x_{12} & x_{13} & x_{14} \\ x_{21} & x_{22} & x_{23} & x_{24} \\ x_{31} & x_{32} & x_{33} & x_{34} \\ x_{41} & x_{42} & x_{43} & x_{44} \end{pmatrix}$$

When its rank is equal to 3, it can be assumed that the fourth row is represented by the first to third row:

$$\begin{aligned} (x_{41}\ x_{42}\ x_{43}\ x_{44}) = p&(x_{11}\ x_{12}\ x_{13}\ x_{14}) \\ +q&(x_{21}\ x_{22}\ x_{23}\ x_{24}) \\ +r&(x_{31}\ x_{32}\ x_{33}\ x_{34}) \end{aligned} \tag{31}$$

Then, the following three types of equations are obtained by simple calculation.

$$\begin{aligned}
\Delta(1,j) &= (1+q)\left\{x_{1j}\sum_{k=1}^{4} x_{2k} - x_{2j}\sum_{k=1}^{4} x_{1k}\right\} \\
&\quad +(1+r)\left\{x_{1j}\sum_{k=1}^{4} x_{3k} - x_{3j}\sum_{k=1}^{4} x_{1k}\right\} \\
\Delta(2,j) &= (1+p)\left\{x_{2j}\sum_{k=1}^{4} x_{1k} - x_{1j}\sum_{k=1}^{4} x_{2k}\right\} \\
&\quad +(1+r)\left\{x_{2j}\sum_{k=1}^{4} x_{3k} - x_{3j}\sum_{k=1}^{4} x_{2k}\right\} \\
\Delta(3,j) &= (1+p)\left\{x_{2j}\sum_{k=1}^{4} x_{1k} - x_{1j}\sum_{k=1}^{4} x_{2k}\right\} \\
&\quad +(1+q)\left\{x_{1j}\sum_{k=1}^{4} x_{2k} - x_{2j}\sum_{k=1}^{4} x_{1k}\right\} \\
\Delta(4,j) &= (p-q)\left\{x_{1j}\sum_{k=1}^{4} x_{2k} - x_{2j}\sum_{k=1}^{4} x_{1k}\right\} \\
&\quad +(r-p)\left\{x_{3j}\sum_{k=1}^{4} x_{2k} - x_{1j}\sum_{k=1}^{4} x_{1k}\right\} \\
&\quad +(q-r)\left\{x_{2j}\sum_{k=1}^{4} x_{3k} - x_{3j}\sum_{k=1}^{4} x_{2k}\right\}
\end{aligned}$$

Thus, the following theorem is obtained.

Theorem 11. *The fourth row represened by a linear combination of first to third rows (basis) will satisfy the condition of statistical independence if and only if* $\Delta(4, j) = 0$.

Unfortunately, the condition is not simpler than Theorem 10. It is notable $\Delta(4, j) = 0$ is a diophatine equation whose trivial solution is $p = q = r$. That is, the solution space includes not only $p = q = r$, but other solutions. Thus,

Corollary 2. *If* $p = q = r$, *then the fourth row satisfies the condition of statistical independence.*

The converse is not true.

Example 3. Let us consider the following matrix:

$$E = \begin{pmatrix} 1 & 1 & 2 & 2 \\ 2 & 2 & 3 & 3 \\ 4 & 4 & 5 & 5 \\ x_{41} & x_{42} & x_{43} & x_{44} \end{pmatrix}.$$

The question is when the fourth row represented by the other rows satisfies the condition of statistical independence. Since $x_{1j} \sum_{k=1}^{4} x_{2k} - x_{2j} \sum_{k=1}^{4} x_{1k} = -2$, $x_{1j} \sum_{k=1}^{4} x_{3k} - x_{3j} \sum_{k=1}^{4} x_{1k} = 6$ and $x_{2j} \sum_{k=1}^{4} x_{1k} - x_{1j} \sum_{k=1}^{4} x_{2k} = -4$, $\Delta(4, j)$ is equal to: $-2(p-q) + 6(r-p) - 4(q-r) = -8p - 2q + 10r$.

Thus, the set of solutions is $\{(p, q, r) | 10r = 8p + 2q\}$, where $p = q = r$ is included.

It is notable that the characteristics of solutions will be characterized by a diophantine equation $10r = 8p + 2q$ and a contingency table given by a tripule (p, q, r) may be represented by another tripule. For example, $(3, 3, 3)$ gives the same contingency table as $(1, 6, 2)$:

$$\begin{pmatrix} 1 & 1 & 2 & 2 \\ 2 & 2 & 3 & 3 \\ 4 & 4 & 5 & 5 \\ 21 & 21 & 30 & 30 \end{pmatrix}.$$

It will be our future work to investigate the general characteristics of the solution space.

8.3 Four-Way Contingency Table (Rank: 2)

When its rank is equal to 2, it can be assumed that the third and fourth rows are represented by the first to third row:

$$\begin{aligned}(x_{41}\ x_{42}\ x_{43}\ x_{44}) = {} & p(x_{11}\ x_{12}\ x_{13}\ x_{14}) \\ & + q(x_{21}\ x_{22}\ x_{23}\ x_{24}) \end{aligned} \qquad (32)$$

$$\begin{aligned}(x_{31}\ x_{32}\ x_{33}\ x_{34}) = {} & r(x_{11}\ x_{12}\ x_{13}\ x_{14}) \\ & + s(x_{21}\ x_{22}\ x_{23}\ x_{24}) \end{aligned} \qquad (33)$$

$$\Delta(1,j) = (1+q+s)\left\{x_{1j}\sum_{k=1}^{4}x_{2k} - x_{2j}\sum_{k=1}^{4}x_{1k}\right\}$$
$$\Delta(2,j) = (1+p+r)\left\{x_{2j}\sum_{k=1}^{4}x_{1k} - x_{1j}\sum_{k=1}^{4}x_{2k}\right\}$$
$$\Delta(3,j) = (p-q+ps-qr) \times \left\{x_{2j}\sum_{k=1}^{4}x_{1k} - x_{1j}\sum_{k=1}^{4}x_{2k}\right\}$$
$$\Delta(4,j) = (r-s+qr-ps) \times \left\{x_{1j}\sum_{k=1}^{4}x_{2k} - x_{2j}\sum_{k=1}^{4}x_{2k}\right\}$$

Since $p-q+ps-qr=0$ and $r-s+qr-ps=0$ gives the only reasonable solution $p=q$ and $r=s$, the following theorem is obtained.

Theorem 12. *The third and fourth rows represened by a linear combination of first and second rows (basis) will satisfy the condition of statistical independence if and only if* $p=q$ *and* $r=w$.

9 Pseudo Statiatical Independence

Now, we will generalize the results shown in Section 8. Let us consider the $n\times m$ contingency table whose r rows (columns) are described by $n-s$ rows (columns). Thus, we assume a corresponding matrix with the following equations.

$$\begin{pmatrix} x_{11} & x_{12} & \cdots & x_{1n} \\ x_{21} & x_{22} & \cdots & x_{2n} \\ \vdots & \vdots & \ddots & \vdots \\ x_{m1} & x_{m2} & \cdots & x_{mn} \end{pmatrix}$$

$$(x_{n-s+p,1}\ x_{n-s+p,2}\ \cdots\ x_{n-s+p,m}) = \sum_{i=1}^{n-s} k_{pi}(x_{i1}\ x_{i2}\ \cdots\ x_{im})$$
$$(1\le s\le n-1, 1\le p\le s) \quad (34)$$

Then, the following theorem about $\Delta(u,v)$ is obtained.

Theorem 13. *For a contingency table with size $n \times m$:*

$$\Delta(u,v) = \begin{cases} \sum_{i=1}^{n-s}(1+\sum_{p=1}^{n-s} k_{pi}) \\ \quad \times \left\{ x_{uv}(\sum_{j=1}^{m} x_{ij}) - x_{iv}(\sum_{j=1}^{m} x_{uj}) \right\} \\ \quad (1 \leq u \leq n-s,\ 1 \leq v \leq m) \\ \sum_{i=1}^{n-s}\sum_{j=1}^{m}\sum_{q=1}^{n-s} x_{q1}x_{ij} \\ \quad \times \{(k_{uq} - k_{ui}) \\ \quad + k_{uq}\sum_{p=1}^{n-s} k_{pi} - k_{ui}\sum_{p=1}^{n-s} k_{pq} \} \\ \quad (n-s+1 \leq u \leq n,\ 1 \leq v \leq m) \end{cases} \tag{35}$$

Thus, from the above theorem, if and only if $\Delta(u,v) = 0$ for all v, then the u-th row will satisfy the condition of statistically independence. Especially, the following theorem is obtained.

Theorem 14. *If the following equation holds for all $v (1 \leq v \leq m)$, then the condition of statistical independence will hold for the u-th row in a contingency table.*

$$\sum_{i=1}^{n-s}\sum_{j=1}^{m}\sum_{q=1}^{n-s} \left\{ (k_{uq} - k_{ui}) + k_{uq}\sum_{p=1}^{n-s} k_{pi} - k_{ui}\sum_{p=1}^{n-s} k_{pq} \right\} = 0 \tag{36}$$

It is notable that the above equations give diophatine equations which can check whether each row (column) will satisfy the condition of statistical independence. As a corollary,

Corollary 3. *If k_{ui} is equal for all $i = 1, \cdots, n-s$, then the u-th satisfies the condition of statistical independence.*

The converse is not true.

Example 4. Let us consider the following matrix:

$$F = \begin{pmatrix} 1 & 1 & 2 \\ 2 & 2 & 3 \\ 4 & 4 & 5 \\ x_{41} & x_{42} & x_{43} \\ x_{51} & x_{52} & x_{53} \end{pmatrix},$$

where the last two rows are represented by the first three columns. That is, the rank of a matrix is equal to 3. Then, according to Theorem 14, the following equations are obtained:

$$(5k_{53} - k_{52} - 4k_{51}) \times \{k_{41} - 2k_{43} + (k_{51} - 2k_{53} - 1\} = 0 \quad (37)$$

$$(5k_{43} - k_{42} - 4k_{41}) \times \{k_{41} - 2k_{43} + (k_{51} - 2k_{53} - 1\} = 0 \quad (38)$$

In case of $k_{41} - 2k_{43} + (k_{51} - 2k_{53} - 1) = 0$, simple calculations give several equations for those coefficients.

$$k_{41} + k_{51} = 2(k_{43} + k_{53}) + 1$$
$$k_{42} + k_{52} = -3(k_{43} + k_{53})$$

The solutions of these two equations give examples of pseudo statistical independence. □

10 Conclusion

In this paper, a contingency table is interpreted from the viewpoint of granular computing and statistical independence. From the definition of statistical independence, statistical independence in a contingency table will holds when the equations of collinearity(Equation 14) are satisfied. In other words, statistical independence can be viewed as linear dependence. Then, the correspondence between contingency table and matrix, gives the theorem where the rank of the contingency matrix of a given contingency table is equal to 1 if two attributes are statistical independent. That is, all the rows of contingency table can be described by one row with the coefficient given by a marginal distribution. If the rank is maximum, then two attributes are dependent. Otherwise, some probabilistic structure can be found within attribute -value pairs in a given attribute, which we call contextual independence. Moreover, from the characteristics of statistical independence, a contingency table may be composed of statistical independent and dependent parts, which we call pseudo statistical dependence. In such cases, if we merge several rows or columns, then we will obtain a new contingency table with statistical independence, whose rank of its corresponding matrix is equal to 1.0. Especially, we obtain Diophatine equations for a pseudo statistical dependence. Thus, matrix algebra and elementary number theory are the key methods of the analysis of a contingency table and the degree of independence, where its rank and the structure of linear dependence as Diophatine equations play very important roles in determining the nature of a given table.

References

1. Maimon, O., Rokach, L. (eds.): The Data Mining and Knowledge Discovery Handbook. Springer, Heidelberg (2005)
2. Joe, H.: Multivariate Models and Dependence Concepts. CRC/Chapman & Hall (1997)
3. Zadeh, L.: Toward a theory of fuzzy information granulation and its certainty in human reasoning and fuzzy logic. Fuzzy Sets and Systems 90, 111–127 (1997)
4. Lin, T.Y., Liau, C.J.: Granular computing and rough sets. In: [1], pp. 535–561
5. Skowron, A., Grzymala-Busse, J.: From rough set theory to evidence theory. In: Yager, R., Fedrizzi, M., Kacprzyk, J. (eds.) Advances in the Dempster-Shafer Theory of Evidence, pp. 193–236. John Wiley & Sons, New York (1994)
6. Yao, Y., Wong, S.: A decision theoretic framework for approximating concepts. International Journal of Man-machine Studies 37, 793–809 (1992)
7. Yao, Y., Zhong, N.: An analysis of quantitative measures associated with rules. In: Zhong, N., Zhou, L. (eds.) PAKDD 1999. LNCS (LNAI), vol. 1574, pp. 479–488. Springer, Heidelberg (1999)
8. Tsumoto, S.: Automated induction of medical expert system rules from clinical databases based on rough set theory. Information Sciences 112, 67–84 (1998)
9. Coxeter, H. (ed.): Projective Geometry, 2nd edn. Springer, New York (1987)

Applying Rough Sets to Information Tables Containing Possibilistic Values

Michinori Nakata[1] and Hiroshi Sakai[2]

[1] Faculty of Management and Information Science,
Josai International University
1 Gumyo, Togane, Chiba, 283-8555, Japan
nakatam@ieee.org

[2] Department of Mathematics and Computer Aided Sciences,
Faculty of Engineering, Kyushu Institute of Technology,
Tobata, Kitakyushu, 804-8550, Japan
sakai@mns.kyutech.ac.jp

Abstract. Rough sets are applied to information tables containing imprecise values that are expressed in a normal possibility distribution. A method of weighted equivalence classes is proposed, where each equivalence class is accompanied by a possibilistic degree to which it is an actual one. By using a family of weighted equivalence classes, we derive lower and upper approximations. The lower and upper approximations coincide with ones obtained from methods of possible worlds. Therefore, the method of weighted equivalence classes is justified. When this method is applied to missing values interpreted possibilistically, it creates the same relation for indiscernibility as the method of Kryszkiewicz that gave an assumption for indiscernibility of missing values. Using weighted equivalence classes correctly derives a lower approximation from the viewpoint of possible worlds, although using a class of objects that is not an equivalence class does not always derive a lower approximation.

Keywords: Rough sets, Imprecise value, Missing value, Possibility distribution, Weighted equivalence class, Lower and upper approximations.

1 Introduction

Rough sets play a significant role in the field of knowledge discovery and data mining since the first paper published by Pawlak [29]. Methods of rough sets are originally constructed under the premise that information tables consisting of precise information are obtained, which are called the traditional methods of rough sets. However, information tables actually obtained contain imprecise data such as missing values in many cases. Furthermore, there ubiquitously exists imperfect information containing imprecision and uncertainty in the real world [28]. Under these circumstances, it has been investigated to apply rough sets to information tables containing imprecise information represented by a missing value, an or-set, a possibility distribution, a probability distribution,

M.L. Gavrilova et al. (Eds.): Trans. on Comput. Sci. II, LNCS 5150, pp. 180–204, 2008.

etc [5,6,7,11,14,15,16,17,20,21,22,23,24,25,26,31,32,33,35,37,38]. The methods are broadly separated into three ways.

The first method is one based on possible worlds, which is called a method of possible worlds [27, 31, 32, 33, 34]. The method creates an extended set of possible tables from an information table. All possible tables consist of precise values. Each possible table is dealt with in terms of the traditional methods of applying rough sets to information tables not containing imprecise values, and then the results from possible tables are aggregated. In other words, the methods that are already established are applied to each possible table. Therefore, There is no doubt about the correctness of the method of possible worlds in the sense that the method is based on results obtained from the established methods. However, the method has difficulties for knowledge discovery at the level of a set of possible values, although it is suitable for finding knowledge at the level of possible values. This is because the number of possible tables exponentially increases as the number of imprecise attribute values increases.

The second method is to use assumptions on indiscernibility of missing values [5, 6, 11, 14, 15, 16, 17, 20, 37, 38]. The assumption that Kryszkiewicz used is that a missing value is indiscernible with any value. Stefanowski and Tsoukiàs pointed out that the assumption creates quite poor results for lower approximations [37]. To improve the situation, another assumption was proposed. The assumption is that indiscernibility is directional [5, 6, 37, 38]. A missing value is indiscernible with any precise value when viewed from the missing value, whereas any precise value is not indiscernible with a missing value when viewed from the precise value. Under the assumptions, we can obtain a binary relation for indiscernibility between objects. To the binary relation, rough sets are applied by using a class of objects which is not an equivalence class; for instance, a tolerance class. In the method, it is not clarified why the assumptions are valid to real data sets.

The third method directly deals with imprecise values, without using any assumptions on indiscernibility, under extending the traditional method of applying rough sets to information tables not containing imprecise values [21,22,23,38]. In the method, imprecise values are dealt with probabilistically or possibilistically and the traditional methods are probabilistically or possibilistically extended. A binary relation for indiscernibility is constructed by calculating a degree for indiscernibility between objects. A criterion is proposed to check the correctness of extended methods [21, 22, 23]. The correctness criterion is that any extended method has to give the same results as the method of possible worlds [21]. This criterion is commonly used in the field of databases handling imprecise information [1, 2, 3, 4, 12, 13, 39].

Stefanowski and Tsoukiàs obtained a lower approximation by using implication operators to calculate an inclusion degree between tolerance classes [38]. Nakata and Sakai have shown that the results in terms of implication operators do not satisfy the correctness criterion and has proposed the method that satisfies the correctness criterion [21, 22, 23]. However, the proposed method has difficulties for definability, because rough approximations are defined by constructing sets from singletons. Therefore, we propose a method using equivalence classes, called a method of weighted equivalence classes. In this paper, we show how weighted equivalence classes are applied to information tables containing possibilistic values expressed in a normal possibility distribution.[1]

In Section 2, we briefly address the traditional methods of applying rough sets to information tables not containing imprecise values. In Section 3, the method of possible worlds is mentioned. The extended set of possible tables is created from an information table containing imprecise values. The traditional methods of applying rough sets to precise information deal with each possible table and then the results from possible tables are aggregated. In Section 4, a method of applying rough sets to information tables containing imprecise values expressed in a normal possibility distribution is described in terms of weighted equivalence classes. In Section 5, the method is applied to information tables containing missing values under possibilistic interpretation. Section 6 presents conclusions.

2 Rough Sets under Precise Information

A data set is represented as a table, called an information table, where each row represents an object and each column represents an attribute. The information table is expressed in (U, AT), where U is a non-empty finite set of objects called the universe and AT is a non-empty finite set of attributes such that $\forall a \in AT : U \to V_a$. V_a is called the domain of attribute a. In information table T consisting of set AT of attributes, binary relation $IND(\Psi_A)$ for indiscernibility of objects in subset $\Psi \subseteq U$ on subset $A \subseteq AT$ of attributes is,

$$IND(\Psi_A) = \{(o, o') \in \Psi \times \Psi \mid \forall a \in A \;\; a(o) = a(o')\}, \tag{1}$$

where o and o' denotes objects, (o, o') a pair of o and o', $a(o)$ and $a(o')$ attribute values of o and o' on a. This relation is called an indiscernibility relation. Obviously, $IND(\Psi_A)$ is an equivalence relation. From the indiscernibility relation, the equivalence class containing object o on A, denoted by $E(\Psi_A)_o (= \{o' \mid (o, o') \in IND(\Psi_A)\})$, is obtained. This is also the set of objects that is indiscernible with object o on A, called the indiscernible class on A for object o. Finally, a family of equivalence classes on A, denoted by $\Psi/IND(\Psi_A)$ $(= \{E(\Psi_A)_o \mid o \in \Psi\})$, is derived from the indiscernibility relation. All equivalence classes obtained from the indiscernibility relation do not intersect with each other. This means that the objects are partitioned.

Example 1
Let the following information table T_1 be obtained:

[1] See references [24, 26] for information tables containing probabilistic information.

T_1

O	a_1	a_2	a_3
1	x	1	a
2	x	2	b
3	y	2	b
4	z	1	b
5	x	1	a
6	y	2	b

Mark O denotes the object identity and $U = \{o_1, o_2, o_3, o_4, o_5, o_6\}$. For set $\Psi(= \{o_2, o_3, o_4, o_5, o_6\})$ of objects, we obtain the following binary relation for indiscernibility on attribute a_1:

$$IND(\Psi_{a_1}) = \{(o_2, o_2), (o_2, o_5), (o_3, o_3), (o_3, o_6), (o_4, o_4), (o_5, o_2), (o_5, o_5), (o_6, o_3), (o_6, o_6)\}.$$

From $IND(\Psi_{a_1})$, we obtain the following family of equivalence classes:

$$\Psi/IND(\Psi_{a_1}) = \{\{o_4\}, \{o_2, o_5\}, \{o_3, o_6\}\}.$$

Similarly, for $\Phi = \{o_1, o_2, o_3, o_5, o_6\}$ on attribute a_3,

$$\Phi/IND(\Phi_{a_3}) = \{\{o_1, o_5\}, \{o_2, o_3, o_6\}\}.$$

Using equivalence classes, lower approximation $\underline{Apr}(\Phi_B, \Psi_A)$ and upper approximation $\overline{Apr}(\Phi_B, \Psi_A)$ of $\Phi/IND(\Phi_B)$ by $\Psi/IND(\Psi_A)$ are,

$$\underline{Apr}(\Phi_B, \Psi_A) = \{E(\Psi_A) \mid \exists E(\Phi_B)\ E(\Psi_A) \subseteq E(\Phi_B)\}, \quad (2)$$
$$\overline{Apr}(\Phi_B, \Psi_A) = \{E(\Psi_A) \mid \exists E(\Phi_B)\ E(\Psi_A) \cap E(\Phi_B) \neq \emptyset\}. \quad (3)$$

where $E(\Psi_A) \in \Psi/IND(\Psi_A)$ and $E(\Phi_B) \in \Phi/IND(\Phi_B)$ are equivalence classes for sets Ψ and Φ of objects on sets A and B of attributes, respectively. These formulas are expressed in terms of equivalence classes. For lower and upper approximations in terms of objects, the following expressions are used:

$$\underline{apr}(\Phi_B, \Psi_A) = \{o \mid o \in E(\Psi_A) \wedge \exists E(\Phi_B)\ E(\Psi_A) \subseteq E(\Phi_B)\}, \quad (4)$$
$$\overline{apr}(\Phi_B, \Psi_A) = \{o \mid o \in E(\Psi_A) \wedge \exists E(\Phi_B)\ E(\Psi_A) \cap E(\Phi_B) \neq \emptyset\}. \quad (5)$$

Example 2

We check equivalence classes comprising families $\Psi/IND(\Psi_{a_1})$ and $\Phi/IND(\Phi_{a_3})$ in Example 1. For inclusion and intersection between equivalence classes, $\{o_4\} \cap \{o_1, o_5\} = \emptyset$, $\{o_4\} \cap \{o_2, o_3, o_6\} = \emptyset$, $\{o_2, o_5\} \not\subseteq \{o_1, o_5\}$, $\{o_2, o_5\} \cap \{o_1, o_5\} \neq \emptyset$, $\{o_2, o_5\} \not\subseteq \{o_2, o_3, o_6\}$, and $\{o_3, o_6\} \subset \{o_2, o_3, o_6\}$. Thus, for lower and upper approximations in terms of equivalence classes,

$$\underline{Apr}(\Phi_{a_3}, \Psi_{a_1}) = \{\{o_3, o_6\}\},$$
$$\overline{Apr}(\Phi_{a_3}, \Psi_{a_1}) = \{\{o_2, o_5\}, \{o_3, o_6\}\}.$$

For the expressions in terms of objects,

$$\underline{apr}(\Phi_{a_3}, \Psi_{a_1}) = \{o_3, o_6\},$$
$$\overline{apr}(\Phi_{a_3}, \Psi_{a_1}) = \{o_2, o_3, o_5, o_6\}.$$

3 Methods of Possible Worlds

In methods of possible worlds, the traditional methods addressed in the previous section are applied to each possible table, and then the results from possible tables are aggregated. We suppose that every imprecise value is expressed in a normal possibility distribution where an element has the maximum possibilistic degree 1. When imprecise values are contained in information table T, we obtain the following extended set $rep(T)$ of possible tables:

$$rep(T) = \{(pt_1, \mu(pt_1)), \ldots, (pt_n, \mu(pt_n))\}, \tag{6}$$

where pt_i and $\mu(pt_i)$ denote a possible table and the possibilistic degree to which pt_i is the actual one, n is equal to $\Pi_{i=1,m} l_i$, m is the number of imprecise attribute values that are expressed in a normal possibility distribution having $l_i(i = 1, m))$ elements. A possible table is a table such that each imprecise value expressed in a normal possibility distribution is replaced by an element comprising the normal possibility distribution. When replaced values in pt_i are expressed in terms of elements v_{i1}, v_{i2}, ..., v_{im},

$$\mu(pt_i) = \min_{k=1,m} \pi(v_{ik}), \tag{7}$$

where $\pi(v_{ik})$ is the possibilistic degree of v_{ik} and comes from normal possibility distribution π expressing the imprecise attribute value in which v_{ik} is an element.

Example 3
Let the following information table T_2 be obtained:

T_2

O	a_1	a_2	a_3
1	x	1	a
2	$\{(x,1),(y,0.8)\}_p$	2	b
3	y	2	$\{(a,0.4),(b,1)\}_p$
4	$\{(x,1),(y,0.3)\}_p$	1	a

$U = \{o_1, o_2, o_3, o_4\}$ and subscript p of $\{(x,1),(y,0.8)\}_p$ denotes a normal possibility distribution. The following extended set $rep(T_2)$ of possible tables is obtained:

$$rep(T_2) = \{(pt_1, \mu(pt_1)), \cdots, (pt_8, \mu(pt_8))\},$$

where pt_i and $\mu(pt_i)$ are:

pt_1

O	a_1	a_2	a_3
1	x	1	a
2	x	2	b
3	y	2	a
4	x	1	a

pt_2

O	a_1	a_2	a_3
1	x	1	a
2	x	2	b
3	y	2	b
4	x	1	a

pt_3

O	a_1	a_2	a_3
1	x	1	a
2	x	2	b
3	y	2	a
4	y	1	a

pt_4

O	a_1	a_2	a_3
1	x	1	a
2	x	2	b
3	y	2	b
4	y	1	a

pt_5

O	a_1	a_2	a_3
1	x	1	a
2	y	2	b
3	y	2	a
4	x	1	a

pt_6

O	a_1	a_2	a_3
1	x	1	a
2	y	2	b
3	y	2	b
4	x	1	a

pt_7

O	a_1	a_2	a_3
1	x	1	a
2	y	2	b
3	y	2	a
4	y	1	a

pt_8

O	a_1	a_2	a_3
1	x	1	a
2	y	2	b
3	y	2	b
4	y	1	a

$$\begin{aligned}
\mu(pt_1) &= \min(1,1,0.4) = 0.4,\\
\mu(pt_2) &= \min(1,1,1) = 1,\\
\mu(pt_3) &= \min(1,0.3,0.4) = 0.3,\\
\mu(pt_4) &= \min(1,0.3,1) = 0.3,\\
\mu(pt_5) &= \min(0.8,1,0.4) = 0.4,\\
\mu(pt_6) &= \min(0.8,1,1) = 0.8,\\
\mu(pt_7) &= \min(0.8,0.3,0.4) = 0.3,\\
\mu(pt_8) &= \min(0.8,0.3,1) = 0.3.
\end{aligned}$$

Each possible table consists of precise values. Possible table pt_i is accompanied by possibilistic degree $\mu(pt_i)$ to which it is the actual information table. Thus, the family of equivalence classes accompanied by a possibilistic degree is obtained for each possible table, which is denoted by $(U/IND(U_A)_{pt_i}, \mu(pt_i))$.[2] $(U/IND(U_A)_{pt_i}, \mu(pt_i))$ is defined by,

$$(U/IND(U_A)_{pt_i}, \mu(pt_i)) = \{(E(U_A), \mu(pt_i)) \mid E(U_A) \in U/IND(U_A)_{pt_i}\}. \quad (8)$$

$U/IND(U_A)$ is the union of $(U/IND(U_A)_{pt_i}, \mu(pt_i))$,

$$U/IND(U_A) = \cup_i (U/IND(U_A)_{pt_i}, \mu(pt_i)). \quad (9)$$

In union $\cup_i$, the maximum possibilistic degree is taken if there are the same elements accompanied by a possibilistic degree. Thus,

$$\begin{aligned}
U/IND(U_A) = \{(E(U_A), \kappa(E(U_A) \in U/IND(U_A))) \mid \\
\kappa(E(U_A) \in U/IND(U_A)) = \max_{E(U_A) \in U/IND(U_A)_{pt_i}} \mu(pt_i)\}, \quad (10)
\end{aligned}$$

where $E(U_A)$ is an equivalence class on A and $\kappa(E(U_A) \in U/IND(U_A))$ is the possibilistic degree to which $E(U_A)$ is contained in $U/IND(U_A)$.

[2] U is used in place of a set when it is the universe.

Example 4
Binary relations $IND(U_{a_1})_{pt_1}$ and $IND(U_{a_3})_{pt_1}$ for indiscernibility on attributes a_1 and a_3 in possible table pt_1 of Example 3 are,

$$\begin{aligned} IND(U_{a_1})_{pt_1} = \{&(o_1,o_1),(o_1,o_2),(o_1,o_4),(o_2,o_1),(o_2,o_2),(o_2,o_4),(o_3,o_3),\\ &(o_4,o_1),(o_4,o_2),(o_4,o_4)\},\\ IND(U_{a_3})_{pt_1} = \{&(o_1,o_1),(o_1,o_3),(o_1,o_4),(o_2,o_2),(o_3,o_1),(o_3,o_3),\\ &(o_3,o_4),(o_4,o_1),(o_4,o_3),(o_4,o_4)\}. \end{aligned}$$

Families $U/IND(U_{a_1})_{pt_1}$ and $U/IND(U_{a_3})_{pt_1}$ of equivalence classes on attributes a_1 and a_3 are,

$$\begin{aligned} U/IND(U_{a_1})_{pt_1} &= \{\{o_3\},\{o_1,o_2,o_4\}\},\\ U/IND(U_{a_3})_{pt_1} &= \{\{o_2\},\{o_1,o_3,o_4\}\}. \end{aligned}$$

Possible table pt_1 is accompanied by possibilistic degree $\mu(pt_1)$ to which it is the actual one. Thus, families $(U/IND(U_{a_1})_{pt_1}, \mu(pt_1))$ and $(U/IND(U_{a_3})_{pt_1}, \mu(pt_1))$ of equivalence classes accompanied by possibilistic degree $\mu(pt_1)(= 0.4)$ are,

$$\begin{aligned} (U/IND(U_{a_1})_{pt_1}, \mu(pt_1)) &= \{(\{o_3\},0.4),(\{o_1,o_2,o_4\},0.4)\},\\ (U/IND(U_{a_3})_{pt_1}, \mu(pt_1)) &= \{(\{o_2\},0.4),(\{o_1,o_3,o_4\},0.4)\}. \end{aligned}$$

Similarly, for the other possible tables,

$$\begin{aligned} (U/IND(U_{a_1})_{pt_2}, \mu(pt_2)) &= \{(\{o_3\},1),(\{o_1,o_2,o_4\},1)\},\\ (U/IND(U_{a_3})_{pt_2}, \mu(pt_2)) &= \{(\{o_1,o_4\},1),(\{o_2,o_3\},1)\},\\ (U/IND(U_{a_1})_{pt_3}, \mu(pt_3)) &= \{(\{o_1,o_2\},0.3),(\{o_3,o_4\},0.3)\},\\ (U/IND(U_{a_3})_{pt_3}, \mu(pt_3)) &= \{(\{o_2\},0.3),(\{o_1,o_3,o_4\},0.3)\},\\ (U/IND(U_{a_1})_{pt_4}, \mu(pt_4)) &= \{(\{o_1,o_2\},0.3),(\{o_3,o_4\},0.3)\},\\ (U/IND(U_{a_3})_{pt_4}, \mu(pt_4)) &= \{(\{o_1,o_4\},0.3),(\{o_2,o_3\},0.3)\},\\ (U/IND(U_{a_1})_{pt_5}, \mu(pt_5)) &= \{(\{o_1,o_4\},0.4),(\{o_2,o_3\},0.4)\},\\ (U/IND(U_{a_3})_{pt_5}, \mu(pt_5)) &= \{(\{o_2\},0.4),(\{o_1,o_3,o_4\},0.4)\},\\ (U/IND(U_{a_1})_{pt_6}, \mu(pt_6)) &= \{(\{o_1,o_4\},0.8),(\{o_2,o_3\},0.8)\},\\ (U/IND(U_{a_3})_{pt_6}, \mu(pt_6)) &= \{(\{o_1,o_4\},0.8),(\{o_2,o_3\},0.8)\},\\ (U/IND(U_{a_1})_{pt_7}, \mu(pt_7)) &= \{(\{o_1\},0.3),(\{o_2,o_3,o_4\},0.3)\},\\ (U/IND(U_{a_3})_{pt_7}, \mu(pt_7)) &= \{(\{o_2\},0.3),(\{o_1,o_3,o_4\},0.3)\},\\ (U/IND(U_{a_1})_{pt_8}, \mu(pt_8)) &= \{(\{o_1\},0.3),(\{o_2,o_3,o_4\},0.3)\},\\ (U/IND(U_{a_3})_{pt_8}, \mu(pt_8)) &= \{(\{o_1,o_4\},0.3),(\{o_2,o_3\},0.3)\}. \end{aligned}$$

The possibilistic degree to which $\{o_1\}$ is an actual equivalence class on a_1 is,

$$\kappa(\{o_1\} \in U/IND(U_{a_1})) = \max(0.3, 0.3) = 0.3.$$

Similarly, for the other equivalence classes,

$$\begin{aligned}
\kappa(\{o_3\} \in U/IND(U_{a_1})) &= \max(0.4, 1) = 1,\\
\kappa(\{o_1, o_2\} \in U/IND(U_{a_1})) &= \max(0.3, 0.3) = 0.3,\\
\kappa(\{o_1, o_4\} \in U/IND(U_{a_1})) &= \max(0.4, 0.8) = 0.8,\\
\kappa(\{o_2, o_3\} \in U/IND(U_{a_1})) &= \max(0.4, 0.8) = 0.8,\\
\kappa(\{o_3, o_4\} \in U/IND(U_{a_1})) &= \max(0.3, 0.3) = 0.3,\\
\kappa(\{o_1, o_2, o_4\} \in U/IND(U_{a_1})) &= \max(0.4, 1) = 1,\\
\kappa(\{o_2, o_3, o_4\} \in U/IND(U_{a_1})) &= \max(0.3, 0.3) = 0.3.
\end{aligned}$$

Finally,

$$\begin{aligned}
U/IND(U_{a_1}) = \{&(\{o_1\}, 0.3), (\{o_3\}, 1), (\{o_1, o_2\}, 0.3), (\{o_1, o_4\}, 0.8),\\
&(\{o_2, o_3\}, 0.8), (\{o_3, o_4\}, 0.3), (\{o_1, o_2, o_4\}, 1), (\{o_2, o_3, o_4\}, 0.3)\}.
\end{aligned}$$

Similarly,

$$U/IND(U_{a_3}) = \{(\{o_2\}, 0.4), (\{o_1, o_4\}, 1), (\{o_2, o_3\}, 1), (\{o_1, o_3, o_4\}, 0.4)\}.$$

To obtain lower and upper approximations, the traditional methods addressed in the previous section are applied to possible tables. Let $\underline{Apr}(U_B, U_A)_{pt_i}$ and $\overline{Apr}(U_B, U_A)_{pt_i}$ denote the lower approximation and the upper approximation of $U/IND(U_B)_{pt_i}$ by $U/IND(U_A)_{pt_i}$ in possible table pt_i having possibilistic degree $\mu(pt_i)$. $\underline{Apr}(U_B, U_A)_{pt_i}$ and $\overline{Apr}(U_B, U_A)_{pt_i}$ are accompanied by possibilistic degree $\mu(pt_i)$, which are denoted by $(\underline{Apr}(U_B, U_A)_{pt_i}, \mu(pt_i))$ and $(\overline{Apr}(U_B, U_A)_{pt_i}, \mu(pt_i))$, respectively.

$$(\underline{Apr}(U_B, U_A)_{pt_i}, \mu(pt_i)) = \{(E(U_A), \mu(pt_i)) \mid E(U_A) \in \underline{Apr}(U_B, U_A)_{pt_i}\}, \quad (11)$$
$$(\overline{Apr}(U_B, U_A)_{pt_i}, \mu(pt_i)) = \{(E(U_A), \mu(pt_i)) \mid E(U_A) \in \overline{Apr}(U_B, U_A)_{pt_i}\}. \quad (12)$$

$(\underline{Apr}(U_B, U_A))$ and $(\overline{Apr}(U_B, U_A))$ are the union of $(\underline{Apr}(U_B, U_A)_{pt_i}, \mu(pt_i))$ and $(\overline{Apr}(U_B, U_A)_{pt_i}, \mu(pt_i))$, respectively,

$$\underline{Apr}(U_B, U_A) = \cup_i (\underline{Apr}(U_B, U_A)_{pt_i}, \mu(pt_i)), \quad (13)$$
$$\overline{Apr}(U_B, U_A) = \cup_i (\overline{Apr}(U_B, U_A)_{pt_i}, \mu(pt_i)). \quad (14)$$

Considering the same equivalence classes accompanied by a possibilistic degree,

$$\begin{aligned}
\underline{Apr}(U_B, U_A) = \{&(E(U_A), \kappa(E(U_A) \in \underline{Apr}(U_B, U_A))) \mid\\
&\kappa(E(U_A) \in \underline{Apr}(U_B, U_A)) = \max_{E(U_A) \in \underline{Apr}(U_B, U_A))_{pt_i}} \mu(pt_i)\}, \quad (15)
\end{aligned}$$

$$\begin{aligned}
\overline{Apr}(U_B, U_A) = \{&(E(U_A), \kappa(E(U_A) \in \overline{Apr}(U_B, U_A))) \mid\\
&\kappa(E(U_A) \in \overline{Apr}(U_B, U_A)) = \max_{E(U_A) \in \overline{Apr}(U_B, U_A))_{pt_i}} \mu(pt_i)\}, \quad (16)
\end{aligned}$$

where $\kappa(E(U_A) \in \underline{Apr}(U_B, U_A))$ and $\kappa(E(U_A) \in \overline{Apr}(U_B, U_A))$ are possibilistic degrees to which $E(U_A)$ is contained in $\underline{Apr}(U_B, U_A)$ and $\overline{Apr}(U_B, U_A)$, respectively. These formulas show that the maximum of the possibilistic degrees of possible tables where $E(U_A)$ is contained in the approximations is equal to the possibilistic degree for $E(U_A)$.

Proposition 1
When $(E(U_A), \kappa(E(U_A) \in \underline{Apr}(U_B, U_A)))$ is an element of $\underline{Apr}(U_B, U_A)$ in an information table, there exists possible table pt_i where $\underline{Apr}(U_B, U_A)_{pt_i}$ contains $E(U_A)$ and $\mu(pt_i)$ is equal to $\kappa(E(U_A) \in \underline{Apr}(U_B, U_A))$.

Proof
Let $(E(U_A), \kappa(E(U_A) \in \underline{Apr}(U_B, U_A)))$ be an element of $\underline{Apr}(U_B, U_A)$. From formula (15), there are possible tables where a lower approximation contains $E(U_A)$ and $\kappa(E(U_A) \in \underline{Apr}(U_B, U_A))$ is equal to the maximum of possibilistic degrees that the possible tables have. Thus, Proposition 1 holds.

Proposition 2
When $(E(U_A), \kappa(E(U_A) \in \overline{Apr}(U_B, U_A)))$ is an element of $\overline{Apr}(U_B, U_A)$ in an information table, there exists possible table pt_i where $\overline{Apr}(U_B, U_A)_{pt_i}$ contains $E(U_A)$ and $\mu(pt_i)$ is equal to $\kappa(E(U_A) \in \overline{Apr}(U_B, U_A))$.

Proof
The proof is similar to that of Proposition 1.

When lower and upper approximations are expressed in terms of objects,

$$\underline{apr}(U_B, U_A) = \{(o, \kappa(o \in \underline{apr}(U_B, U_A))) \mid \kappa(o \in \underline{apr}(U_B, U_A)) > 0\}, \quad (17)$$
$$\overline{apr}(U_B, U_A) = \{(o, \kappa(o \in \overline{apr}(U_B, U_A))) \mid \kappa(o \in \overline{apr}(U_B, U_A)) > 0\}, \quad (18)$$

where

$$\kappa(o \in \underline{apr}(U_B, U_A)) = \max_{E(U_A) \ni o} \kappa(E(U_A) \in \underline{Apr}(U_B, U_A)), \quad (19)$$
$$\kappa(o \in \overline{apr}(U_B, U_A)) = \max_{E(U_A) \ni o} \kappa(E(U_A) \in \overline{Apr}(U_B, U_A)). \quad (20)$$

Example 5
For inclusion and intersection of equivalence classes on attributes a_1 and a_3 of possible table pt_1 in Example 4, $\{o_1, o_2, o_4\} \not\subseteq \{o_1, o_3, o_4\}$, $\{o_1, o_2, o_4\} \not\subseteq \{o_2\}$, $\{o_3\} \subseteq \{o_1, o_3, o_4\}$, $\{o_1, o_2, o_4\} \cap \{o_1, o_3, o_4\} \neq \emptyset$, and $\{o_3\} \cap \{o_1, o_3, o_4\} \neq \emptyset$. Thus,

$$(\underline{Apr}(U_{a_3}, U_{a_1})_{pt_1}, \mu(pt_1)) = \{(\{o_3\}, 0.4)\},$$
$$(\overline{Apr}(U_{a_3}, U_{a_1})_{pt_1}, \mu(pt_1)) = \{(\{o_3\}, 0.4), (\{o_1, o_2, o_4\}, 0.4)\}.$$

Similarly for the other possible tables,

$$(\underline{Apr}(U_{a_3}, U_{a_1})_{pt_2}, \mu(pt_2)) = \{(\{o_3\}, 1)\},$$

$$
\begin{aligned}
(\overline{Apr}(U_{a_3}, U_{a_1})_{pt_2}, \mu(pt_2)) &= \{(\{o_3\}, 1), (\{o_1, o_2, o_4\}, 1)\},\\
(\underline{Apr}(U_{a_3}, U_{a_1})_{pt_3}, \mu(pt_3)) &= \{(\{o_3, o_4\}, 0.3)\},\\
(\overline{Apr}(U_{a_3}, U_{a_1})_{pt_3}, \mu(pt_3)) &= \{(\{o_1, o_2\}, 0.3), (\{o_3, o_4\}, 0.3)\},\\
(\underline{Apr}(U_{a_3}, U_{a_1})_{pt_4}, \mu(pt_4)) &= \{(\{\emptyset\}, 0.3)\},\\
(\overline{Apr}(U_{a_3}, U_{a_1})_{pt_4}, \mu(pt_4)) &= \{(\{o_1, o_2\}, 0.3), (\{o_3, o_4\}, 0.3)\},\\
(\underline{Apr}(U_{a_3}, U_{a_1})_{pt_5}, \mu(pt_5)) &= \{(\{o_1, o_4\}, 0.4)\},\\
(\overline{Apr}(U_{a_3}, U_{a_1})_{pt_5}, \mu(pt_5)) &= \{(\{o_1, o_4\}, 0.4), (\{o_2, o_3\}, 0.4)\},\\
(\underline{Apr}(U_{a_3}, U_{a_1})_{pt_6}, \mu(pt_6)) &= \{(\{o_1, o_4\}, 0.8), (\{o_2, o_3\}, 0.8)\},\\
(\overline{Apr}(U_{a_3}, U_{a_1})_{pt_6}, \mu(pt_6)) &= \{(\{o_1, o_4\}, 0.8), (\{o_2, o_3\}, 0.8)\},\\
(\underline{Apr}(U_{a_3}, U_{a_1})_{pt_7}, \mu(pt_7)) &= \{(\{o_1\}, 0.3)\},\\
(\overline{Apr}(U_{a_3}, U_{a_1})_{pt_7}, \mu(pt_7)) &= \{(\{o_1\}, 0.3), (\{o_2, o_3, o_4\}, 0.3)\},\\
(\underline{Apr}(U_{a_3}, U_{a_1})_{pt_8}, \mu(pt_8)) &= \{(\{o_1\}, 0.3)\},\\
(\overline{Apr}(U_{a_3}, U_{a_1})_{pt_8}, \mu(pt_8)) &= \{(\{o_1\}, 0.3), (\{o_2, o_3, o_4\}, 0.3)\}.
\end{aligned}
$$

We aggregate the results obtained from possible tables. The union of the results from possible tables is made. For lower approximation $\underline{Apr}(U_{a_3}, U_{a_1})$, the equivalence classes that satisfy $\kappa(E(U_{a_1}) \in \underline{Apr}(U_{a_3}, U_{a_1})) > 0$ are $\{o_1\}$, $\{o_3\}$, $\{o_1, o_4\}$, $\{o_2, o_3\}$, and $\{o_3, o_4\}$.

$$
\begin{aligned}
\kappa(\{o_1\} \in \underline{Apr}(U_{a_3}, U_{a_1})) &= \max(0.3, 0.3) = 0.3,\\
\kappa(\{o_3\} \in \underline{Apr}(U_{a_3}, U_{a_1})) &= \max(0.4, 1) = 1,\\
\kappa(\{o_1, o_4\} \in \underline{Apr}(U_{a_3}, U_{a_1})) &= \max(0.4, 0.8) = 0.8,\\
\kappa(\{o_2, o_3\} \in \underline{Apr}(U_{a_3}, U_{a_1})) &= 0.8,\\
\kappa(\{o_3, o_4\} \in \underline{Apr}(U_{a_3}, U_{a_1})) &= 0.3.
\end{aligned}
$$

Finally,

$$
\begin{aligned}
\underline{Apr}(U_{a_3}, U_{a_1}) = \{&(\{o_1\}, 0.3), (\{o_3\}, 1), (\{o_1, o_4\}, 0.8), (\{o_2, o_3\}, 0.8),\\
&(\{o_3, o_4\}, 0.3)\}.
\end{aligned}
$$

Similarly, for upper approximation $\overline{Apr}(U_{a_3}, U_{a_1})$, the equivalence classes that satisfy $\kappa(E(U_{a_1}) \in \overline{Apr}(U_{a_3}, U_{a_1})) > 0$ are $\{o_1\}$, $\{o_3\}$, $\{o_1, o_2\}$, $\{o_1, o_4\}$, $\{o_2, o_3\}$, $\{o_3, o_4\}$, $\{o_1, o_2, o_4\}$, $\{o_2, o_3, o_4\}$.

$$
\begin{aligned}
\overline{Apr}(U_{a_3}, U_{a_1}) = \{&(\{o_1\}, 0.3), (\{o_3\}, 1), (\{o_1, o_2\}, 0.3), (\{o_1, o_4\}, 0.8),\\
&(\{o_2, o_3\}, 0.8), (\{o_3, o_4\}, 0.3), (\{o_1, o_2, o_4\}, 1), (\{o_2, o_3, o_4\}, 0.3)\}.
\end{aligned}
$$

When the lower approximation is expressed in terms of objects,

$$
\begin{aligned}
\kappa(o_1 \in \underline{apr}(U_{a_3}, U_{a_1})) &= \max(0.3, 0.8) = 0.8,\\
\kappa(o_2 \in \underline{apr}(U_{a_3}, U_{a_1})) &= 0.8,\\
\kappa(o_3 \in \underline{apr}(U_{a_3}, U_{a_1})) &= \max(1, 0.8, 0.3) = 1,\\
\kappa(o_4 \in \underline{apr}(U_{a_3}, U_{a_1})) &= \max(0.8, 0.3) = 0.8.
\end{aligned}
$$

Similarly, for the upper approximation,

$$\kappa(o_1 \in \underline{apr}(U_{a_3}, U_{a_1})) = 1,$$
$$\kappa(o_2 \in \underline{apr}(U_{a_3}, U_{a_1})) = 1,$$
$$\kappa(o_3 \in \underline{apr}(U_{a_3}, U_{a_1})) = 1,$$
$$\kappa(o_4 \in \underline{apr}(U_{a_3}, U_{a_1})) = 1.$$

Thus,

$$\underline{apr}(U_{a_3}, U_{a_1}) = \{(o_1, 0.8), (o_2, 0.8), (o_3, 1), (o_4, 0.8)\},$$
$$\overline{apr}(U_{a_3}, U_{a_1}) = \{(o_1, 1), (o_2, 1), (o_3, 1), (o_4, 1)\}.$$

We adopt results from the method of possible worlds as a correctness criterion of extended methods of applying rough sets to information tables containing imprecise values. This is commonly used in the field of databases handling imprecise information [1, 2, 3, 4, 12, 13, 39].

Correctness criterion
Results obtained from applying an extended method to an information table containing imprecise information are the same as ones obtained from applying the corresponding traditional method to every possible table derived from that information table and aggregating the results created in the possible tables.

This is formulated as follows:
Suppose that operator rep creates extended set rep(T) of possible tables derived from information table T containing imprecise values. Let q be an extended method directly applied to T and the corresponding method q′ be applied to rep(T) in the method of possible worlds. The two results is the same; namely,

$$q(T) = q'(rep(T)).$$

This condition is schematized in Figure 1.

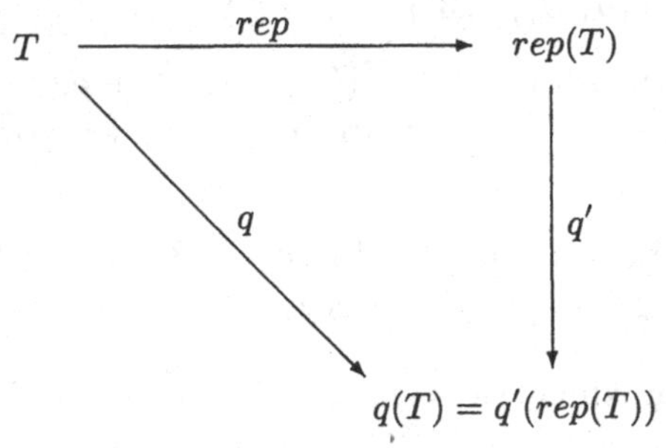

Fig. 1. Correctness criterion of extended method q

When this condition is valid, extended method q gives correct results at the level of possible values. This correctness criterion is checked as follows:

- Derive the extended set of possible tables from an information table containing imprecise values.
- Apply the traditional methods to each possible table.
- Aggregate the results obtained from possible tables.
- Apply the extended method to the original information table.
- Compare the aggregated results with ones obtained from the extended method.

4 Applying Rough Sets to Information Tables Containing Possibilistic Values

When object o takes imprecise values on attributes, we calculate the degree to which the attribute values are the same as those of another object o'. The degree is the indiscernibility degree of objects o and o' on the attributes. In this case, a binary relation for indiscernibility on set A of attributes is,

$$IND(U_A) = \{((o,o'), \kappa(A(o) = A(o'))) \mid (\kappa(A(o) = A(o')) \neq 0) \wedge (o \neq o')\} \cup \{((o,o),1)\}, \quad (21)$$

where $\kappa(A(o) = A(o'))$ denotes the indiscernibility degree of objects o and o' on set A of attributes and is equal to degree $\kappa((o,o') \in IND(U_A))$ to which (o,o') is included in $IND(U_A)$.

$$\kappa(A(o) = A(o')) = \bigotimes_{a \in A} \kappa(a(o) = a(o')), \quad (22)$$

where operator $\bigotimes$ depends on properties of imprecise attribute values. When the imprecise attribute values are expressed in a normal possibility distribution, the operator is min.

$$\kappa(a(o) = a(o')) = \max_{u,v \in V_a} \min(\mu_=(u,v), \pi_{a(o)}(u), \pi_{a(o')}(v)),$$

where $\pi_{a(o)}(u)$ and $\pi_{a(o')}(v)$ are possibilistic degrees to which attribute values $a(o)$ and $a(o')$ are equal to u and v, respectively, and,

$$\mu_=(u,v) = \begin{cases} 1 \text{ if } u = v, \\ 0 \text{ otherwise.} \end{cases} \quad (23)$$

From $IND(U_A)$, family $U/IND(U_A)$ of weighted equivalence classes is obtained via indiscernible classes. Among the elements of $IND(U_A)$, set $S(U_A)_o$ of objects that are paired with object o, called the indiscernible class on A for o, is,

$$S(U_A)_o = \{o' \mid \kappa((o,o') \in IND(U_A)) > 0\}. \quad (24)$$

$S(U_A)_o$ is the greatest one among equivalence classes containing objects o, when o has a precise value on every attribute in A. Let $PS(U_A)_o$ denote the power set of

$S(U_A)_o$. From $PS(U_A)_o$, family $Can(U/IND(U_A))_o$ of candidates of equivalence classes containing o is obtained:

$$Can(U/IND(U_{U_A}))_o = \{E(U_A) \mid E(U_A) \in PS(U_A)_o \wedge o \in E(U_A)\}. \quad (25)$$

Whole family $Can(U/IND(U_A))$ of candidates of equivalence classes is,

$$Can(U/IND(U_A)) = \cup_o Can(U/IND(U_A))_o. \quad (26)$$

Possibilistic degree $\kappa(E(U_A) \in U/IND(U_A))$ to which $E(U_A) \in Can(U/IND(U_A))$ is an actual equivalence class is,

$$\kappa(E(U_A) \in U/IND(U_A)) = \kappa(\wedge_{o \in E(U_A) \text{ and } o' \in E(U_A)}(A(o) = A(o')) \\ \wedge_{o'' \in E(U_A) \text{ and } o''' \notin E(U_A)}(A(o'') \neq A(o'''))), \quad (27)$$

where $o \neq o'$, $\kappa(f)$ is the possibilistic degree to which formula f is satisfied, and $\kappa(f) = 1$ when there exists no f. When an information table contains k objects and $E(U_A)$ consists of l objects,

$$\kappa(E(U_A) \in U/IND(U_A)) = \\ \max_{(u,v_1,\cdots,v_{k-l})} \min(\min_{o_i \in E(U_A)} (\pi_{A(o_1)}(u), \pi_{A(o_2)}(u), \ldots, \pi_{A(o_l)}(u)), \\ \min_{o'_i \notin E(U_A)} (\pi_{A(o'_1)}(v_1), \pi_{A(o'_2)}(v_2), \ldots, \pi_{A(o'_{k-l})}(v_{k-l}))), \quad (28)$$

where

$$\pi_{A(o_i)}(u) = \min_{j=1,m} \pi_{a_j(o_i)}(u_j), \quad (29)$$

$$\pi_{A(o'_i)}(v_i) = \min_{j=1,m} \pi_{a_j(o'_i)}(v_{ij}), \quad (30)$$

where different values u and v_i on set $A(= \{a_1, a_2, \ldots, a_m\})$ of attributes are expressed in $(u_1, \cdots, u_m)$ and $(v_{i1}, \cdots, v_{im})$, respectively. Finally, family $U/IND(U_A)$ of weighted equivalence classes is,

$$U/IND(U_A) = \\ \{(E(U_A), \kappa(E(U_A) \in U/IND(U_A))) \mid \kappa(E(U_A) \in U/IND(U_A)) > 0\}. \quad (31)$$

Example 6
From applying formula (21) to information table T_2 in Example 3,

$$U/IND(U_{a_1}) = \{((o_1, o_1), 1), ((o_1, o_2), 1), ((o_1, o_4), 1), ((o_2, o_1), 1), \\ ((o_2, o_2), 1), ((o_2, o_3), 0.8), ((o_2, o_4), 1), ((o_3, o_2), 0.8), \\ ((o_3, o_3), 1), ((o_3, o_4), 0.3), ((o_4, o_1), 1), ((o_4, o_2), 1), \\ ((o_4, o_3), 0.3), ((o_4, o_4), 1)\}.$$

For the binary relation for indiscernibility, each indiscernible class on attribute a_1 for object o_i is, respectively,

$$\begin{aligned} S(U_{a_1})_{o_1} &= \{o_1, o_2, o_4\}, \\ S(U_{a_1})_{o_2} &= \{o_1, o_2, o_3, o_4\}, \\ S(U_{a_1})_{o_3} &= \{o_2, o_3, o_4\}, \\ S(U_{a_1})_{o_4} &= \{o_1, o_3, o_3, o_4\}. \end{aligned}$$

Each power set of these sets is, respectively,

$$\begin{aligned} PS(U_{a_1})_{o_1} &= \{\emptyset, \{o_1\}, \{o_2\}, \{o_4\}, \{o_1, o_2\}, \{o_1, o_4\}, \{o_2, o_4\}, \{o_1, o_2, o_4\}\}, \\ PS(U_{a_1})_{o_2} &= PS(U_{a_1})_{o_4} \\ &= \{\emptyset, \{o_1\}, \{o_2\}, \{o_3\}, \{o_4\}, \{o_1, o_2\}, \{o_1, o_3\}, \{o_1, o_4\}, \{o_2, o_3\}, \\ &\quad \{o_2, o_4\}, \{o_3, o_4\}, \{o_1, o_2, o_3\}, \{o_1, o_2, o_4\}, \{o_1, o_3, o_4\}, \\ &\quad \{o_2, o_3, o_4\}, \{o_1, o_2, o_3, o_4\}\}, \\ PS(U_{a_1})_{o_3} &= \{\emptyset, \{o_2\}, \{o_3\}, \{o_4\}, \{o_2, o_3\}, \{o_2, o_4\}, \{o_3, o_4\}, \{o_2, o_3, o_4\}\}. \end{aligned}$$

Each family of candidates of equivalence classes containing o_i is, respectively,

$$\begin{aligned} Can(U/IND(U_{a_1}))_{o_1} &= \{\{o_1\}, \{o_1, o_2\}, \{o_1, o_4\}, \{o_1, o_2, o_4\}\}, \\ Can(U/IND(U_{a_1}))_{o_2} &= \{\{o_2\}, \{o_1, o_2\}, \{o_2, o_3\}, \{o_2, o_4\}, \{o_1, o_2, o_3\}, \\ &\quad \{o_1, o_2, o_4\}, \{o_2, o_3, o_4\}, \{o_1, o_2, o_3, o_4\}\}, \\ Can(U/IND(U_{a_1}))_{o_3} &= \{\{o_3\}, \{o_2, o_3\}, \{o_3, o_4\}, \{o_2, o_3, o_4\}\}, \\ Can(U/IND(U_{a_1}))_{o_4} &= \{\{o_4\}, \{o_1, o_4\}, \{o_2, o_4\}, \{o_3, o_4\}, \{o_1, o_2, o_4\}, \\ &\quad \{o_1, o_3, o_4\}, \{o_2, o_3, o_4\}, \{o_1, o_2, o_3, o_4\}\}. \end{aligned}$$

The whole family of candidates of equivalence classes is,

$$\begin{aligned} Can(U/IND(U_{a_1})) = \{&\{o_1\}, \{o_2\}, \{o_3\}, \{o_4\}, \{o_1, o_2\}, \{o_1, o_3\}, \{o_1, o_4\}, \{o_2, o_3\}, \\ &\{o_2, o_4\}\{o_3, o_4\}, \{o_1, o_2, o_3\}, \{o_1, o_2, o_4\}, \{o_1, o_3, o_4\}, \\ &\{o_2, o_3, o_4\}, \{o_1, o_2, o_3, o_4\}\}. \end{aligned}$$

Possibilistic degree $\kappa(\{o_1\} \in U/IND(U_{a_1}))$ to which $\{o_1\}$ is an actual equivalence class is,

$$\begin{aligned} \kappa(\{o_1\} \in U/IND(U_{a_1})) &= \kappa((a_1(o_1) \neq a_1(o_2)) \wedge (a_1(o_1) \neq a_1(o_3)) \wedge \\ &\quad (a_1(o_1) \neq a_1(o_4)) \\ &= \min(0.8, 1, 0.3) \\ &= 0.3. \end{aligned}$$

Similarly,

$$\kappa(\{o_2\} \in U/IND(U_{a_1})) = 0,$$

$$\kappa(\{o_3\} \in U/IND(U_{a_1})) = 1,$$
$$\kappa(\{o_4\} \in U/IND(U_{a_1})) = 0,$$
$$\kappa(\{o_1, o_2\} \in U/IND(U_{a_1})) = 0.3,$$
$$\kappa(\{o_1, o_3\} \in U/IND(U_{a_1})) = 0,$$
$$\kappa(\{o_1, o_4\} \in U/IND(U_{a_1})) = 0.8,$$
$$\kappa(\{o_2, o_3\} \in U/IND(U_{a_1})) = 0.8,$$
$$\kappa(\{o_2, o_4\} \in U/IND(U_{a_1})) = 0,$$
$$\kappa(\{o_3, o_4\} \in U/IND(U_{a_1})) = 0.3,$$
$$\kappa(\{o_1, o_2, o_3\} \in U/IND(U_{a_1})) = 0,$$
$$\kappa(\{o_1, o_2, o_4\} \in U/IND(U_{a_1})) = 1,$$
$$\kappa(\{o_1, o_3, o_4\} \in U/IND(U_{a_1})) = 0,$$
$$\kappa(\{o_2, o_3, o_4\} \in U/IND(U_{a_1})) = 0.3,$$
$$\kappa(\{o_1, o_2, o_3, o_4\} \in U/IND(U_{a_1})) = 0.$$

Thus, the family of weighted equivalence classes on attribute a_1 is,

$$U/IND(U_{a_1}) = \{(\{o_1\}, 0.3), (\{o_3\}, 1), (\{o_1, o_2\}, 0.3), (\{o_1, o_4\}, 0.8), (\{o_2, o_3\}, 0.8), (\{o_3, o_4\}, 0.3), (\{o_1, o_2, o_4\}, 1), (\{o_2, o_3, o_4\}, 0.3)\}.$$

Similarly, the family of weighted equivalence classes on attribute a_3 is,

$$U/IND(U_{a_3}) = \{(\{o_2\}, 0.4), (\{o_1, o_4\}, 1), (\{o_2, o_3\}, 1), (\{o_1, o_3, o_4\}, 0.4)\}.$$

Proposition 3
When $(E(U_A), \kappa(E(U_A) \in U/IND(U_A)))$ is an element of $U/IND(U_A)$ in an information table, there exists possible table pt_i where $U/IND(U_A)_{pt_i}$ contains $E(U_A)$ and $\mu(pt_i)$ is equal to $\kappa(E(U_A) \in U/IND(U_A))$.

Proof
A calculated possibilistic degree $\kappa(E(U_A) \in U/IND(U_A)))$ by using formulas (27) – (30) is equal to the maximum possibilistic degree to which each object $o \in E(U_A)$ takes an equal possible value, denoted by u, on A and each object $o'_i \notin E(U_A)$ takes a possible value, denoted by v_i, different from u on A. Let u and v_i be $(u_1, \cdots, u_m)$ and $(v_{i1}, \cdots, v_{im})$ on $A (= \{a_1, a_2, \ldots, a_m\})$, respectively. The possibilistic degree is equal to that of the possible table where each object $o \in E(U_A)$ takes an equal possible value u_j and each object $o'_i \notin E(U_A)$ takes possible value v_{ij} as the value of attribute a_j for $j = 1, m$ and all objects take a possible value with the maximum degree 1 on the attributes not included in A. Thus, Proposition 3 holds.

Example 7
We check whether or not each element of $U/IND(U_{a_1})$ in Example 6 exists in families of equivalence classes obtained in Example 4. $(\{o_1\}, 0.3)$ is an element of $(U/IND(U_{a_1})_{pt_7}, \mu(pt_7))$ and $(U/IND(U_{a_1})_{pt_8}, \mu(pt_8))$. Indeed, there exist possible tables pt_7 and pt_8 where $U/IND(U_A)_{pt_7}$ and $U/IND(U_A)_{pt_8}$ contain $\{o_1\}$ and $\mu(pt_7)$ and $\mu(pt_8)$ is equal to 0.3. Similarly, for the other elements $(\{o_3\}, 1)$, $(\{o_1, o_2\}, 0.3)$, $(\{o_1, o_4\}, 0.8)$, $(\{o_2, o_3\}, 0.8)$, $(\{o_3, o_4\}, 0.3)$, $(\{o_1, o_2, o_4\}, 1)$, and $(\{o_2, o_3, o_4\}, 0.3)$, there exist the corresponding possible tables pt_2, pt_3 and pt_4, pt_6, pt_6, pt_3 and pt_4, pt_2, and pt_7 and pt_8.

Proposition 4
$U/IND(U_A)$ in an information table is equal to one obtained from the union of the families of equivalence classes accompanied by a possibilistic degree, where each family of equivalence classes is obtained from a possible table created from the information table.

Proof
From Proposition 3 and the proof, if $(E(U_A), \kappa(E(U_A) \in U/IND(U_A)))$ is an element of $U/IND(U_A)$, there exist possible tables having the family of equivalence classes containing $E(U_A)$. $\kappa(E(U_A) \in U/IND(U_A)))$ is equal to the maximum of possibilistic degrees that the possible tables have. The maximum degree is taken as the possibilistic degree, as is shown in formula (10), when more than one equivalence class accompanied by a possibilistic degree is obtained in the union operation. The possibilistic degree by which an equivalence class is accompanied in a possible table is equal to one that the possible table has. So, $(E(U_A), \kappa(E(U_A) \in U/IND(U_A)))$ is equal to one obtained from the union of the families of equivalence classes.

Proposition 5
For any object o,

$$\max_{E(U_A) \ni o} \kappa(E(U_A) \in U/IND(U_A)) = 1. \tag{32}$$

Proof
From Proposition 4, $U/IND(U_A)$ in an information table is equal to the union of the families of equivalence classes accompanied by a possibilistic degree, where each family of equivalence classes is obtained from a possible table created from the information table. Every imprecise value is expressed in a normal possibility distribution where an element has the maximum possibilistic degree 1. So, there is a possible table where all imprecise attribute values are replaced by an element having the maximum possibilistic degree 1 for any information table. The possibilistic degree to which the possible table is the actual one is equal to 1. Each object belongs to either of the equivalence classes obtained in the possible table. Thus, the above formula holds.

Example 8
From Example 6, $\kappa(\{o_3\} \in U/IND(U_{a_1})) = 1$ and $\kappa(\{o_1, o_2, o_4\} \in U/IND(U_{a_1})) = 1$. Indeed, for any object o, $\max_{E(U_{a_1}) \ni o} \kappa(E(U_{a_1}) \in U/IND(U_{a_1})) = 1$.

Using families of weighted equivalence classes, we can obtain lower approximation $\underline{Apr}(U_B, U_A)$ and upper approximation $\overline{Apr}(U_B, U_A)$ of $U/IND(U_B)$ by $U/IND(U_A)$. For the lower approximation,

$$\underline{Apr}(U_B, U_A) = \\ \{(E(U_A), \kappa(E(U_A) \in \underline{Apr}(U_B, U_A))) \mid \kappa(E(U_A) \in \underline{Apr}(U_B, U_A)) > 0\}, \quad (33)$$

$$\kappa(E(U_A) \in \underline{Apr}(U_B, U_A)) = \max_{E(U_B)} \min(\kappa(E(U_A) \subseteq E(U_B)), \\ \kappa(E(U_A) \in U/IND(U_A)), \kappa(E(U_B) \in U/IND(U_B))), \quad (34)$$

where

$$\kappa(E(U_A) \subseteq E(U_B)) = \begin{cases} 1 \text{ if } E(U_A) \subseteq E(U_B), \\ 0 \text{ otherwise.} \end{cases} \quad (35)$$

Proposition 6
If $(E(U_A), \kappa(E(U_A) \in \underline{Apr}(U_B, U_A)))$ is an element of $\underline{Apr}(U_B, U_A)$ in an information table, there exists possible table pt_i where $\underline{Apr}(U_B, U_A)_{pt_i}$ contains $E(U_A)$ and $\mu(pt_i)$ is equal to $\kappa(E(U_A) \in \underline{Apr}(U_B, U_A))$.

Proof
From formulas (33) and (34), we suppose to obtain $E(U_A) \subseteq E(U_B)$, $\kappa(E(U_A) \in U/IND(U_A)) > 0$, and $\kappa(E(U_B) \in U/IND(U_B)) > 0$ that gives $\kappa(E(U_A) \in \underline{Apr}(U_B, U_A)) > 0$. From Proposition 3, there is possible tables pt_j and pt_k accompanied by $\mu(pt_j)$ and $\mu(pt_k)$ equal to $\kappa(E(U_A) \in U/IND(U_A))$ and $\kappa(E(U_B) \in U/IND(U_B))$, respectively. In possible table pt_j we suppose that every $o \in E(U_A)$ takes the same possible value x_j on A and every $o' \notin E(U_A)$ takes x'_j different from x_j. All objects take a possible value with the maximum degree 1 on the attributes not included in A. And similarly in possible table pt_k we suppose that every $o \in E(U_B)$ takes the same possible value y_k on B and every $o' \notin E(U_B)$ takes y'_k different from y_k. All objects take a possible value with the maximum degree 1 on the attributes not included in B. Clearly, there is possible table pt_i where every $o \in E(U_A)$ takes the same possible value x_j on A and every $o' \notin E(U_A)$ takes x'_j different from x_j and every $o \in E(U_B)$ takes the same possible value y_k on B and every $o' \notin E(U_B)$ takes y'_k different from y_k and all objects take a possible value with the maximum degree 1 on the attributes not included in $A \cup B$. And in pt_i $\underline{Apr}(U_B, U_A)_{pt_i}$ contains $E(U_A)$ and pt_i is accompanied by possibilistic degree $\mu(pt_i)$ equal to $\min(\mu(pt_j), \mu(pt_k))(= \kappa(E(U_A) \in \underline{Apr}(U_B, U_A)))$. Thus, Proposition 6 holds.

For the upper approximation,

$$\overline{Apr}(U_B, U_A) = \{(E(U_A), \kappa(o \in \overline{Apr}(U_B, U_A))) \mid \kappa(E(U_A) \in \overline{Apr}(U_B, U_A)) > 0\}, \quad (36)$$

$$\kappa(E(U_A) \in \overline{Apr}(U_B, U_A)) = \max_{E(U_B)} \min(\kappa(E(U_A) \cap E(U_B) \neq \emptyset), \kappa(E(U_A) \in U/IND(U_A)), \kappa(E(U_B) \in U/IND(U_B))), \quad (37)$$

where

$$\kappa(E(U_A) \cap E(U_B) \neq \emptyset) = \begin{cases} 1 \text{ if } E(U_A) \cap E(U_B) \neq \emptyset, \\ 0 \text{ otherwise.} \end{cases} \quad (38)$$

Proposition 7
If $(E(U_A), \kappa(E(U_A) \in \overline{Apr}(U_B, U_A)))$ is an element of $\overline{Apr}(U_B, U_A)$ in an information table, there exists possible table pt_i where $\overline{Apr}(U_B, U_A)_{pt_i}$ contains $E(U_A)$ and $\mu(pt_i)$ is equal to $\kappa(E(U_A) \in \overline{Apr}(U_B, U_A))$.

Proof
The proof is similar to that of Proposition 6.

For expressions in terms of a set of objects, the same expressions as in Section 3 are used.

Proposition 8
The lower and upper approximations that are obtained by the method of weighted equivalence classes coincide with ones obtained by the method of possible worlds.

Proof
This proposition is proved by showing that the lower and upper approximations that are obtained by the method of weighted equivalence classes are equal to the union of ones obtained from possible tables. For the lower approximation, from Proposition 6, if $(E(U_A), \kappa(E(U_A) \in \underline{Apr}(U_B, U_A))$ is an element of $\underline{Apr}(U_B, U_A)$, there exist possible tables having the lower approximation containing $E(U_A)$. Each possible table is accompanied by a possibilistic degree. The possibilistic degree is also one by which $E(U_A)$ is accompanied for the lower approximation in the possible table. $\kappa(E(U_A) \in \underline{Apr}(U_B, U_A))$ is equal to the maximum degree among the possibilistic degrees. The maximum degree is taken as the possibilistic degree, as is shown in formula (15), when more than one equivalence class accompanied by a possibilistic degree is obtained in the union operation. So, $(E(U_A), \kappa(E(U_A) \in \underline{Apr}(U_B, U_A))$ is equal to one obtained from the union of the lower approximations obtained from possible tables. For the upper approximation, the proof is similar to that of the lower approximation.

Example 9

Using the families of weighted equivalence classes in Example 6, we derive the lower and upper approximations of $U/IND(U_{a_3})$ by $U/IND(U_{a_1})$. For the lower approximation, the possibilistic degree to which equivalence class $\{o_1\}$ in $U/IND(U_{a_1})$ is contained in $\underline{Apr}(U_{a_3}, U_{a_1})$ is,

$$\kappa(\{o_1\} \in \underline{Apr}(U_{a_3}, U_{a_1})) = \max(\min(1, 0.3, 1), \min(1, 0.3, 0.4)) = 0.3.$$

Similarly, for the other equivalence classes,

$$\begin{aligned}
\kappa(\{o_3\} \in \underline{Apr}(U_{a_3}, U_{a_1})) &= \max(\min(1, 1, 1), \min(1, 1, 0.4)) = 1,\\
\kappa(\{o_1, o_2\} \in \underline{Apr}(U_{a_3}, U_{a_1})) &= 0,\\
\kappa(\{o_1, o_4\} \in \underline{Apr}(U_{a_3}, U_{a_1})) &= \max(\min(1, 0.8, 1), \min(1, 0.8, 0.4)) = 0.8,\\
\kappa(\{o_2, o_3\} \in \underline{Apr}(U_{a_3}, U_{a_1})) &= \min(1, 0.8, 1) = 0.8,\\
\kappa(\{o_3, o_4\} \in \underline{Apr}(U_{a_3}, U_{a_1})) &= \min(1, 0.3, 0.4) = 0.3,\\
\kappa(\{o_1, o_2, o_4\} \in \underline{Apr}(U_{a_3}, U_{a_1})) &= 0,\\
\kappa(\{o_2, o_3, o_4\} \in \underline{Apr}(U_{a_3}, U_{a_1})) &= 0.
\end{aligned}$$

Thus,

$$\begin{aligned}
\underline{Apr}(U_{a_3}, U_{a_1}) = \{&(\{o_1\}, 0.3), (\{o_3\}, 1), (\{o_1, o_4\}, 0.8), (\{o_2, o_3\}, 0.8),\\
&(\{o_3, o_4\}, 0.3)\}.
\end{aligned}$$

For all elements of $\underline{Apr}(U_{a_3}, U_{a_1})$, there exist corresponding possible tables in Example 5. For element $(\{o_1\}, 0.3)$, the element exists in $(\underline{Apr}(U_{a_3}, U_{a_1})_{pt_7}, \mu(pt_7))$ and $\underline{Apr}(U_{a_3}, U_{a_1})_{pt_8}, \mu(pt_8))$. Thus, there exists corresponding possible tables pt_7 and pt_8 to this element. Similarly, for the other elements $(\{o_3\}, 1)$, $(\{o_1, o_4\}, 0.8)$, $(\{o_2, o_3\}, 0.8)$, $(\{o_3, o_4\}, 0.3)$, there exists corresponding possible tables pt_2, pt_6, pt_6, and pt_3, respectively. Thus, Proposition 6 holds.

From $\underline{Apr}(U_{a_3}, U_{a_1})$, the possibilistic degree to which each object is contained in the lower approximation is, respectively,

$$\begin{aligned}
\kappa(o_1 \in \underline{apr}(U_{a_3}, U_{a_1})) &= \max(0.3, 0.8) = 0.8,\\
\kappa(o_2 \in \underline{apr}(U_{a_3}, U_{a_1})) &= 0.8,\\
\kappa(o_3 \in \underline{apr}(U_{a_3}, U_{a_1})) &= \max(1, 0.8, 0.3) = 1,\\
\kappa(o_4 \in \underline{apr}(U_{a_3}, U_{a_1})) &= \max(0.8, 0.3) = 0.8.
\end{aligned}$$

Thus,

$$\underline{apr}(U_{a_3}, U_{a_1}) = \{(o_1, 0.8), (o_2, 0.8), (o_3, 1), (o_4, 0.8)\}.$$

Similarly, for the upper approximation,

$$\begin{aligned}
\overline{Apr}(U_{a_3}, U_{a_1}) = \{&(\{o_1\}, 0.3), (\{o_3\}, 1), (\{o_1, o_2\}, 0.3), (\{o_1, o_4\}, 0.8),\\
&(\{o_2, o_3\}, 0.8), (\{o_3, o_4\}, 0.3), (\{o_1, o_2, o_4\}, 1), (\{o_2, o_3, o_4\}, 0.3)\},\\
\overline{apr}(U_{a_3}, U_{a_1}) = \{&(o_1, 1), (o_2, 1), (o_3, 1), (o_4, 1)\}.
\end{aligned}$$

For all elements of $\overline{Apr}(U_{a_3}, U_{a_1})$, there exist corresponding possible tables in Example 5. For element $(\{o_1\}, 0.3)$, the element exists in $(\overline{Apr}(U_{a_3}, U_{a_1})_{pt_7}, \mu(pt_7))$ and $\overline{Apr}(U_{a_3}, U_{a_1})_{pt_8}, \mu(pt_8))$. Thus, there exists corresponding possible tables pt_7 and pt_8 to this element. Similarly, for the other elements $(\{o_3\}, 1)$, $(\{o_1, o_2\}, 0.3)$, $(\{o_1, o_4\}, 0.8)$, $(\{o_2, o_3\}, 0.8)$, $(\{o_3, o_4\}, 0.3)$, $(\{o_1, o_2, o_4\}, 1)$, $(\{o_2, o_3, o_4\}, 0.3)$, there exist corresponding possible tables pt_2, pt_3 and pt_4, pt_6, pt_6, pt_3 and pt_4, pt_2, and pt_7 and pt_8, respectively. Thus, Proposition 7 holds.

Indeed, the lower and upper approximations coincide with ones obtained from the method of possible worlds in Example 5.

5 Information Tables Containing Missing Values

We apply the method of weighted equivalence classes to information tables containing missing values. We briefly compare the method that Kryszkiewicz used with the method of weighted equivalence classes.

When missing values are contained in an information table, Kryszkiewicz defined binary relation $TOR(U_A)$ for indiscernibility between objects on set A of attributes as follows [14, 16]:

$$TOR(U_A) = \{(o, o') \in U \times U \mid \forall a \in A, a(o) = a(o') \vee a(o) = * \vee a(o') = *\}, \quad (39)$$

where $*$ denotes a missing value. This relation for indiscernibility is a tolerance relation. When an object has a missing value as an attribute value, the object may have the same properties as another object on the attribute. Then, the tolerance relation treats two objects as indiscernible. This corresponds to "do not care" semantics of missing values addressed by Grzymala-Busse [9, 10], where missing values are replaced by all domain elements of the attribute [8]. Indeed, the above definition means that an object having a missing value on every attribute in A is indiscernible with any object.

By using tolerance classes obtained from $TOR(U_A)$, Kryszkiewicz expressed lower and upper approximations of set Φ of objects as follows:

$$\underline{apr}(\Phi, U_A) = \{o \in U \mid T(U_A)_o \subseteq \Phi\}, \quad (40)$$

$$\overline{apr}(\Phi, U_A) = \{o \in U \mid T(U_A)_o \cap \Phi \neq \emptyset\}, \quad (41)$$

where $T(U_A)_o (= \{o' \mid (o, o') \in TOR(U_A)\})$ denotes the tolerance class for object o.

When a missing value in an attribute is possibilistically interpreted, every element in the domain of the attribute has the same possibilistic degree 1 to which the element is the actual value. In other words, the missing value is equal to the possibilistic value expressed in the uniform possibility distribution where every element over the domain has the maximum possibilistic degree 1. When attribute value $a(o)$ of o is a missing value,

$$\kappa(a(o) = a(o')) = \max_{u, v \in V_a} \min(\mu_{=}(u, v), \pi_{a(o)}(u), \pi_{a(o')}(v)) = 1,$$

where $o \neq o'$ and $\pi_{a(o)}(u)$ and $\pi_{a(o')}(u)$ denote normal possibility distributions expressing attribute values $a(o)$ and $a(o')$,[3] respectively. This shows that the indiscernibility degree of an object taking a missing value with the other objects is equal to 1; namely, the object is indiscernible with any object. This is equivalent to adopting the assumption that Kryszkiewicz used for indiscernibility of missing values [14, 16] and is equivalent to "do not care" semantics of missing values addressed by Grzymala-Busse [9, 10]. We express lower and upper approximations by using weighted equivalence classes, as is shown in the previous section, although Kryszkiewicz used tolerance classes that are not an equivalence class. This difference is clarified in the following example.

Example 10
Let the following information table T_3 containing missing values be given:

T_3

O	a_1	a_2	a_3
1	x	1	a
2	y	2	b
3	$*$	2	b
4	$*$	3	c

Let domains V_{a_1}, V_{a_2}, and V_{a_3} of attributes a_1, a_2, and a_3 be $\{x, y\}$, $\{1, 2, 3\}$ and $\{a, b\}$, respectively. Tolerance classes on a_1 for each object are,

$$T(U_{a_1})_{o_1} = \{o_1, o_3, o_4\},$$
$$T(U_{a_1})_{o_2} = \{o_2, o_3, o_4\},$$
$$T(U_{a_1})_{o_3} = \{o_1, o_2, o_3, o_4\},$$
$$T(U_{a_1})_{o_4} = \{o_1, o_3, o_3, o_4\}.$$

We suppose that $\Phi = \{o_2, o_3\}$. Because of $\{o_1, o_3, o_4\} \not\subseteq \{o_2, o_3\}$, $\{o_2, o_3, o_4\} \not\subseteq \{o_2, o_3\}$, $\{o_1, o_2, o_3, o_4\} \not\subseteq \{o_2, o_3\}$, $\{o_1, o_3, o_4\} \cap \{o_2, o_3\} \neq \emptyset$, $\{o_2, o_3, o_4\} \cap \{o_2, o_3\} \neq \emptyset$, and $\{o_1, o_2, o_3, o_4\} \cap \{o_2, o_3\} \neq \emptyset$,

$$\underline{apr}(\Phi, U_{a_1}) = \emptyset,$$
$$\overline{apr}(\Phi, U_{a_1}) = \{o_1, o_2, o_3, o_4, o_5\}.$$

Note that we do not obtain any information for the lower approximation. This is true for different expressions proposed by several authors, where equivalence classes are not used [10, 11, 20]. On the other hand, when we use the method of weighted equivalence classes, applying formulas (24) – (31) to information table T_3, family $U/IND(U_{a_1})$ of weighted equivalence classes is,

$$U/IND(U_{a_1}) = \{(\{o_1\}, 1), (\{o_2\}, 1), (\{o_1, o_3\}, 1), (\{o_1, o_4\}, 1), (\{o_2, o_3\}, 1), (\{o_2, o_4\}, 1), (\{o_1, o_3, o_4\}, 1), (\{o_2, o_3, o_4\}, 1)\}.$$

[3] When $a(o')$ is a precise value; for example, $a(o') = x$, normal possibility distribution $\pi_{a(o')}$ is expressed in $\{(x, 1)\}_p$.

Applying formulas (33) – (38),

$$\underline{Apr}(\Phi, U_{a_1}) = \{(\{o_2\}, 1), (\{o_2, o_3\}, 1)\}.$$
$$\overline{Apr}(\Phi, U_{a_1}) = \{(\{o_2\}, 1), (\{o_1, o_3\}, 1), (\{o_2, o_3\}, 1), (\{o_2, o_4\}, 1), (\{o_1, o_3, o_4\}, 1), (\{o_2, o_3, o_4\}, 1)\}.$$

Using formulas (17) – (20),

$$\underline{apr}(\Phi, U_{a_1}) = \{(o_2, 1), (o_3, 1)\}.$$
$$\overline{apr}(\Phi, U_{a_1}) = \{(o_1, 1), (o_2, 1), (o_3, 1), (o_4, 1)\}.$$

As is shown in this example, the method of weighted equivalence classes gives some information for the lower approximation, but the method of Kryszkiewicz using tolerance classes does not any information.

To examine the reason why such a difference is created, we show the following simple example.

Example 11

Let the following information table T_4 be obtained:

T_4

O	a_1	a_2	a_3
1	x	1	a
2	x	1	a
3	x	1	a
4	x	1	a
5	$*$	2	b

pt_1

O	a_1	a_2	a_3
1	x	1	a
2	x	1	a
3	x	1	a
4	x	1	a
5	x	2	b

pt_2

O	a_1	a_2	a_3
1	x	1	a
2	x	1	a
3	x	1	a
4	x	1	a
5	y	2	b

Let domains V_{a_1}, V_{a_2}, and V_{a_3} of attributes a_1, a_2, and a_3 be $\{x, y\}$, $\{1, 2\}$, and $\{a, b\}$, respectively. For tolerance classes on a_1 for each objects, which are derived from the binary relation for indiscernibility obtained from applying formula (39) to T_4,

$$T(U_{a_1})_{o_1} = T(U_{a_1})_{o_2} = T(U_{a_1})_{o_3} = T(U_{a_1})_{o_4} = T(U_{a_1})_{o_5} = \{o_1, o_2, o_3, o_4, o_5\}.$$

We suppose that $\Phi = \{o_1, o_2, o_3, o_4\}$ for simplicity. We focus on lower approximation $\underline{apr}(\Phi, U_{a_1})$, because the upper approximation is trivial in this case. For the method of Kryszkiewicz using formula (40), because of $\{o_1, o_2, o_3, o_4, o_5\} \not\subseteq \{o_1, o_2, o_3, o_4\}$,

$$\underline{apr}(\Phi, U_{a_1}) = \emptyset$$

This shows that we do not obtain any information for the lower approximation. On the other hand, the method of possible worlds creates different results. We obtain two possible tables pt_1 and pt_2 from T_4, because missing value $*$ of object o_5 is replaced by x or y, which comprise the domain of a_1. The family of equivalence classes on a_1 in pt_1 is,

$$U/IND(U_{a_1}) = \{o_1, o_2, o_3, o_4, o_5\}.$$

The family of equivalence classes on a_1 in pt_2 is,

$$U/IND(U_{a_1}) = \{\{o_1, o_2, o_3, o_4\}, \{o_5\}\}.$$

From $\{o_1, o_2, o_3, o_4, o_5\} \not\subseteq \Phi$, pt_1 has $\underline{apr}(\Phi, U_{a_1}) = \emptyset$. On the other hand, pt_2 has $\underline{apr}(\Phi, U_{a_1}) = \{o_1, o_2, o_3, o_4\}$ from $\{o_1, o_2, o_3, o_4\} \subseteq \Phi$.

In the above example, possible table pt_1 corresponds to the case that object o_5 is indiscernible with the other objects whereas pt_2 does to the case that o_5 is discernible with the other objects. The method of Kryszkiewicz deals with only the situation that corresponds to pt_1. The reason why the method of Kryszkiewicz creates the empty set to the lower approximation is due to that discernibility of missing values is not considered, although indiscernibility of missing values is considered [25]. On the other hand, the method of weighted equivalence classes deals with not only indiscernibility but also discernibility of missing values.

6 Conclusions

We have proposed a method, where weighted equivalence classes are used, to deal with imprecise information expressed in a normal possibility distribution. The lower and upper approximations by the method of weighted equivalence classes coincide with ones by the method of possible worlds. In other words, this method satisfies the correctness criterion that is used in the field of incomplete databases. This is justification of the method of weighted equivalence classes.

We have applied the method of weighted equivalence classes to information tables containing missing values under possibilistic interpretation. A binary relation for indiscernibility is the same as one obtained from the assumption that Kryszkiewicz used. We obtain correct results for rough approximations when weighted equivalence classes are used, although we do not obtain any results for the lower approximation when tolerance classes are used.

Acknowledgment. This research has been partially supported by the Grant-in-Aid for Scientific Research (C), Japan Society for the Promotion of Science, No. 18500214.

References

1. Abiteboul, S., Hull, R., Vianu, V.: Foundations of Databases. Addison-Wesley Publishing Company, Reading (1995)
2. Bosc, P., Duval, L., Pivert, O.: An Initial Approach to the Evaluation of Possibilistic Queries Addressed to Possibilistic Databases. Fuzzy Sets and systems 140, 151–166 (2003)
3. Bosc, P., Liétard, N., Pivert, O.: About the Processing of Possibilistic Queries Involving a Difference Operation. Fuzzy Sets and systems 157, 1622–1640 (2006)
4. Grahne, G.: The Problem of Incomplete Information in Relational Databases. LNCS, vol. 554. Springer, Heidelberg (1991)

5. Greco, S., Matarazzo, B., Slowinski, R.: Handling Missing Values in Rough Set Analysis of Multi-attribute and Multi-criteria Decision Problem. In: Zhong, N., Skowron, A., Ohsuga, S. (eds.) RSFDGrC 1999. LNCS (LNAI), vol. 1711, pp. 146–157. Springer, Heidelberg (1999)
6. Greco, S., Matarazzo, B., Slowinski, R.: Rough Sets Theory for Multicriteria Decision Analysis. European Journal of Operational Research 129, 1–47 (2001)
7. Grzymala-Busse, J.W.: On the Unknown Attribute Values in Learning from Examples. LNCS (LNAI), vol. 542, pp. 368–377. Springer, Heidelberg (1991)
8. Grzymala-Busse, J.W.: MLEM2: A New Algorithm for Rule Induction from Imperfect Data. In: Proceedings of the IPMU 2002, 9th International Conference on Information Processing and Management of Uncertainty in Knowledge-Based Systems, Annecy, France, pp. 243–250 (2002)
9. Grzymala-Busse, J.W.: Characteristic Relations for Incomplete Data: A Generalization of the Indiscernibility Relation. In: Tsumoto, S., Słowiński, R., Komorowski, J., Grzymała-Busse, J.W. (eds.) RSCTC 2004. LNCS (LNAI), vol. 3066, pp. 244–253. Springer, Heidelberg (2004)
10. Grzymala-Busse, J.W.: Incomplete Data and Generalization of Indiscernibility Relation, Definability, and Approximation. In: Ślęzak, D., Wang, G., Szczuka, M.S., Düntsch, I., Yao, Y. (eds.) RSFDGrC 2005. LNCS (LNAI), vol. 3641, pp. 244–253. Springer, Heidelberg (2005)
11. Guan, Y.-Y., Wang, H.-K.: Set-valued Information Systems. Information Sciences 176, 2507–2525 (2006)
12. Imielinski, T.: Incomplete Information in Logical Databases. Data Engineering 12, 93–104 (1989)
13. Imielinski, T., Lipski, W.: Incomplete Information in Relational Databases. Journal of the ACM 31(4), 761–791 (1984)
14. Kryszkiewicz, M.: Rough Set Approach to Incomplete Information Systems. Information Sciences 112, 39–49 (1998)
15. Kryszkiewicz, M.: Properties of Incomplete Information Systems in the framework of Rough Sets. In: Polkowski, L., Skowron, A. (eds.) Rough Set in Knowledge Discovery 1: Methodology and Applications, Studies in Fuzziness and Soft Computing, vol. 18, pp. 422–450. Physica Verlag (1998)
16. Kryszkiewicz, M.: Rules in Incomplete Information Systems. Information Sciences 113, 271–292 (1999)
17. Kryszkiewicz, M., Rybiński, H.: Data Mining in Incomplete Information Systems from Rough Set Perspective. In: Polkowski, L., Tsumoto, S., Lin, T.Y. (eds.) Rough Set Methods and Applications, Studies in Fuzziness and Soft Computing, vol. 56, pp. 568–580. Physica Verlag (2000)
18. Latkowski, R.: On Decomposition for Incomplete Data. Fundamenta Informaticae 54, 1–16 (2003)
19. Latkowski, R.: Flexible Indiscernibility Relations for Missing Values. Fundamenta Informaticae 67, 131–147 (2005)
20. Leung, Y., Li, D.: Maximum Consistent Techniques for Rule Acquisition in Incomplete Information Systems. Information Sciences 153, 85–106 (2003)
21. Nakata, N., Sakai, H.: Rough-set-based Approaches to Data Containing Incomplete Information: Possibility-based Cases. In: Nakamatsu, K., Abe, J.M. (eds.) Advances in Logic Based Intelligent Systems. Frontiers in Artificial Intelligence and Applications, vol. 132, pp. 234–241. IOS Press, Amsterdam (2005)

22. Nakata, N., Sakai, H.: Checking Whether or Not Rough-Set-Based Methods to Incomplete Data Satisfy a Correctness Criterion. In: Torra, V., Narukawa, Y., Miyamoto, S. (eds.) MDAI 2005. LNCS (LNAI), vol. 3558, pp. 227–239. Springer, Heidelberg (2005)
23. Nakata, N., Sakai, H.: Rough Sets Handling Missing Values Probabilistically Interpreted. In: Ślęzak, D., Wang, G., Szczuka, M.S., Düntsch, I., Yao, Y. (eds.) RSFDGrC 2005. LNCS (LNAI), vol. 3641, pp. 325–334. Springer, Heidelberg (2005)
24. Nakata, N., Sakai, H.: Applying Rough Sets to Data Table to Data Tables Containing Imprecise Information under Probabilistic Interpretation. In: Greco, S., Hata, Y., Hirano, S., Inuiguchi, M., Miyamoto, S., Nguyen, H.S., Słowiński, R. (eds.) RSCTC 2006. LNCS (LNAI), vol. 4259, pp. 213–223. Springer, Heidelberg (2006)
25. Nakata, N., Sakai, H.: Applying Rough Sets to Data Table Containing Missing Values. In: Kryszkiewicz, M., Peters, J.F., Rybinski, H., Skowron, A. (eds.) RSEISP 2007. LNCS (LNAI), vol. 4585, pp. 181–191. Springer, Heidelberg (2007)
26. Nakata, N., Sakai, H.: Applying Rough Sets to Information Tables Containing Probabilistic Values. In: Torra, V., Narukawa, Y., Yoshida, Y. (eds.) MDAI 2007. LNCS (LNAI), vol. 4617, pp. 282–294. Springer, Heidelberg (2007)
27. Orłowska, E., Pawlak, Z.: Representation of Nondeterministic Information. Theoretical Computer Science 29, 313–324 (1984)
28. Parsons, S.: Current Approaches to Handling Imperfect Information in Data and Knowledge Bases. IEEE Transactions on Knowledge and Data Engineering 8(3), 353–372 (1996)
29. Pawlak, Z.: Rough Sets. International Journal of Computer and Information Sciences 11, 341–356 (1982)
30. Pawlak, Z.: Rough Sets: Theoretical Aspects of Reasoning about Data. Kluwer Academic Publishers, Dordrecht (1991)
31. Sakai, H.: Some Issues on Nondeterministic Knowledge Bases with Incomplete Information. In: Polkowski, L., Skowron, A. (eds.) RSCTC 1998. LNCS (LNAI), vol. 1424, pp. 424–431. Springer, Heidelberg (1998)
32. Sakai, H.: Effective Procedures for Handling Possible Equivalence Relation in Nondeterministic Information Systems. Fundamenta Informaticae 48, 343–362 (2001)
33. Sakai, H., Nakata, M.: An Application of Discernibility Functions to Generating Minimal Rules in Non-deterministic Information Systems. Journal of Advanced Computational Intelligence and Intelligent Informatics 10, 695–702 (2006)
34. Sakai, H., Okuma, A.: Basic Algorithms and Tools for Rough Non-deterministic Information Systems. Transactions on Rough Sets 1, 209–231 (2004)
35. Słowiński, R., Stefanowski, J.: Rough Classification in Incomplete Information Systems. Mathematical and Computer Modelling 12(10/11), 1347–1357 (1989)
36. Slowiński, R., Vanderpooten, D.: A Generalized Definition of Rough Approximations Based on Similarity. IEEE Transactions on Knowledge and Data Engineering 12(2), 331–336 (2000)
37. Stefanowski, J., Tsoukiàs, A.: On the Extension of Rough Sets under Incomplete Information. In: Zhong, N., Skowron, A., Ohsuga, S. (eds.) RSFDGrC 1999. LNCS (LNAI), vol. 1711, pp. 73–81. Springer, Heidelberg (1999)
38. Stefanowski, J., Tsoukiàs, A.: Incomplete Information Tables and Rough Classification. Computational Intelligence 17(3), 545–566 (2001)
39. Zimányi, E., Pirotte, A.: Imperfect Information in Relational Databases. In: Motro, A., Smets, P. (eds.) Uncertainty Management in Information Systems: From Needs to Solutions, pp. 35–87. Kluwer Academic Publishers, Dordrecht (1997)

Toward a Generic Mathematical Model of Abstract Game Theories

Yingxu Wang

Theoretical and Empirical Software Engineering Research Centre (TESERC)
International Center for Cognitive Informatics (ICfCI)
Dept. of Electrical and Computer Engineering
Schulich School of Engineering, University of Calgary
2500 University Drive, NW, Calgary, Alberta, Canada T2N 1N4
Tel.: (403) 220 6141; Fax: (403) 282 6855
yingxu@ucalgary.ca

Abstract. Games are a complex mathematical structure for modeling dynamic decision processes under competition where opponent players compete for the maximum gain or toward a success state in the same environment according to the same rules of the game. Games are conventionally dealt with payoff tables based on random strategies, which are found inadequate to describe the dynamic behaviors of games and to rigorously predict the outcomes of games. This paper presents an abstract game theory, which enables a formal treatment of games by a set of mathematical models for both the layouts and behaviors of games. A generic mathematical model of abstract games is introduced, based on which the properties of games in terms of decision strategies and serial matches are described. A wide range of generic zero-sum and nonzero-sum games are formally modeled and analyzed using the generic mathematical models of abstract games.

Keywords: Cognitive informatics, abstract games, game theory, mathematical models, static layout, dynamic behaviors, properties, layoff tables, utilities, decision making, zero-sum games, nonzero-sum games, serial matches, decision grids.

1 Introduction

As a complex mathematical structure, games and game theories are a typical paradigm of decision theories. The study on decision making is interested in multiple disciplines, such as cognitive informatics, computer science, computational intelligence, cognitive psychology, management science, operational theories, economics, sociology, political science, and statistics [5], [10], [14], [17], [20], [24], [25], [30], [33], [34], [35], [36], [38].

Decision theories can be categorized into two paradigms: the *descriptive* and *normative* theories. The former are based on empirical observation and on experimental studies of choice behaviors; and the latter assume a rational decision-maker who

M.L. Gavrilova et al. (Eds.): Trans. on Comput. Sci. II, LNCS 5150, pp. 205–223, 2008.

follows well-defined preferences that obey certain axioms of rational behaviors. Typical normative theories are the expected utility paradigm [12], [17], [21], and the Bayesian theory [5], [33]. W. Edwards and B. Fasolo proposed a 19-step decision making process [10] by integrating Bayesian and multi-attribute utility theories. W. Zachary and his colleagues [42] perceived that there are three constituents in decision making known as the *decision situation*, the *decision maker*, and the *decision process*. Although the cognitive capacities of decision makers may be greatly varying, the core cognitive processes of the human brain share similar and recursive characteristics and mechanisms [35], [39], [40].

An overview of the taxonomy and classification of decision theories and related rational strategies can be illustrated as shown in Fig. 1, which may be used as a guideline for studying the whole framework of decision theories. Most of the decision making strategies [5], [6], [10], [14], [24], [33], [42] can be classified into static decision-making strategies, because the changes of environments of decision makers are independent of the decision makers' activities. Also, different decision strategies may be selected in the same situation or environment based on the decision makers' values and attitudes towards risk and their prediction on future outcomes. In classic decision and operations theories [6], [33], although the states of nature or environment may be both deterministic or nondeterministic, its state of nature as an outcome of the environment will not be changed or affected by the decision maker's actions. In other words, there are natural rules but no adaptive competitors in the static decision making processes.

However, when the environment of a decision maker is interactive with one's decisions or the environment changes according to the decision makers' activities and the decision strategies and rules are predetermined, this category of decision making needs are classified into the category of dynamic decisions, such as games [32], [33] and decision grids [36], [37].

Definition 1. The *dynamic strategies and criteria* of decision making are those that all alternatives and criteria are dependent on both the environment and the effect of the historical decisions made by the decision maker.

Classic dynamic decision making methods are decision trees [6], [12], [17], [21]. A new theory of decision grids is developed in [36], [37] for serial decision makings. Decision making under interactive events and competition is commonly modeled by games [12], [17], [32]. According to Fig. 1, games are used to deal with the most complicated decision problems, which are dynamic, interactive, and under uncontrollable competitions.

Definition 2. A *game* is a decision process under competition where opponent players or opponent groups of players compete for the maximum gain or toward a success state in the same environment according to the same predetermined rules and constraints of the game.

Games traditionally deal with probability-based static payoff tables [3], [4], [12], [16], [19], [26]. However, the conventional approach is found inadequate to deal with the dynamic behaviors of games and to rigorously determine the outcomes of games. This

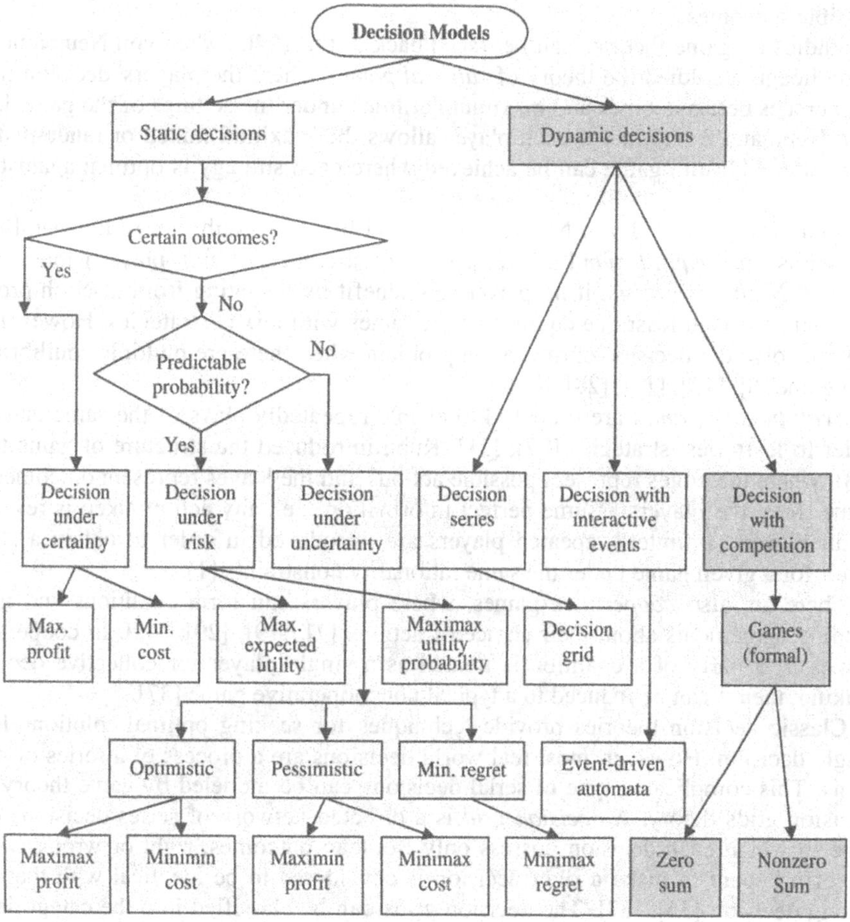

Fig. 1. An overview of decisions and strategies

paper presents a formal model of abstract games, which rigorously describes the architecture or layout of abstract games and their dynamic behaviors with a set of mathematical models. Section 2 reviews related work of various game theories in literature, which indicates a need of a unified model of abstract games. Section 3 develops a formal model of abstract games and describes their properties. Section 4 analyzes the behaviors of abstract games embodied by sets of matches, particularly the zero-sum games and nonzero-sum games. Section 5 discusses strategies of decision making in games such as the maximin and maximum utility strategies.

2 Related Work

Game theory provides a mathematical structure for analyzing the interaction between multiple parties whose decisions affect each other. A game encompasses a finite set of players, a set of courses of actions available to them, and their preferences over the possible outcomes.

Studies on game theories can be traced back to the 1940s when von Neumann and his colleagues studied the theory of *rational games* where the players' decision making pursues best outcomes and maximum utilities under the settings of the game [32]. Von Neumann found that if each player allows the maximin mixed or random strategy, an equilibrium game can be achieved where each strategy is optimal against the other.

John Nash extended von Neumann's rational binary equilibrium to a general case known as *Nash equilibrium* [22], [23]. A set of strategies of an n-player game is said to be in *Nash equilibrium* if no player can benefit by deviating from it. Nash proved that there exists at least one equilibrium in games with mixed strategies. However, he did not solve the decision optimization problem when there are multiple equilibriums in a game [8], [13], [15], [28], [41].

Evolutionary games are proposed to enable repeatedly plays of the same game in order to learn best strategies [27], [31]. Kuhn introduced the structure of game trees [18], where the edges represent possible actions and the leaves represent outcomes. In game trees, the players assume perfect information, i.e., any action taken is revealed to all players. Infinitely repeated players are introduced in order to obtain a stable result for a given game under the same rationality constraints [1].

There are also *cooperative* games, where players can form coalitions and make binding agreements about their choice of actions [7], [19], [29], [32]. In cooperative games, if a party of a coalition is treated as a single player for collective decision making, then it can be reduced to a typical noncooperative game [37].

Classic decision theories provide techniques for seeking optimal solutions for a single decision. However, most real world decisions are a process of a series of decisions. This complicated type of serial decisions can be modeled by game theory and decision grids theory. A *decision grid* is a directed network of series decisions over time in which each decision possess only Boolean outcomes, right or wrong, where the effort spent to make a right decision is considered to be identical with that of a wrong decision [36], [37]. The decision grids can be classified into the categories of unlimited and limited grids according to the scope of allowable trials. When the allowable number of trials t in a decision grid is infinitive, the decision grid is called an *unlimited decision grid*; otherwise, it is a *limited decision grid*. The unlimited decision grid is a suitable model for the series of decisions toward a success state no matter how many trials are needed, such as an experimental process, a research project, or a person's pursuit towards a goal in life. The limited decision grid is a serial decision model for a short period of trials, such as a student towards a degree, an assessment process, or a deadline-specific process.

3 The Generic Mathematical Model of Abstract Games

Although games are usually represented by layoff tables, the lack of a generic mathematical model of games in game theories has greatly limited the exploration of the modeling, properties, and dynamic behaviors of games. This section develops a mathematical model of a general abstract game, based on which the layouts and behaviors of any concrete game may be treated as a derived instance.

3.1 The Formal Model of Abstract Games

The architecture of an abstract game can be formally described by the following definition, where the behaviors of the game will be modeled by a series of matches between the players of the game.

Definition 3. An *abstract game G* is a 4-tuple, i.e.:

$$G = (P, D, M, S) \tag{1}$$

where

- P is a finite nonempty set of *players* $P = \{p_1, p_2, \ldots, p_n\}$, and n denotes the number of players, $n = \#P$, $n \geq 2$.
- D is a finite nonempty set of *decisions* for certain *moves*, $D = \{d_1, d_2, \ldots, d_k\}$, $k \geq 1$, and all players have the same number of alternative decisions in rational games.
- M is a nonempty finite set of *matches* between players, $M = \{m_1, m_2, \ldots, m_q\}$, $q \geq 1$.
- S is a nonempty finite set of *scores* for each player after a match or a series of matches, $S = \{s_1, s_2, \ldots, s_n\}$.

For a generic abstract game, the matches, which represent the behaviors of the game, can be further described below.

Definition 4. A *match* $m \in M$ of an abstract game $G = (P, D, M, S)$ is a function that maps a set of n decisions made by each player of G into a set of n scores S for each of the players, i.e.:

$$m = f_m : D \times D \times \ldots \times D \rightarrow S \tag{2}$$

A match is an individual block given in the payoff table of the game. A set of matches in the given game is constrained by a set of certain rules in order to be rational.

Lemma 1. In an abstract game $G = (P, D, M, S)$, the following rules for matches yield rational, stable, and predictable behaviors and scores:

- **Rule (*a*):** All players are supposed to pursue the maximum gains on the basis of the same predefined payoff table.
- **Rule (*b*):** Whenever the first player initiates a move in a specific set of matches, the remaining moves (actions) of all players in the set of matches are determined according to Rule (a).
- **Rule (*c*):** Each match preset in the payoff table may only be used once in the set of matches.

The rules given in Lemma 1 form the basic constraints of rational games and make a game to be deterministic and its outcomes of all sets of matches are predictable. In lemma 1, Rules (a) and (b) guarantee that all matches of a game are determinable on the basis of the given payoff table. Rule (c) assurances that a set of matches in the game is finite and determinable, although it may force a player to take an unused strategy that would be unfavorable in a particular mach when it is the only strategy left in the setting of the given game.

Lemma 2. The *number of individual matches* n_m in the set of matches of a given game $G = (P, D, M, S)$ is determinable, i.e.:

$$n_m = k^n \tag{3}$$

where n is the number of players in a game, and k is the number of alternative decisions (moves) defined in the game for each player.

Example 1. An $n \times k = 2 \times 2$ game $G_1 = (P, D, M, S)$ can be formally described according to Definition 3 as follows:

- *Players* $P = \{a, b\}, n = 2.$
- *Decisions* $D = \{d_1, d_2\}, k = 2$, i.e. $D_a = \{a_1, a_2\}$, or $D_b = \{b_1, b_2\}$.
- *Scores* $S = \{s_a, s_b\}$.
- *Matches* $M = \{m_{11}, m_{12}, m_{21}, m_{22}\}$, which is determined by Lemma 2, i.e., $n_m = k^n = 2^2 = 4$.

In $G_1 = (P, D, M, S)$, let a_1 and a_2 be the alternative decisions of player A, and b_1 and b_2 the alternative decisions of player B, then the four matches in M can be formally described as follows:

$$m_{11} = a_1 : b_1 \rightarrow s_a : s_b = 0 : 0$$
$$m_{12} = a_1 : b_2 \rightarrow -1 : 1$$
$$m_{21} = a_2 : b_1 \rightarrow -2 : 2$$
$$m_{22} = a_2 : b_2 \rightarrow 3 : -3$$

The above matches can be represented by a payoff table as shown in Table 1.

Table 1. The Payoff Table of M = {m_{11}, m_{12}, m_{21}, m_{22}}

	b_1	b_2
a_1	0 : 0	-1 : 1
a_2	-2 : 2	3 : -3

This is the static architecture or layout of game G_1. Its dynamic behaviors on the basis of the layout will be discussed in the following subsections.

3.2 Properties of Abstract Games

Properties of games are basic characteristics possessed by them. The properties of abstract games are number of alternative strategies (moves), number of sets of matches, and the number of matches. The properties of abstract games can be used to predicate possible outcomes of games and to select optimal strategies or moves in games.

Definition 5. A *set of matches* in an abstract game $G = (P, D, M, S)$ is a series of matches in which all players may use each of their alternative strategies only once determined according to the current move of opponent and the rule of the maximum gains based on the given layout of the game.

Lemma 3. The *number of set of matches* n_s in an abstract game $G = (P, D, M, S)$ is proportional to both the number of alternative strategies (moves) k, and the number of players n, i.e.:

$$n_s = n \bullet k \tag{4}$$

Lemma 4. The *number of matches* q of an abstract game $G = (P, D, M, S)$ is determined by a product of the number of sets of matches n_s and number of matches in each set n_m, i.e.:

$$\begin{aligned} q &= n_s \bullet n_m \\ &= nk \bullet k^n \\ &= n \bullet k^{n+1} \end{aligned} \tag{5}$$

It is noteworthy that Lemmas 1 through 4 provide a set of generic theories for determining the properties of any given game including those that are beyond the process power of conventional game theories. The attributes of some typical games can be predicated as shown in Table 2.

Table 2. Attributes of Arbitrary and Typical Games

n	$k \Rightarrow (n_m = k^n \mid n_s = n \bullet k \mid q = n_s \bullet n_m = nk^{n+1})$											
	$k = 1$			$k = 2$			$k = 3$			$k = 4$		
2	1	2	2	4	4	16	9	6	54	16	8	128
3	1	3	3	8	6	48	27	9	243	64	12	768
4	1	4	4	16	8	128	81	12	972	256	16	4096
5	1	5	5	32	10	320	243	15	3645	1024	20	20480
...												
100	1	100	100	2^{100}	200	$100 \bullet 2^{101}$	3^{100}	200	$100 \bullet 3^{101}$	4^{100}	400	$100 \bullet 4^{101}$

According to Table 2, the complexity of games is explosively increasing proportional to the numbers of both players n and strategies k. This explains why games are so complicated and difficult to be modeled and formally treated in conventional game theory [6], [32]. For example, when the number of players $n = 5$ and the number of alternative strategies of each player $k = 4$, the total number of matches of the game may easily reach as high as 20,480. That is why conventional empirical game theories may only deal with small and simple games with a few of players and alternative strategies.

However, the abstract game theory presented so far is able to analyze any games no matter how large n and k would be based on the generic mathematical model of formal games and their instances.

Since games with multiple players can be divided into a number of pairwise games, the following sections will focus on the analyses of binary game properties as shaded in Table 2.

Definition 6. A *binary game* $G = (P_2, D, M, S_2)$ is a game with only two players $n = 2$, where $P_2 = \{p_1, p_2\}$ and $S_2 = \{s_1, s_2\}$, simply called a game.

Example 2. A well known binary game is the Prisoner's dilemma, where two conspirators in prison may receive a sentence for either two years or eight years for both remaining silent or confess, respectively. They are also given the opportunity to confess in return for a reduced prison sentence of half a year. The payoffs correspond to numbers of years in prison are given in Table 3.

Table 3. The Payoff Table of the Prisoner's Dilemma

	b_1 (silent)	b_2 (confess)
a_1 (silent)	-2 : -2	-0.5 : -10
a_2 (confess)	-10 : -0.5	-8 : -8

It is noteworthy that according to Nash equilibrium, the utility of the above game is (-8, -8), rather than (-2,-2) [2], [22]. However, according to the abstract game theory, the average score of the above game is -20.5 : -20.5, i.e., there is no winner according to the payoff table of the Prisoner's dilemma.

When a game $G = (P, D, M, S)$ is set according to Definitions 3 and 4, the properties of G, such as the number of matches, the number of sets of matches, and the winner are determined. Game theory may be used to predict and select the optimal combination of individual strategies. However, the score for any individual strategy in G has already fixed according to the payoff table.

Theorem 1. The *properties of games* state that an abstract game $G = (P, D, M, S)$ is *deterministic* and *conservative*. Once the game G is set, the properties of G are determined, predictable, and unchangeable to all players in the game.

According to Theorem 1, game theory may be used to predict and select the optimal combinations of individual strategies for a player in a given game G. However, the optimal strategies may not necessarily result in a win situation rather than a minimal loss in some cases, because the scores for individual moves and their combination strategies in G are determined by the settings of the game.

Corollary 1. The outcomes of a formal game $G = (P, D, M, S)$ are constrained by the settings of the game. Although an individual strategy may result in the maximum gain, the final score of a player in the whole set of games is fixed by the payoff table in a particular match, which may not necessarily result in a win situation for all players.

The objective of decision makers in a game is to make the score of a player to the maximum. However, according to Corollary 1, $\max(s_i)$ may not mean a winning score due to the settings of a given game.

4 Behaviors of Abstract Games

There are zero-sum and nonzero-sum games. Each of them has different properties and dynamic behaviors as described in the following subsections.

4.1 Behaviors of Zero-Sum Games

Definition 7. A *zero-sum game* is a type of abstract games where the total score of all players in the game remains zero, i.e.:

$$\sum_{i=1}^{n} s_i = 0 \qquad (6)$$

In the case of a binary game, Eq. 6 can be expressed as follows:

$$s_1 = - s_2 \qquad (7)$$

where Eq. 7 models a decision making situation that one player's gain is always another's loss.

Lemma 5. The *condition for a zero-sum game* is that all n_m individual matches are zero-sum, i.e.:

$$\sum_{i=1}^{n_m} s_i = 0 \qquad (8)$$

Example 3. The game $G_1 = (P, D, M, S)$ as given in Example 1 and Table 1 is a zero-sum game. The properties and behaviors of G_1 can be formally analyzed below.

The properties of $G_1 = (P, D, M, S)$ are:

- *Number of sets of matches:* $n_s = n \bullet k = 2 \bullet 2 = 4$
- *Number of matches in a set:* $n_m = k^n = 2^2 = 4$
- *Total number of matches in the game:*
  $n_m = n_s \bullet n_m = n \bullet k^{n+1} = 2 \bullet 2^3 = 16$

The four sets of matches each with a series of four individual matches can be illustrated in Fig. 2.

	$s_a : s_b$
Set 1: $a_1 \xrightarrow{-1:1} b_2 \xrightarrow{3:-3} a_2 \xrightarrow{-2:2} b_1 \xrightarrow{0:0} a_1$	$\Rightarrow 0:0$
Set 2: $a_2 \xrightarrow{-2:2} b_1 \xrightarrow{0:0} a_1 \xrightarrow{-1:1} b_2 \xrightarrow{3:-3} a_2$	$\Rightarrow 0:0$
Set 3: $b_1 \xrightarrow{0:0} a_1 \xrightarrow{-1:1} b_2 \xrightarrow{3:-3} a_2 \xrightarrow{-2:2} b_1$	$\Rightarrow 0:0$
Set 4: $b_2 \xrightarrow{3:-3} a_2 \xrightarrow{-2:2} b_1 \xrightarrow{0:0} a_1 \xrightarrow{-1:1} b_2$	$\Rightarrow 0:0$

Fig. 2. Sets of matches in the zero-sum game G_1

Lemma 6. The *final scores of all sets of matches* of an abstract games *G* are the same, no matter who moves first and which strategy (decision alternative) is selected for the first move.

Theorem 2. The *scores* of a $2 \times k$ abstract game, $s_a : s_b$, is predetermined by the settings of the payoff table, i.e.:

$$\begin{aligned} s_a : s_b &= (\sum_{i=1}^{k}\sum_{j=1}^{k} s_{ij}^a) : (\sum_{i=1}^{k}\sum_{j=1}^{k} s_{ij}^b) \\ &= (\sum_{i=1}^{k}\sum_{j=1}^{k} s_{ij}^a) : (-\sum_{i=1}^{k}\sum_{j=1}^{k} s_{ij}^a) \end{aligned} \tag{9}$$

where *k* is the number of alternative decision strategies and *k* is identical for all players.

According to Theorem 2, the results of all possible sets of matches for a given zero-sum game can be predicated using Eq. 9. For instance, the final score of Example 3 can be calculated according Eq. 9 as follows:

$$\begin{aligned} s_a : s_b &= (\sum_{i=1}^{k}\sum_{j=1}^{k} s_{ij}^a) : (\sum_{i=1}^{k}\sum_{j=1}^{k} s_{ij}^b) \\ &= (s_{11}^a + s_{12}^a + s_{21}^a + s_{22}^a) : \\ &\quad (s_{11}^b + s_{12}^b + s_{21}^b + s_{22}^b) \\ &= (0 - 1 - 2 + 3) : \\ &\quad (0 + 1 + 2 - 3) \\ &= 0 : 0 \end{aligned}$$

Example 4. For a 2×3 game $G_2 = (P, D, M, S)$ with the following payoff table, try to determine its properties and behaviors.

Table 4. The Payoff Table of $G_2 = (P, D, M, S)$

	b_1	b_2	b_3
A_1	0 : 0	100 : -100	200 : -200
A_2	-300 : 300	0 : 0	-100 : 100
A_3	500 : -500	-200 : 200	0 : 0

The properties of $G_2 = (P, D, M, S)$ are:

- *Number of sets of matches:* $n_s = n \bullet k = 2 \bullet 3 = 6$
- *Number of matches in a set:* $n_m = k^n = 3^2 = 9$
- *Total number of matches in the game:*
  $n_m = n_s \bullet n_m = n \bullet k^{n+1} = 2 \bullet 3^3 = 54$

According to Theorem 2, the final scores of $G_2 = (P, D, M, S)$ are as follows:

$$
\begin{aligned}
s_a : s_b &= (\sum_{i=1}^{k}\sum_{j=1}^{k} s_{ij}^a) : (\sum_{i=1}^{k}\sum_{j=1}^{k} s_{ij}^b) \\
&= (s_{11}^a + s_{12}^a + s_{13}^a + s_{21}^a + s_{22}^a + s_{23}^a + s_{31}^a + s_{32}^a + s_{33}^a) : \\
&\quad (s_{11}^b + s_{12}^b + s_{13}^b + s_{21}^b + s_{22}^b + s_{23}^b + s_{31}^b + s_{32}^b + s_{33}^b) \\
&= (0 + 100 + 200 - 300 + 0 - 100 + 500 - 200 + 0) : \\
&\quad (0 - 100 - 200 + 300 + 0 + 100 - 500 + 200 + 0) \\
&= 200 : \text{-}200
\end{aligned}
$$

The behaviors of $G_2 = (P, D, M, S)$ can be modeled by 54 detailed matches in 6 sets as shown in Fig. 3.

Corollary 2. The *outcomes* of a given game G is determined according to the scores $s_a : s_b$ as follows:

$$
\begin{cases}
s_a > s_b\text{: Player A won} \\
s_a = s_b\text{: Tie} \\
s_a < s_b\text{: Player B won}
\end{cases}
\tag{10}
$$

Therefore, the final score of Example 1, $s_a : s_b = 0 : 0$, shows a tie game; while Example 3, $s_a : s_b = 200 : \text{-}200$, indicates that Player A will always win.

4.2 Behaviors of Nonzero-Sum Games

A more general type of games is nonzero-sum games where all players involved share a certain pie with a fixed size. From this view, the zero-sum game discussed in Section 4.1 is a special case of nonzero-sum games where the size of the pie is zero.

$$s_a : s_b$$

$$\text{Set 1: } a_1 \xrightarrow{0:0} b_1 \xrightarrow{500:-500} a_3 \xrightarrow{-200:200} b_2 \xrightarrow{100:-100} a_1$$
$$\xrightarrow{200:-200} b_3 \xrightarrow{0:0} a_3,\ a_2 \xrightarrow{-300:300} b_1,$$
$$a_2 \xrightarrow{0:0} b_2,\ a_2 \xrightarrow{-100:100} b_3 \quad \Rightarrow 200 : -200$$

$$\text{Set 2: } a_2 \xrightarrow{-300:300} b_1 \xrightarrow{500:-500} a_3 \xrightarrow{-200:200} b_2 \xrightarrow{100:-100} a_1$$
$$\xrightarrow{0:0} b_1, a_1 \xrightarrow{200:-200} b_3,\ a_2 \xrightarrow{0:0} b_2,$$
$$a_2 \xrightarrow{-100:100} b_3,\ a_3 \xrightarrow{0:0} b_3 \quad \Rightarrow 200 : -200$$

$$\text{Set 3: } a_3 \xrightarrow{-200:200} b_2 \xrightarrow{100:-100} a_1 \xrightarrow{0:0} b_1 \xrightarrow{500:-500} a_3$$
$$\xrightarrow{0:0} b_3 \xrightarrow{200:-200} a_1,\ a_2 \xrightarrow{-300:300} b_1,$$
$$a_2 \xrightarrow{0:0} b_2,\ a_2 \xrightarrow{-100:100} b_3 \quad \Rightarrow 200 : -200$$

$$\text{Set 4: } b_1 \xrightarrow{500:-500} a_3 \xrightarrow{-200:200} b_2 \xrightarrow{100:-100} a_1 \xrightarrow{0:0} b_1$$
$$\xrightarrow{-300:300} a_2 \xrightarrow{-100:100} b_3,\ b_2 \xrightarrow{0:0} a_2,$$
$$b_3 \xrightarrow{200:-200} a_1,\ b_3 \xrightarrow{0:0} a_3 \quad \Rightarrow 200 : -200$$

$$\text{Set 5: } b_2 \xrightarrow{100:-100} a_1 \xrightarrow{0:0} b_1 \xrightarrow{500:-500} a_3 \xrightarrow{-200:200} b_2$$
$$\xrightarrow{0:0} a_2 \xrightarrow{-300:300} b_1,\ b_3 \xrightarrow{200:-200} a_1,$$
$$b_3 \xrightarrow{-100:100} a_2,\ b_3 \xrightarrow{0:0} a_3 \quad \Rightarrow 200 : -200$$

$$\text{Set 6: } b_3 \xrightarrow{200:-200} a_1 \xrightarrow{0:0} b_1 \xrightarrow{500:-500} a_3 \xrightarrow{-200:200} b_2$$
$$\xrightarrow{100:-100} a_1, a_3 \xrightarrow{0:0} b_3,\ b_1 \xrightarrow{-300:300} a_2,$$
$$b_2 \xrightarrow{0:0} a_2,\ b_3 \xrightarrow{-100:100} a_2 \quad \Rightarrow 200 : -200$$

Fig. 3. Sets of matches of the 2×3 zero-sum game

Definition 8. A *nonzero-sum game* is a game where the total scores of all players in the game is a nonzero positive value, i.e.:

$$\sum_{i=1}^{n} s_i > 0 \tag{11}$$

A group on a common project or a set of partners bidding for a contract is a typical example of nonzero-sum games.

The most interesting property of decision making in nonzero-sum game is that there is an ideal state of result known as the win-win situation.

Definition 9. A *win-win game* is a nonzero-sum game in which all players gain a certain score constrained by Eq. 11.

Lemma 7. A win-win game can only exist in nonzero-sum games.

According to Lemma 7, if a number of competitive players in a nonzero-sum game are coordinated, i.e., a superset of partnership is established in the game, every party may be benefit.

Theorem 3. A *win-win decision* can be achieved in a nonzero-sum game when the following condition is satisfied:

$$\sigma \geq \frac{1}{n_s} \sum_{i=1}^{n_s} s_i \qquad (12)$$

where σ is the sum of the game that is a nonzero positive constant, s_i is the expected score of player *i*, and n_s is the number of sets of matches in the game.

According to Theorem 3, a win-win game may satisfy all coordinative players when the constant sum σ is large enough as determined by Eq. 12.

Example 5. Given a 2 × 2 nonzero-sum game $G_3 = (P, D, M, S)$ with the following payoff table and $\sigma = 100$, try to determine its properties and behaviors.

Table 5. The Payoff Table of $G_3 = (P, D, M, S)$

	b_1	b_2
a_1	70 : 30	20 : 80
a_2	60 : 40	90 : 10

The properties of $G_3 = (P, D, M, S)$ are:

- *Number of sets of matches:* $n_s = n \bullet k = 2 \bullet 2 = 4$
- *Number of matches in a set:* $n_m = k^n = 2^2 = 4$
- *Total number of matches in the game:* $n_m = n_s \bullet n_m = n \bullet k^{n+1} = 2 \bullet 2^3 = 16$

According to Theorem 2, the final scores of $G_3 = (P, D, M, S)$ are as follows:

$$\begin{aligned} s_a : s_b &= (\sum_{i=1}^{k}\sum_{j=1}^{k} s_{ij}^a) : (\sum_{i=1}^{k}\sum_{j=1}^{k} s_{ij}^b) \\ &= (s_{11}^a + s_{12}^a + s_{21}^a + s_{22}^a) : (s_{11}^b + s_{12}^b + s_{21}^b + s_{22}^b) \\ &= (70 + 20 + 60 + 90) : (30 + 80 + 40 + 10) \\ &= 240 : 160 \end{aligned}$$

This result indicates that the four sets of matches defined by G_3 will result in an average score in each match as 60 : 40, in which Players A and B share $\sigma = 100$. This can be proved by the following four sets of matches as shown in Fig. 4.

	$s_a : s_b$
Set 1: $a_1 \xrightarrow{20:80} b_2 \xrightarrow{90:10} a_2 \xrightarrow{60:40} b_1 \xrightarrow{70:30} a_1$	$\Rightarrow 240 : 160$
Set 2: $a_2 \xrightarrow{60:40} b_1 \xrightarrow{70:30} a_1 \xrightarrow{20:80} b_2 \xrightarrow{90:10} a_2$	$\Rightarrow 240 : 160$
Set 3: $b_1 \xrightarrow{70:30} a_1 \xrightarrow{20:80} b_2 \xrightarrow{90:10} a_2 \xrightarrow{60:40} b_1$	$\Rightarrow 240 : 160$
Set 4: $b_2 \xrightarrow{90:10} a_2 \xrightarrow{60:40} b_1 \xrightarrow{70:30} a_1 \xrightarrow{20:80} b_2$	$\Rightarrow 240 : 160$

Fig. 4. Sets of matches of the 2×2 nonzero-sum game G_3

It may be observed that for a given game in a certain context, it would appear to be competitive between conflict interests of players. However, at a higher level of an enlarged scope of the given game, it can be perceived differently as a cooperative game for all parties involved. This leads to the following corollary for management attitude and skills in decision making.

Corollary 3. The *art of management*, to a certain extent, is to create a win-win environment for all members, partners, and parent organizations involved in a game context.

5 Strategies of Decision Making in Games

According to Lemma 4, a certain layout of a game implies a whole set of $q = nk^{n+1}$ match outcomes. Therefore, the choice of decision strategies in games is crucial. The following subsections discuss typical decision making strategies in games known as those of maximin and maximum utility.

5.1 The Maximin Strategy in Games

A conservative strategy to try to gain the maximum utility or to spend the minimum cost under uncertainty is known as the maximin strategy in games [9], [37].

Definition 10. A *conservative decision making under uncertainty* $d_{maximin}$ or $d_{minimax}$ yields a decision with the *maximum-minimum* strategy for utility or a *minimum-maximum* strategy for cost, i.e.:

$$\begin{aligned} d_{maximin} &= f\colon \mathcal{A} \times C \to \mathcal{A} \\ &= \{a_i \mid max\,(min\,(u_{ij} \mid 1 \le i \le n) \mid 1 \le j \le k)\} \end{aligned} \tag{13a}$$

or

$$\begin{aligned} d_{minimax} &= f\colon \mathcal{A} \times C \to \mathcal{A} \\ &= \{a_i \mid min\,(max\,(u_{ij} \mid 1 \le i \le n) \mid 1 \le j \le k)\} \end{aligned} \tag{13b}$$

where $\mathcal{A}$ is a set of given alternative strategies, and C a set of decision-making criteria.

Example 6. Consider the following engineering project where a maximin or a pessimistic uncertainty decision can be made based on the project gains for different architecture-result combinations as shown in Table 6.

Table 6. Maximin Decision Making for an Engineering Project

Alternative ($\mathcal{A}$)	Situation (S) Result 1 (s_1)	Result 2 (s_2)	Result 3 (s_3)	Result 4 (s_4)	Criterion (maximin utility)
Architecture (a_1)	100	10	40	60	u_{12} = \$10k
Architecture (a_2)	- 10	50	200	30	
Architecture (a_3)	50	20	5	130	

According to Eq. 13a, the maximin decision under uncertainty is as follows:

$$\begin{aligned} d_{maximax} &= f\colon \mathcal{A} \times C \to \mathcal{A} \\ &= \{a_i \mid max\ (min\ (u_{ij} \mid 1 \leq i \leq 3) \mid 1 \leq j \leq 4)\} \\ &= \{a_i \mid max\ (u_{12}, u_{21}, u_{33})\} \\ &= \{a_1 \mid u_{12} = 10\} \end{aligned}$$

The solution indicates that the conservative decision for this given project with the maximin criterion is (a_1, s_2), which will result in a maximin project gain $u_{max} = u_{12} =$ \$10,000. It is noteworthy that, by choosing this solution, there is a chance to lose the opportunity gain of u_{23} =\$200,000 if the uncertain outcomes turn out to be Result 3 constrained by the other party of the game. However, in any case, this decision can prevent the project from a negative result as that of u_{21} = -\$10,000.

5.2 The Maximum Utility Strategy in Games

Subsection 5.1 deals with decision strategies where the probabilities of the opponent party are uncertain. When the opponent strategies in a game are individually predictable, i.e., the probabilities or likelihoods are known, the risk for a decision can be better estimated. In this case, decision making process will be directed based on the weights of probabilities for each payoff.

Definition 11. A *decision making under risk* is a selection of an alternative a_i among $\mathcal{A}$ that meets a given criterion C, when the likelihood or probability of each possible situation is known or can be predicated.

The criterion for a decision making under risk can be based on the maximum expected utility of alternatives.

Definition 12. An *expected utility EU* is a weighted sum of all utilities u_j for each decision alternative based on known probabilities for each possible situation p_j, i.e.:

$$EU_i = \sum_{j=1}^{k} u_{ij} \bullet p_j,\ 1 \leq i \leq n \tag{14}$$

Definition 13. A *decision making under risk with maximum expected utility* d_{maxEU} yields a decision with the *maximum expected utilities* of all alternatives, i.e.:

$$\begin{aligned} d_{maxEU} &= f\!: \mathcal{A} \times C \to \mathcal{A} \\ &= \{a_i \mid max\,(EU_i \mid 1 \leq i \leq n)\} \end{aligned} \tag{15}$$

Example 7. Consider the same layout given in Example 6. A decision under risk with maximum expected utility can be made based on the *EUs* determined by Eq. 14 for different decision alternatives as shown in Table 7.

After the expected utilities for all three alternatives are obtained as shown in Table 7, the best decision with the maximum expected utility can be determined according to Eq. 15 as follows:

$$\begin{aligned} d_{maxEU} &= f\!: \mathcal{A} \times C \to \mathcal{A} \\ &= \{a_i \mid max\,(EU_i \mid 1 \leq i \leq n)\} \\ &= \{a_i \mid man\,(EU_1, EU_2, EU_3)\} \\ &= \{a_2 \mid EU_2 = 66\} \end{aligned}$$

Table 7. Decision Making based on the Maximum Expected Utility for an Engineering Project

Alternative ($\mathcal{A}$)	Situation (S)				Expected Utility (EU)	Criterion (Maximum EU)
	Result 1 (s_1) [$p_1 = 0.2$]	Result 2 (s_2) [$p_2 = 0.5$]	Result 3 (s_3) [$p_3 = 0.2$]	Result 4 (s_4) [$p_4 = 0.1$]		
Architecture (a_1)	100	10	40	60	$EU_1 = 39$	
Architecture (a_2)	- 10	50	200	30	$EU_2 = 66$	$EU_{max} = 66$
Architecture (a_3)	50	20	5	130	$EU_3 = 34$	

The solution indicates that the decision under risk for this given project with the maximum expected utility criterion is Architecture a_2 that will result in a maximum weighted sum EU_2 = \$66,000.

Decision making under risk with the maximum expected utility d_{maxEU} can be described by a backward-inducted *decision tree* as shown in Fig. 5. The decision tree provides another approach to derive the maximum expected utility in two steps [11]. First, the individual weighted utilities of all the alternatives are calculated according to Eq. 14, which yields EU_i, $1 \leq i \leq 3$, represented by the three middle nodes. Then, the maximum utility EU_{max} is selected from these three middle nodes according to Eq. 15, which yields node *A* represented by decision d_2 with $EU_{max} = 66$.

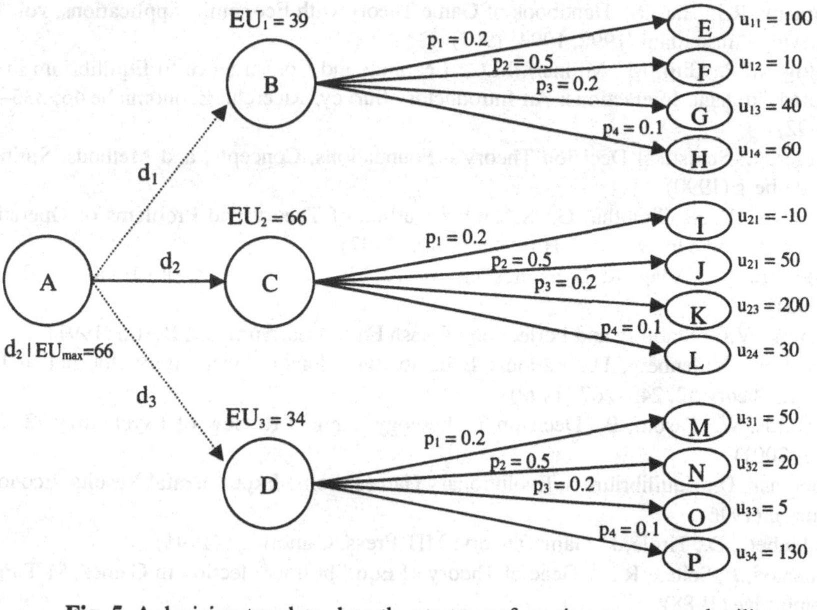

Fig. 5. A decision tree based on the strategy of maximum expected utility

6 Conclusions

This paper has presented a generic mathematical model of abstract games. Based on the abstract game theory, properties of games have been analyzed, and the predictability of games has been derived. This paper has demonstrated a formal treatment of games by a set of mathematical models on both of the layout and behaviors of abstract games in terms of serial matches constrained by the generic rules. Then, all specific games has been treated as particular instances. On the basis of the generic game theory, zero-sum and nonzero-sum games have been rigorously analyzed and their properties and relations have been formally described. A set of decision strategies of games such as the maximin and maximum utility has been explained. The generic abstract game theories and mathematical models can be applied in the design and implementation of game systems as well as autonomous agent systems.

Acknowledgement. This work is partially sponsored by Natural Sciences and Engineering Research Council of Canada (NSERC). The author would like to thank the anonymous reviewers for their valuable suggestions and comments on this work.

References

1. Abreu, D., Rubinstein, A.: The Structure of Nash Equilibria in Repeated Games with Finite Automata. Econometrica 56, 1259–1281 (1992)
2. Aumann, R.: Subjectivity and Correlation in Randomized Strategies. Journal of Mathematical Economics 1, 67–96 (1974)

3. Aumann, R.J., Hart, S.: Handbook of Game Theory with Economic Applications, vol. 1–3. Elsevier, Amsterdam (1992, 1994, 1997)
4. Battigalli, P., Gilli, M., Milinari, M.C.: Learning and Convergence to Equilibrium in Repeated Strategic Interactions: An Introductory Survey. Ricerche Economiche 46, 335–377 (1992)
5. Berger, J.: Statistical Decision Theory – Foundations, Concepts, and Methods. Springer, Heidelberg (1990)
6. Bronson, R., Naadimuthu, G.: Schaum's Outline of Theory and Problems of Operations Research, 2nd edn. McGraw-Hill, New York (1997)
7. Crawford, V.: Adaptive Dynamics in Coordination Games. Econometrica 63, 103–158 (1995)
8. Damme, V.E.: Stability and Perfection of Nash Equilibria. Springer, Berlin (1991)
9. Dekel, E., Fudenberg, D.: Rational Behavior with Payoff Uncertainty. Journal of Economic Theory 52, 243–267 (1990)
10. Edwards, W., Fasolo, B.: Decision Technology. Annual Review of Psychology 52, 581–606 (2001)
11. Friedman, D.: Equilibrium in Evolutionary Games: Some Experimental Results. Economic Journal (1996)
12. Fudenberg, D., Tirole, J.: Game Theory. MIT Press, Cambridge (1991)
13. Harsanyi, J., Selten, R.: A General Theory of Equilibrium Selection in Games. MIT Press, Cambridge (1988)
14. Hastie, R.: Problems for Judgment and Decision Making. Annual Review of Psychology 52, 653–683 (2001)
15. Jordan, J.: Three Problems in Learning Mixed-Strategy Equilibria. Games and Economic Behavior 5, 368–386 (1993)
16. Koller, D., Pfeffer, A.: Representations and Solutions for Game-Theoretic Problems. Artificial Intelligence (1997)
17. Koller, D.: Game Theory. In: Wilson, R.A., Keil, F.C. (eds.) The MIT Encyclopedia of the Cognitive Sciences. MIT Press, Cambridge (2001)
18. Kuhn, H.W.: Extensive Games and the Problem of Information. In: Kuhn, H.W., Tucker, A.W. (eds.) Contributions to the Theory of Games II, pp. 193–216. Princeton University Press, Princeton (1953)
19. Luce, R.D., Raiffa, H.: Games and Decisions—Introduction and Critical Survey. Wiley, Chichester (1957)
20. Matlin, M.W.: Cognition, 4th edn. Harcourt Brace College Publishers, Orlando (1998)
21. Myerson, R.: Game Theory. Harvard University Press, Cambridge (1991)
22. Nash, J.F.: Equilibrium Points in n-Person Games. Proc. National Academy of Sciences 36, 48–49 (1950)
23. Nash, J.F.: Non-Cooperative Games. Annals of Mathematics 54, 286–295 (1951)
24. Payne, D.G., Wenger, M.J.: Cognitive Psychology. Houghton Mifflin Co., New York (1998)
25. Pinel, J.P.J.: Biopsychology, 3rd edn. Allyn and Bacon, Needham Heights (1997)
26. Rosenschein, J.S., Zlotkin, G.: Consenting Agents: Designing Conventions for Automated Negotiation. AI Magazine 15(3), 29–46 (1994)
27. Roth, A., Er'ev, I.: Learning in Extensive Form Games: Experimental Data and Simple Dynamic Models in the Intermediate Run. Games and Economic Behavior 6, 164–212 (1995)
28. Selten, R.: Reexamination of the Perfectness Concept for Equilibrium Points in Extensive Games. International Journal of Game Theory 4, 25–55 (1975)

29. Shapley, L.S., Shubik, M.: Solutions of n-Person Games with Ordinal Utilities. Econometrica 21, 348–349 (1953)
30. Simon, H.A.: The New Science of Management Decision. Harper & Row, NY (1960)
31. Smith, M.J.: Evolution and the Theory of Games. Cambridge University Press, Cambridge (1982)
32. von Neumann, J., Morgenstern, O.: Theory of Games and Economic Behavior. Princeton Univ. Press, Princeton (1944)
33. Wald, A.: Statistical Decision Functions. John Wiley & Sons, Chichester (1950)
34. Wang, Y.: Keynote: On Cognitive Informatics. In: Proc. 1st IEEE International Conference on Cognitive Informatics (ICCI 2002), Calgary, Canada, pp. 34–42. IEEE CS Press, Los Alamitos (2002)
35. Wang, Y.: On Cognitive Informatics. Brain and Mind: A Transdisciplinary Journal of Neuroscience and Neurophilosophy 4(3), 151–167 (2003)
36. Wang, Y.: A Novel Decision Grid Theory for Dynamic Decision Making. In: Proc. 4th IEEE International Conference on Cognitive Informatics (ICCI 2005), Irvin, CA (August 2005)
37. Wang, Y.: Software Engineering Foundations: A Software Science Perspective. CRC Series in Software Engineering, vol. II. Auerbach Publications, NY, USA (2007)
38. Wang, Y.: The Theoretical Framework of Cognitive Informatics. International Journal of Cognitive Informatics and Natural Intelligence 1(1), 1–27 (2007)
39. Wang, Y., Ruhe, G.: The Cognitive Process of Decision Making. International Journal of Cognitive Informatics and Natural Intelligence 1(2), 73–85 (2007)
40. Wang, Y., Wang, Y., Patel, S., Patel, D.: A Layered Reference Model of the Brain (LRMB). IEEE Transactions on Systems, Man, and Cybernetics (C) 36(2), 124–133 (2006)
41. Wilson, R.: Computing Equilibria of n-Person Games. SIAM Journal of Applied Mathematics 21, 80–87 (1971)
42. Zachary, W., Wherry, R., Glenn, F., Hopson, J.: Decision Situations, Decision Processes, and Decision Functions: Towards a Theory-Based Framework for Decision-Aid Design. In: Proc. Conference on Human Factors in Computing Systems (1982)

A Comparative Study of STOPA and RTPA*

Natalia Lopez[1], Manuel Núñez[1], and Fernando L. Pelayo[2]

[1] Dept. Sistemas Informáticos y Programación
Facultad de Informática
Universidad Complutense de Madrid, 28040 Madrid, Spain
{natalia, mn}@sip.ucm.es
[2] Dept. de Sistemas Informáticos
Escuela Politécnica Superior
Universidad Castilla-La Mancha, 02071 Albacete, Spain
fpelayo@dsi.uclm.es

Abstract. During the last years it has been widely recognized that formal semantic frameworks improve the capability to represent cognitive processes. In this line, process algebras have been introduced as formal frameworks to represent this kind of processes. In this paper we compare two process algebras oriented towards the specification of cognitive processes: RTPA (*Real Time Process Algebra*) and STOPA (*Stochastic Process Algebra*). These two formal languages share a common characteristic: Both of them include a notion of time. Thus, when comparing the two languages we will concentrate on the different treatment of time. In order to illustrate how these two languages work we specify a cognitive model of the memorizing process both in RTPA and in STOPA. In order to represent the memory, we follow the classical memory classification (*sensory buffer*, *short-term*, and *long-term* memories) where we also consider the so-called *action buffer* memory.

1 Introduction

Cognitive informatics [34,36,46] is emerging as a new, separate, interdisciplinary branch of research. Quoting from [41], we can describe this field as:

> Cognitive Informatics is a transdisciplinary enquiry of cognitive and information sciences that investigates the internal information processing mechanisms and processes of the brain and natural intelligence, and their engineering applications via an interdisciplinary approach.

Since this quite new research line is in a preliminary phase, we are still in need of having good formalisms to represent the different agents involved in the behaviour of cognitive processes. In order to show that this is a truly interdisciplinary area, researchers in cognitive informatics very often take advantage of developments in other areas. Thus, they often adapt implementation and representation languages, theoretical models, and algorithms as well as consider the best practices obtained in these areas.

* Research partially supported by the Spanish MCYT project WEST TIN2006-15578-C02.

M.L. Gavrilova et al. (Eds.): Trans. on Comput. Sci. II, LNCS 5150, pp. 224–245, 2008.

In this line, *descriptive mathematics for cognitive informatics* introduced as a representation language an adaption of classical process algebras [15,21,2]. It is worth to note that even though process algebras were not originally created to describe cognitive processes, they are a very suitable mechanism to do it since they are very appropriate to define systems where concurrency plays a fundamental role. In fact, variants of process algebras have been already used several times to represent *human* processes (see, for example, [26,19,37,44,27,29,18]).

The first *cognitive process algebra* is RTPA [35,38,41,40,42]. By conveniently putting together the ideas underlying the definition of classical process algebras, RTPA is a new mathematical framework to represent cognitive processes and systems. Following RTPA, the process algebra STOPA [17] represents an advance since the notion of time is further developed to consider the representation of *stochastic time*.

In order to understand why we consider stochastic time, it is worth to briefly review the main milestones in the development of process algebras (see [3] for a good overview of the current research topics in the field). Process algebras are very suitable to formally specify systems where concurrency is an essential key. This is so because they allow to model these systems in a compositional manner. The first work on process algebras settled an important theoretical background for the study of concurrent and distributed systems. In fact, this work was very significant, mainly to shed light on concepts and to open research methodologies. However, due to the abstraction of the complicated features, models were still far from real systems. Therefore, some of the solutions were not specific enough, for instance, those related to real time systems.

Thus, researchers in process algebras have tried to bridge the gap between formal models described by process algebras and real systems. In particular, features which were abstracted before have been introduced in the models. Thus, they allow the design of systems where not only functional requirements but also performance ones are included. The most significant of these additions are related to notions such as time (e.g. [31,22,47,10,1]) and probabilities (e.g. [11,25,24,8,7,23]). An attempt to integrate time and probabilistic information has been given by introducing *stochastic process algebras* (e.g. [12,14,4,28,16,20]). The idea underlying the definition of stochastic time is that the actual amount of time is given by taking into account different probabilities. Thus, we will be able to specify not only *fix* delays (e.g. the process will last 2 milliseconds) but delays that can vary according to a probability distribution function. For example, we may specify a process finishing before 1 millisecond with probability $\frac{1}{3}$, finishing before 2 milliseconds with probability $\frac{1}{2}$, finishing before 4 milliseconds with probability $\frac{4}{5}$, and so on.

Most stochastic process algebras work exclusively with exponential distributions (some exceptions are [5,9,13,6,20]). The problem is that the combination of parallel/concurrency composition operators and general distributions strongly complicates the definition of semantic models. That is why stochastic process algebras are usually based on (semi)-Markov chains. However, this assumption decreases the expressiveness of the language, because it does not allow to properly express some behaviours where timed distributions are not exponential.

The main purpose of this paper is to compare the two cognitive process algebras RTPA and STOPA. As we mentioned before, RTPA [35,38] represents a step forward for formalization of cognitive processes including time information. Our STOchastic Process Algebra builds, on the one hand, on RTPA and, on the other hand, on our work on stochastic and probabilistic process algebras [25,7,23,20]. Thus, STOPA represents a new stochastic process algebra to formally represent cognitive processes containing stochastic information. For example, a process such as $\xi; P$ is delayed an amount of time t with probability $F_\xi(t)$, where F_ξ is the probability distribution function associated with ξ. Let us remark that deterministic delays (as presented in timed process algebras) can be specified by using Dirac distributions. The main improvement of our framework with respect to RTPA is that we may specify timed information given by stochastic delays. As we will see along the paper, the inclusion of stochastic time introduces some additional complexity in the definition of the operational semantics of the language.

In order to assess the usefulness of the two languages and compare how they represent the same process, we formally represent a high level description of the cognitive process of memorizing. We follow contemporary theories of memory classification already stated in [33,32]. We consider *sensory buffer*, *short-term*, and *long-term* memories. Moreover, borrowing from [45], we also consider the so-called *action buffer* memory.

2 The Language RTPA

RTPA is a formal method for specifying software system architectures and also, static and dynamic behaviours. An uploaded version of RTPA, can be found in [43], this version provides 17 meta-processes, 17 process relations and 17 primitive types, on the other hand, the one referred in this paper is almost as complete as the most recent but it is completely worth for our purposes. Quoting from [35]:

Definition 1. *behaviour of a software system is outcomes and effects of computational operations that affect or change the state of a system in a space of input/output events and variables, as well as internal variables and related memory structures.*

Behaviours of software systems can be classified as static and dynamic ones as described below.

Definition 2. *The static behaviour of a software system is a software behaviour that can be determined at design and compile time. The dynamic behaviour of a software system is a software behaviour that can be determined at run-time.*

In software engineering, basic requirements for describing and specifying a software system can be considered in two categories: *architectural* components and *operational* components. System specifications can be described as:

- system architecture,
- system static behaviours,
- system dynamic behaviours.

They can be described by a set of real-time processes in RTPA.

Definition 3. *A process is a basic unit of software system behaviours that represents a transition procedure of a system from one state to another by changing its sets of inputs, outputs, and/or internal variables.*

Table 1. RTPA meta-processes (1/2)

No.	Meta-process	Syntax	Operational Semantics
1.1	System	$\S(SysID_S)$	Represents a system, $SysID$, identified by a string, S
1.2	Assignment	$y_{Type} := x_{Type}$	if $x.type = y.type$ then $x.value \Rightarrow y.value$ else !(@AssigmentTypeError$_S$) where $Type \in$ Meta − Types
1.3	Addressing	$ptrP\hat{} := x_{Type}$	if $prt.type = x.type$ then $x.value \Rightarrow ptr.value$ else !(@AssigmentTypeError$_S$) where $Type \in \{H, Z, P\hat{}\}$
1.4	Input	$\text{Port}(ptrP\hat{})_{Type}\vert > x_{Type}$	if $\text{Port}(ptrP\hat{}).type = x.type$ then $\text{Port}(ptrP\hat{}).value \Rightarrow x.value$ else !(@InputTypeError$_S$) where $Type \in \{B, H\}$, $P\hat{} \in \{H, N, Z\}$
1.5	Output	$x_{Type}\vert < \text{Port}(ptrP\hat{})_{Type}$	if $\text{Port}(ptrP\hat{}).type = x.type$ then $x.value \Rightarrow \text{Port}(ptrP\hat{}).value$ else !(@OutputTypeError$_S$) where $Type \in \{B, H\}$, $P\hat{} \in \{H, N, Z\}$
1.6	Read	$\text{Mem}(ptrP\hat{})_{Type} > x_{Type}$	if $\text{Mem}(ptrP\hat{}).type = x.type$ then $\text{Mem}(ptrP\hat{}).value \Rightarrow x.value$ else !(@ReadTypeError$_S$) where $Type \in \{B, H\}$, $P\hat{} \in \{H, N, Z\}$
1.7	Write	$x_{Type} < \text{Mem}(ptrP\hat{})_{Type}$	if $\text{Mem}(ptrP\hat{}).type = x.type$ then $x.value \Rightarrow \text{Mem}(ptrP\hat{}).value$ else !(@WriteTypeError$_S$) where $Type \in \{B, H\}$, $P\hat{} \in \{H, N, Z\}$
1.8	Timing	a) $@t_{\mathtt{hh:mm:ss:ms}} := \S t_{\mathtt{hh:mm:ss:ms}}$ b) $@t_{\mathtt{yy:MM:dd}} := \S t_{\mathtt{yy:MM:dd}}$ c) $@t_{\mathtt{yy:MM:dd:hh:mm:ss:ms}} := \S t_{\mathtt{yy:MM:dd:hh:mm:ss:ms}}$	if $@t.type = \S t.type$ then $\S t.value \Rightarrow @t.value$ else !(@TimingTypeError$_S$) where $\mathtt{yy} \in \{0, \ldots, 99\}$, $\mathtt{MM} \in \{1, \ldots, 12\}$, $\mathtt{dd} \in \{1, \ldots, 31\}$, $\mathtt{hh} \in \{0, \ldots, 23\}$, $\mathtt{mm}, \mathtt{ss} \in \{0, \ldots, 59\}$, $\mathtt{ms} \in \{0, \ldots, 999\}$
1.9	Duration	$@tn_Z := \S tn_Z + \Delta n_Z$	if $@tn.type = \Delta n.type = \S tn.type = Z$ then $(\S tn.value + \Delta n.value)$ mod MaxValue $\Rightarrow @tn.value$ where MaxValue is the upper bound of the system relative-clock, and the unit of all values is ms
1.10	Memory allocation	AllocateObject (ObjectID$_S$, NofElements$_N$, ElementType$_{RT}$)	$n_N :=$ NofElements; $R^n_{i=1}$(new ObjectID(i_N) : ElementType$_{RT}$); Ⓢ ObjectID.Existed$_{BL}$:= **true**

Table 2. RTPA meta-processes (2/2)

No.	Meta-process	Syntax	Operational Semantics
1.11	Memory release	ReleaseObject(ObjectID$_S$)	delete ObjectID$_S$// System.Garbage Collection( ) ; ObjectID$_S$:= null ; Ⓢ ObjectID.Release$_{BL}$:= `true`
1.12	Increase	$\uparrow (n_{Type})$	if $n.value <$ MaxValue then $n.value + 1 \Rightarrow n.value$ else !(@ValueOutofRange$_S$) where $Type \in \{N, Z, B, H, P\hat{}\}$ MaxValue = min{run-time defined upper bound, nature upper bound of $Type$}
1.13	Decrease	$\downarrow (n_{Type})$	if $n.value > 0$ then $n.value - 1 \Rightarrow n.value$ else !(@ValueOutofRange$_S$) where $Type \in \{N, Z, B, H, P\hat{}\}$
1.14	Exception detection	!(@e_S)	$\uparrow$ (ExceptionLogPtr$_{P\hat{}}$); @$e_S \Rightarrow$ Mem(ExceptionLogPtr$_{P\hat{}}$)$_S$
1.15	Skip	$\emptyset$	Exit a current control structure, such as loop, branch, or switch
1.16	Stop	stop	System stop

A process can be defined as a single *meta-process* or as a complex process based on meta-processes using *process relations*. Thus, RTPA is described by using the following structure:

RTPA ::= Meta-processes
| Primary types
| Abstract data types
| Process relations
| System architectures
| Specification refinement sequences

RTPA distinguishes the concepts of meta-processes from complex processes and complex relations. A meta-process is an elementary process that serves as a basic building block in a software system. Complex processes can be derived from meta-processes according to given process combinatory rules. The syntax and operational semantics of the meta-processes are given in Tables 1 and 2. Each meta-process is a basic operation on one or more operands such as variables, memory elements, or I/O ports. Structures of the operands and their allowable operations are constrained by their types.

The RTPA notation is strongly typed. Every operand in RTPA is assigned with a data type labeled as a suffix. The definition of the primary data types are the meta-types defined from 2.1 to 2.10 in Table 3. The meta-types date/time (2.11) are specially types for continuous real-time systems, where long-range timing manipulation is needed. The

Table 3. RTPA meta-types

No.	Meta-type	Syntax
2.1	Natural number	N
2.2	Integer	Z
2.3	Real	R
2.4	String	S
2.5	Boolean	$BL = \{\texttt{true}, \texttt{false}\}$
2.6	Byte	B
2.7	Hexadecimal	H
2.8	Pointer	$P\hat{}$
2.9	Time	hh : mm : ss : ms where hh $\in \{0, \ldots, 23\}$, mm, ss $\in \{0, \ldots, 59\}$, ms $\in \{0, \ldots, 999\}$
2.10	Date	yy : MM : dd where yy $\in \{0, \ldots, 99\}$, MM $\in \{1, \ldots, 12\}$, dd $\in \{1, \ldots, 31\}$
2.11	Date/Time	yyyy : MM : dd : hh : mm : ss : ms where yyyy $\in \{0, \ldots, 9999\}$, MM $\in \{1, \ldots, 12\}$, dd $\in \{1, \ldots, 31\}$, hh $\in \{0, \ldots, 23\}$, mm, ss $\in \{0, \ldots, 59\}$, ms $\in \{0, \ldots, 999\}$
2.12	Run-time determinable type	RT
2.13	System architectural type	ST
2.14	Event	$@e_S$
2.15	Status	Ⓢs_{BL}

runtime determinable (2.12) is a subset of all the rest meta-types defined, which is designed to support flexible type specification that is unknown at compile-time, but will be instantiated at run-time. The system architectural components (2.13) is a novel and important data type in RTPA that models system architectural components. The event and status types are used to model systems event variables (2.14) as a string type, and system status variables (2.15) as a Boolean type.

In addition to the meta-types for system modeling, a set of typical and frequently used combinational data objects in system architectural modeling, the abstract data types are selected and predefined in RTPA. They are described in Table 4. The interested reader may find in [35] a detailed explanation of all the definitions appearing in Tables 1-4 as well as the complete definitions of *system architectures* and *specification refinement sequences*. Quoting from [35]:

> There are four types of system meta-architectures known as: *parallel, serial, pipeline*, and *nested*. Any complicated system architecture can be represented by a combination of these four meta-architectures between components. It is interesting to find that each of the meta-architectures corresponds to a key RTPA process relation as defined in Table 5.

The combination rules of meta-processes in RTPA are governed by a set of algebraic process relations. In Table 5 the syntax and operational semantics of the process relations are described. The process relations sequential, branch and iteration (4.1 – 4.6)

Table 4. RTPA abstract data types

No.	ADT	Syntax	Designed behaviours
3.1	Stack	Stack:ST	Stack. (Create, Push, Pop, Clear, EmptyTest, FullTest, Release)
3.2	Record	Record:ST	Record. (Create, fieldUpdate, Update, FieldRetrieve, Retrieve, Release)
3.3	Array	Array:ST	Array. (Create, Enqueue, Serve, Clear, EmptyTest, FullTest, Release)
3.4	Queue (FIFO)	Queue:ST	Queue.(Create, Enqueue, Serve, Clear, EmptyTest, FullTest, Release)
3.5	Sequence	Sequence:ST	Sequence.(Create, Retrieve, Append, Clear, EmptyTest, FullTest, Release)
3.6	List	List:ST	List. (Create, FindNext, FindPrior, Findith, FindKey, Retrieve, Update, InsetAfter, InsertBefore, Delete, CurrentPos, FullTest, EmptyTest,SizeTest,Clear,Release)
3.7	Set	Set:ST	Set. (Create, Assign, In, Intersection, Union, Difference, Equal, Subset, Release)
3.8	File (Sequential)	SeqFile:ST	SeqFile. (Create, Reset, Read, Append, Clear, EndTest, Release)
3.9	File (Random)	RandFile:ST	RandFile. (Create, Reset, Read, Write, Clear, EndTest, Release)
3.10	Binary Tree	BTree:ST	BTree. (Create, TRaverse, Insert, DeleteSub, Update, Retrieve, Find, Characteristics, EmptyTest, Clear, Release)

are identified as the basic control structures of software architectures. To represent the modern programming structural concepts seven additional process relations are included: function call, recursion, parallel, concurrency, interleave, pipeline, and jump (4.7 – 4.12, 4.16). The language RTPA extends the process relations with time-driven dispatch, event-driven dispatch, and interrupt (4.13 – 4.15).

3 The Language STOPA

In this section it is presented the language STOPA. The semantic model is strongly based on RTPA. The modifications are given mainly in the presentation of the operational semantics for process relations and the introduction of the stochastic time. In the forthcoming subsection a detailed explanation will be given about all these modifications. In this language, we have included, with respect to RTPA, three choice operators. These operators represent the external, internal and stochastic choices. Besides, some modifications are needed in the description of the parallel operator due to we have to take into account the passing of time. Finally, two process relations of the language RTPA have been omitted, the function call and the jump operator.

Table 5. RTPA process relations

No.	Process relation	Syntax	Operational semantics
4.1	Sequence	$P \to Q$	$P\,;Q$
4.2	Branch	$(?expBL = \texttt{true})\,;P$ $\mid (?expBL = \texttt{false})\,;Q$	if $expBL = \texttt{true}$ then P else Q
4.3	Switch	$?expNUM =$ $0 \to P_0$ $1 \to P_1$ $\ldots$ $n-1 \to P_{n-1}$ $else \to \emptyset$	case $expNUM =$ $0 : P_0$ $1 : P_1$ $\ldots$ $n-1 : P_{n-1}$ $else : exit$
4.4	For-do	$\mathcal{R}_{i=1}^{n} P(i)$	for $i := 1$ to n do $P(i)$
4.5	Repeat	$\mathcal{R}_{\geq 1}^{expBL \neq \texttt{true}} P$	repeat P until $expBL \neq \texttt{true}$
4.6	While-do	$\mathcal{R}_{\geq 0}^{expBL \neq \texttt{true}} P$	while $expBL = \texttt{true}$ do P
4.7	Function call	$P \looparrowright F$	$P' \to F \to P''$ where $P = P' \cup P''$
4.8	Recursion	$\mathrm{P} \circlearrowleft \mathrm{P}$	$P' \to P \to P''$ where $P = P' \cup P''$
4.9	Parallel	$P \parallel Q$	MPSC (multi-processor single clock) internal parallel
4.10	Concurrence	$P \oint Q$	MPMC (multi-processor multi-clock) external parallel
4.11	Interleave	$P \mid\mid\mid Q$	SPSC (single processor single clock) internal virtual parallel
4.12	Pipeline	$P \gg Q$	$P \to Q$ and $\{P_{outputs}\} \Rightarrow \{Q_{inputs}\}$
4.13	Time-driven dispatch	$@ti_{\mathtt{hh:mm:ss:ms}}$ $\looparrowright P_i,\ i \in \{1, \ldots, n\}$	$@t1_{\mathtt{hh:mm:ss:ms}} \looparrowright P_1$ $\mid @t2_{\mathtt{hh:mm:ss:ms}} \looparrowright P_2$ $\ldots$ $\mid @tn_{\mathtt{hh:mm:ss:ms}} \looparrowright P_n$
4.14	Event-driven dispatch	$@ei_S \looparrowright P_i,\ i \in \{1, \ldots, n\}$	$@e1_S \looparrowright P_1$ $\mid @e2_S \looparrowright P_2$ $\mid \ldots$ $\mid @en_S \looparrowright P_n$
4.15	Interrupt	$P \parallel \odot(@e_S \nearrow Q \searrow \odot)$	$P \parallel$ system interrupt capture; if $@e_S$ captured = true then (record interrupt point $\odot$ and variables $\looparrowright Q$ $\to$ recover interrupted variables $\to$ return to the interrupt point $\odot$ and continue P)
4.16	Jump	$P \to Q$	$P \to$ goto $Q \to Q$

As we have already commented in the introduction of this paper, the aim of this language is to add stochastic information in the framework of RTPA. For that reason random variables are included in the syntax of the process relations, that is, a process can be delayed according to a random variable. It is supposed that the sample space (that is, the domain of random variables) is the set of real numbers $\mathbf{R}$ and that random variables take only positive values, that is, given a random variable ξ we have $\mathrm{P}(\xi \leq t) = 0$ for any $t \leq 0$. The reason for this restriction is that random variables are always associated with time distributions.

Definition 4. Let ξ be a random variable. Its *probability distribution function*, denoted by F_ξ, is defined as the function $F_\xi : \mathbf{R} \longrightarrow (0, 1]$ such that $F_\xi(x) = \mathrm{P}(\xi \leq x)$, where $\mathrm{P}(\xi \leq x)$ is the probability that ξ assumes values less than or equal to x. □

In the forthcoming examples as well as in the last section of this paper we will use the following probability distribution functions.

Uniform Distributions. Let $a, b \in \mathbf{R}^+$ such that $a < b$. A random variable ξ follows a uniform distribution in the interval $[a, b]$, denoted by $U(a, b)$, if its associated probability distribution function is:

$$F_\xi(x) = \begin{cases} 0 & \text{if } x \leq a \\ \frac{x-a}{b-a} & \text{if } a < x < b \\ 1 & \text{if } x \geq b \end{cases}$$

These distributions allow us to keep compatibility with time intervals in timed process algebras in the sense that the same *weight* is assigned to all the times in the interval.

Discrete Distributions. Let $P_I = \{(t_i, p_i)\}_{i \in I}$ be a set of pairs such that for any $i \in I$ we have that $t_i \geq 0$, $p_i > 0$, for any $i, j \in I$, if $i \neq j$ then $t_i \neq t_j$, and $\sum p_i = 1$. A random variable ξ follows a discrete distribution with respect to P_I, denoted by $D(P_I)$, if its associated probability distribution function is:

$$F_\xi(x) = \sum_{i \in I} \{\!| p_i \mid x \geq t_i |\!\}$$

Let us note that $\{\!|$ and $|\!\}$ represent multisets. Discrete distributions are important because they allow us to express *passive* actions, that is, actions that are willing, from a certain point of time, to be executed with probability 1.

Exponential Distributions. Let $0 < \lambda \in \mathbf{R}$. A random variable ξ follows an exponential distribution with parameter λ, denoted by $E(\lambda)$ or simply λ, if its associated probability distribution function is:

$$F_\xi(x) = \begin{cases} 1 - e^{-\lambda x} & \text{if } x \geq 0 \\ 0 & \text{if } x < 0 \end{cases}$$

Poisson Distributions. Let $0 < \lambda \in \mathbf{R}$. A random variable ξ follows a Poisson distribution with parameter λ, denoted by $P(\lambda)$, if it takes positives values only in $\mathbf{N}$ and its associated probability distribution function is:

$$F_\xi(x) = \begin{cases} \sum_{t \in \mathbf{N}, t \leq x} \frac{\lambda^t}{t!} e^{-\lambda t} & \text{if } x \geq 0 \\ 0 & \text{if } x < 0 \end{cases}$$

During the rest of the paper, mainly in the examples and when no confusion arises, we will identify random variables with their probability distribution functions.

Example 1. Let us consider the process $U(1,3)\ ;\ P$. This process will behave as P in less than 1 milliseconds with probability 0. In a time less than or equal to t milliseconds, when $1 < t < 3$, with probability $\frac{t-1}{2}$. And it will behave as P in more than 3 milliseconds with probability 1. □

Regarding *communication* actions, they can be divided into *input* and *output actions*. Next, we define our alphabet of actions.

Definition 5. We consider a set of *communication* actions $Act = \texttt{Input} \cup \texttt{Output}$, where we assume $\texttt{Input} \cap \texttt{Output} = \emptyset$. We suppose that there exists a bijection $f : \texttt{Input} \longleftrightarrow \texttt{Output}$. For any *input* action $a? \in \texttt{Input}$, $f(a?)$ is denoted by the *output* action $a! \in \texttt{Output}$. If there exists a message transmission between $a?$ and $a!$ we say that a is the *channel* of the communication ($a, b, c, \cdots$ to range over Act).

We also consider a special action $\tau \notin Act$ that represents internal behaviour of the processes. We denote by Act_τ the set $Act \cup \{\tau\}$ ($\alpha, \beta, \gamma, \cdots$ to range over Act_τ).

Besides, We consider a denumerable set Id of process identifiers. In addition, we denote by V the set of random variables ($\xi, \xi', \psi, \cdots$ to range over V). □

3.1 Process Relations in STOPA

In this section, we define the syntax and operational semantics of our process algebra to describe process relations. Operational semantics is probably the simplest and more intuitive way to give semantics to any process language. In this part of the description of the language, operational behaviours will be defined by means of transitions $P \xrightarrow{\omega} P'$ that each process can execute. These are obtained in a structured way by applying a set of inference rules [30].

The intuitive meaning of a transition as $P \xrightarrow{\omega} P'$ is that the process P may perform the action ω and, once this action is performed, then it behaves as P'. Let us note that labels appearing in these operational transitions have the following types: $a? \in \texttt{Input}$, $a! \in \texttt{Output}$, $a \in Act$, $\alpha \in Act \cup \{\tau\}$, $\omega \in Act_\tau \cup V$, and $\xi \in V$.

The set of process relations, as well as their operational semantics, is given in Tables 6 and 7. Next, we intuitively describe each of the operators.

The **external, internal**, and **stochastic** choice process relations are used to describe the choice among different actions. They are respectively denoted by

$$\sum a_i\ ;\ P_i \qquad \textstyle\sum\!\!\!\!\!\bigcirc\ \tau\ ;\ P_i \qquad \textstyle\sum\!\!\!\!\!\bullet\ \xi_i\ ;\ P_i$$

where $a_i \in Act$ and $\xi_i \in V$. The inference rules describing the behaviour of these process relations are (CHO1), (CHO2), and (CHO3).

The process $\sum a_i\ ;\ P_i$ will perform one of the actions a_i and after that it behaves as P_i. The term $\textstyle\sum\!\!\!\!\!\bigcirc\ \tau\ ;\ P_i$ represents the internal choice among the processes P_i. Once the choice is made, by performing an internal action τ, the process behaves as the chosen process. Finally, the process $\textstyle\sum\!\!\!\!\!\bullet\ \xi_i\ ;\ P_i$ will be delayed by ξ_i a certain amount of time t with probability $p = \mathrm{P}(\xi_i \leq t)$ and after that it will behave as P_i (see Example 1).

Sequence is a process relation in which two processes are consequently performed. This relation is denoted by:

$$P\ ;\ Q$$

The rules (SEQ1), (SEQ2), (SEQ3), and (SEQ4) describe the behaviour of these process relations. Intuitively, P is initially performed. Once P finishes, Q starts its performance. If the process P can perform an action then $P \,;\, Q$ will perform it. If the process P finishes then the process $P \,;\, Q$ will behave as Q. A process will finish its execution if it performs the action $\surd$ (see rule (SEQ4)).

The **branch** and **switch** process relations are denoted by

$$(?\mathrm{expBL} = \mathtt{true}) \,;\, P \,|\, (?\mathrm{expBL} = \mathtt{false}) \,;\, Q$$

and

$$|\,(?\mathrm{expNUM} = i) \,;\, P_i$$

The behaviour of the first operator is described in the rule (BRA). If the boolean expression expBL evaluates to `true` then P is performed; otherwise, Q is performed. In the second operator, if the value of the numerical expression expBL is equal to i then the process P_i is performed (see rule (SWI)).

The **FOR-DO, REPEAT**, and **WHILE-DO** process relations are denoted, respectively, by:

$$\overset{\mathtt{n}}{\underset{i=1}{R}} P \qquad \overset{\mathtt{expBL} \neq \mathtt{true}}{\underset{\geq 1}{R}} P \qquad \overset{\mathtt{expBL} \neq \mathtt{true}}{\underset{\geq 0}{R}} P$$

and their operational semantics are given in the rules (FOR), (REP), (WHI1) and (WHI2).

The *FOR-DO* process relation may be used to describe the performance of a certain process a fix amount of times n. Thus, $R^{n}_{i=1} P$ behaves as P and after that as $R^{n-1}_{i=1} P$.

The *REPEAT* process relation executes a process P at least once. It continues its execution until a certain expression takes the value `true`. Thus, the term $R^{\mathrm{expBL} \neq \mathtt{true}}_{\geq 1} P$ behaves as P, and after that, as a *WHILE-DO* process relation.

The *WHILE-DO* process relation $R^{\mathrm{expBL} \neq \mathtt{true}}_{\geq 0} P$ behaves as P if the boolean expression expBL is `true`, and after that behaves again as $R^{\mathrm{expBL} \neq \mathtt{true}}_{\geq 0} P$. If the boolean expression expBL is `false` then the process finishes its performance.

Recursion is a process relation in which the definition of a process may contain a call to itself. We will use a notation slightly different to that in RTPA

$$X := P$$

where $X \in Id$, that is, a process identifier. The rule (REC1) applies to external and internal actions while (REC2) applies to stochastic actions. Let us also remark that $P[X/X := P]$ represents the substitution of all the free occurrences of X in P by $X := P$.

The **parallel, concurrence**, and **interleaving** process relations are denoted, respectively, by:

$$P \parallel_{tc} Q \qquad P \oint Q \qquad P \,|||\, Q$$

Parallel is a process relation in which two processes are executed simultaneously, synchronized by a common system clock. The parallel process relation is designed to model behaviours of a multi-processor single-clock system. The parameter tc indicates the time of the system clock. It will vary as time passes. If one of the processes of the parallel composition can perform a non-stochastic action then the composition will perform it (see rules (PAR1) and (PAR2)). We suppose that there is an operation $*$ on the set of actions Act such that $(Act, *)$ is a monoid and τ is its identity element. Thus, by rule (PAR3), if we have the parallel composition of P and Q, P may perform a, Q may perform b, and $a * b \neq \tau$ then they will evolve together. Let us suppose that P can perform a stochastic action, and neither external nor internal actions can be performed by the composition. Then $P \|_{tc} Q$ will perform the temporal action and Q will evolve into $\mathrm{cond}(A, \Delta t)$, being Δt the actual time consumed by the stochastic action. That is, $\Delta t = tc' - tc$, where tc is the system time in the moment that the performance of ξ started and tc' is the system time after the execution of ξ.

The definition of the function $\mathrm{cond}(P, \Delta t)$ is given in Table 8. In that table we have included, for the sake of completeness, all the cases of the function. However, the most relevant part of this definition is the one concerning choice process relations. In this case,

$$\mathrm{cond}(P, \Delta t) = \begin{cases} \sum a_i \,;\, \mathrm{cond}(P_i, \Delta t) & \text{if } P = \sum a_i \,;\, P_i \\ \sum \tau \,;\, \mathrm{cond}(P_i, \Delta t) & \text{if } P = \sum \tau \,;\, P_i \\ \sum \xi_i' \,;\, P_i & \text{if } P = \sum \xi_i \,;\, P_i \end{cases}$$

where ξ_i' is the random variable whose probability distribution function is given as a *conditional probability*. We have that $\mathrm{P}(\xi_i' \leq t') = \mathrm{P}(\xi_i \leq t'' \mid \Delta t + t' \leq t'')$, that is, the probability of the random variable ξ_i' to finish before t' units of time have passed is equal to the probability of the original random variable ξ_i to be less than or equal to t'', provided that $\Delta t + t' \geq t''$. Let us remark that the modification of the process of the parallel composition that does not perform the stochastic action is needed. This is so because, in this case, we have a multi processor single-clock system, so the time consumed by the stochastic action has to be taken into account in both sides of the parallel composition.

Example 2. Let $P = \xi_1 \,;\, P_1 \|_{tc} \xi_2 \,;\, P_2$. Let us suppose that we have transitions of the form

$$P \xrightarrow{\xi_1} P_1 \|_{tc'} \xi_2' \,;\, P_2$$

The passage of time would not be reflected in the right hand side of the parallel composition. Nevertheless, the random variable associated with the action b cannot be ξ_2 because some time has passed. That is why we generate the transition $P \xrightarrow{\xi_1} P_1 \|_{tc'} \mathrm{cond}(\xi_2 \,;\, P_2, tc' - tc)$, where $\mathrm{cond}(\xi_2 \,;\, P_2, tc' - tc) = \xi_2' \,;\, P_2$ and ξ_2' is the random variable with probability distribution function defined as:

$$F_{\xi_2'}(t) = \mathrm{P}(\xi_2' \leq t) = \mathrm{P}(\xi_2 \leq t' \mid t + (tc' - tc) \leq t')$$

□

Concurrence is a process relation in which two processes are simultaneously and asynchronously executed, according to separate system clocks. This process relation is designed to model behaviours of a multi-processor multi-clock system. The rules (CON1) and (CON2) indicate that if one of the processes of the composition can perform an action then the composition will asynchronously perform it. However, if one of the processes of the composition can perform an input action and the other can perform the corresponding output action then there is a communication and the process relation will perform it (see rules (CON3) and (CON4)). Since we have in this case a multi-clock system, if one of the processes can perform a stochastic action then the composition will perform it without modifying the other part of the composition (see rules (CON5) and (CON6)).

Interleaving is a process relation in which two processes are simultaneously executed while maintaining by a common system clock. The interleaving process relation is designed to model behaviours of a single-processor single-clock system. If one of the processes can perform a non-stochastic action then the composition will perform it (see rules (INT1) and (INT2)). Regarding stochastic actions, see rules (INT3) and (INT4), we consider that a stochastic action , once it has started, cannot be interrupted. Besides, if one of the processes performs a stochastic action then the other component is not modified. In fact, as we suppose a single processor single-clock system, the other side of the composition is not active, so no time has passed for it.

Pipeline is a process relation in which two processes are interconnected to each other. The second process takes the inputs from the outputs of the first one. This process relation is denoted by:

$$P >> Q$$

Thus, if P can perform an action then $P >> Q$ will perform it (see rules (PIPE1) and (PIPE3)). Once the process P is finished, that is, P can perform the action $\surd$, $P >> Q$ behaves as Q (see rule (PIPE2)).

Time-driven dispatch is a process relation in which the i-th process is triggered by a predefined system time $ti_{\mathtt{hh:mm:ss:ms}}$. It is denoted as follows:

$$@ti_{\mathtt{hh:mm:ss:ms}} \looparrowright P_i, \; i \in \{1, \ldots, n\}$$

According to the rule (TDD), the process P_i is performed when the value of the system time is equal to $ti_{\mathtt{hh:mm:ss:ms}}$.

Event-driven dispatch is a process relation in which the i-th process is triggered by a system event $@ei_S$. It is denoted as follows:

$$@ei_S \looparrowright P_i, \; i \in \{1, \ldots, n\}$$

When the event occurred is ei_S, then the process P_i is performed, rule (EDD).

Interrupt is a process relation in which a process is temporarily held by another with higher priority. The term describing that the process P is interrupted by the process Q when the event $@e_S$ is captured at interrupt point $\odot$ will be represented by

$$P \parallel \odot(@e_S \nearrow Q \searrow \odot)$$

If the event e_S is captured (rule (INTER1)), the process $P \parallel \odot(@e_S \nearrow Q \searrow \odot)$ will behave as Q, and after that the process will continue its execution behaving as P. If the event e_S is not captured and P can perform an action (rules (INTER2) and (INTER3)), the process $P \parallel \odot(@e_S \nearrow Q \searrow \odot)$ will also perform it.

The semantics of both RTPA and STOPA have been described and form this theoretical level, the differences between them are mainly those allowing STOPA to capture more generic notions of time (whatever time provide that we knew its probability distribution function) than RTPA does, and this fact is reflected in the existence of three choice relations in STOPA, one of them a stochastic choice process relation which introduces the ability of capturing stochastic time, and as a natural consequence the semantics of the parallel operator has been modified for dealing with the definition of this time and its soundness. The parallel operator semantics needs the function *cond* to properly capture its semantics.

We want to analyze the empirical differences between this two process algebras, in particular when describing/specifying the same case study, in order to perform this comparison, the cognitive memorizing process is now described.

4 The Memorizing Process

In this section we give a brief explanation on how the memorization process works.

The main goal of this description is to answer the following three basic questions:

- How are memories formed? (*encoding*)
- How are memories retained? (*storage*)
- How are memories recalled? (*retrieval*)

4.1 Encoding Process

Encoding is an active process which requires selective attention to the material to be encoded. Memories may then be affected by the amount or type of attention devoted to the task of encoding the material.

There may be different levels of processing, being some of them *deeper* than others. These processes are *structural encoding* (where emphasis is placed on the physical structural characteristics of the stimulus), *phonemic encoding* (with emphasis on the sounds of the words) and *semantic encoding* (that is, emphasis on the meaning).

The main aspects of encoding fit the OAR-model in the following sense:

1. Relation: Association with other information.
2. Object: Visual imagery of the real entity or concept that can be used to add richness to the material to be remembered. Besides, it also adds more sensory modalities.
3. Attributes: To make the material personally relevant and to add more detailed information about the object.

4.2 Storage Process

Over the years, analogies with available technologies have been made to try and explain the behaviour of the memory. Nowadays, memory theories use a computer-based, or information processing, model. The most accepted model states that there are three stages of memory storage and one more for memory retrieving:

Table 6. Operational semantics of the process relations (1/2)

$$(\mathrm{CHO1})\frac{}{\sum a_i;P_i \xrightarrow{a_i} P_i} \qquad (\mathrm{CHO2})\frac{}{\sum \tau;P_i \xrightarrow{\tau} P_i} \qquad (\mathrm{CHO3})\frac{}{\sum \xi_i;P_i \xrightarrow{\xi_i} P_i}$$

$$(\mathrm{SEQ1})\frac{P \xrightarrow{\alpha} P'}{P;Q \xrightarrow{\alpha} P';Q} \qquad (\mathrm{SEQ2})\frac{P \xrightarrow{\checkmark} P'}{P;Q \xrightarrow{\tau} Q}$$

$$(\mathrm{SEQ3})\frac{P \xrightarrow{\xi} P'}{P;Q \xrightarrow{\xi} P';Q} \qquad (\mathrm{SEQ4})\frac{}{\mathrm{exit} \xrightarrow{\checkmark} \mathrm{stop}}$$

$$(\mathrm{BRA})\frac{expBL{=}\mathtt{true}}{(?expBL{=}\mathtt{true});P \mid (?expBL{=}\mathtt{false});Q \xrightarrow{\tau} P} \qquad (\mathrm{SWI})\frac{expNUM{=}i}{\mid (?expNUM{=}i);P_i \xrightarrow{\tau} P_i}$$

$$(\mathrm{FOR})\frac{}{R^n_{i=1} P \xrightarrow{\tau} P;R^{n-1}_{i=1} P} \qquad (\mathrm{REP})\frac{}{R^{expBL\neq\mathtt{true}}_{\geq 1} P \xrightarrow{\tau} P;R^{expBL\neq\mathtt{true}}_{\geq 0} P}$$

$$(\mathrm{WHI1})\frac{expBL{=}\mathtt{true}}{R^{expBL\neq\mathtt{true}}_{\geq 0} P \xrightarrow{\tau} P;R^{expBL\neq\mathtt{true}}_{\geq 0} P} \qquad (\mathrm{WHI2})\frac{expBL{=}\mathtt{false}}{R^{expBL\neq\mathtt{true}}_{\geq 0} P \xrightarrow{\tau} \mathrm{exit}}$$

$$(\mathrm{REC1})\frac{P[X/X{:=}P] \xrightarrow{\alpha} P'}{X{:=}P \xrightarrow{\alpha} P'} \qquad (\mathrm{REC2})\frac{P[X/X{:=}P] \xrightarrow{\xi} P'}{X{:=}P \xrightarrow{\xi} P'}$$

$$(\mathrm{PAR1})\frac{P \xrightarrow{\alpha} P'}{P\|_{tc}Q \xrightarrow{\alpha} P'\|_{tc}Q} \qquad (\mathrm{PAR2})\frac{Q \xrightarrow{\alpha} Q'}{P\|_{tc}Q \xrightarrow{\alpha} P\|_{tc}Q'}$$

$$(\mathrm{PAR3})\frac{P \xrightarrow{a} P',\ Q \xrightarrow{b} Q',\ a{*}b\neq\tau}{P\|_{tc}Q \xrightarrow{a*b} P'\|_{tc}Q'}$$

$$(\mathrm{PAR4})\frac{P \xrightarrow{\xi} P',\ P\|_{tc}Q \not\xrightarrow{\alpha}}{P\|_{tc}Q \xrightarrow{\xi} P'\|_{tc'}\mathrm{cond}(Q,tc'-tc)} \qquad (\mathrm{PAR5})\frac{Q \xrightarrow{\xi} Q',\ P\|_{tc}Q \not\xrightarrow{\alpha}}{P\|_{tc}Q \xrightarrow{\xi} \mathrm{cond}(P,tc'-tc)\|_{tc'}Q'}$$

$$(\mathrm{CON1})\frac{P \xrightarrow{\alpha} P'}{P\oint Q \xrightarrow{\alpha} P'\oint Q} \qquad (\mathrm{CON2})\frac{Q \xrightarrow{\alpha} Q'}{P\oint Q \xrightarrow{\alpha} P\oint Q'}$$

$$(\mathrm{CON3})\frac{P \xrightarrow{a?} P',\ Q \xrightarrow{a!} Q'}{P\oint Q \xrightarrow{\tau} P'\oint Q'} \qquad (\mathrm{CON4})\frac{P \xrightarrow{a!} P',\ Q \xrightarrow{a?} Q'}{P\oint Q \xrightarrow{\tau} P'\oint Q'}$$

$$(\mathrm{CON5})\frac{P \xrightarrow{\xi} P'}{P\oint Q \xrightarrow{\xi} P'\oint Q} \qquad (\mathrm{CON6})\frac{Q \xrightarrow{\xi} Q'}{P\oint Q \xrightarrow{\xi} P\oint Q'}$$

$$(\mathrm{INT1})\frac{P \xrightarrow{\alpha} P'}{P|||Q \xrightarrow{\alpha} P'|||Q} \qquad (\mathrm{INT2})\frac{Q \xrightarrow{\alpha} Q'}{P|||Q \xrightarrow{\alpha} P|||Q'}$$

$$(\mathrm{INT3})\frac{P \xrightarrow{\xi} P'}{P|||Q \xrightarrow{\xi} P'|||Q} \qquad (\mathrm{INT4})\frac{Q \xrightarrow{\xi} Q'}{P|||Q \xrightarrow{\xi} P|||Q'}$$

Table 7. Operational semantics of the process relations (2/2)

$$(\text{PIPE1})\frac{P \xrightarrow{\alpha} P'}{P \gg Q \xrightarrow{\alpha} P';Q} \qquad (\text{PIPE2})\frac{P \xrightarrow{\surd} P',\ \texttt{Input}(Q)=\texttt{Output}(P)}{P \gg Q \xrightarrow{\tau} Q}$$

$$(\text{PIPE3})\frac{P \xrightarrow{\xi} P'}{P \gg Q \xrightarrow{\xi} pP' \gg Q}$$

$$(\text{TDD})\frac{\text{tsystem}_{\texttt{hh:mm:ss:ms}} = ti_{\texttt{hh:mm:ss:ms}}}{@ti_{\texttt{hh:mm:ss:ms}} \hookrightarrow P_i,\ i\in\{1,\ldots,n\} \xrightarrow{\tau} P_i} \qquad (\text{EDD})\frac{\text{esystem}=ei_S}{@ei_S \hookrightarrow P_i,\ i\in\{1,\ldots,n\} \xrightarrow{\tau} P_i}$$

$$(\text{INTER1})\frac{@e_S\text{captured}=\texttt{true}}{P\|_{tc}\odot(@e_S \nearrow Q \searrow \odot) \xrightarrow{\tau} Q;P}$$

$$(\text{INTER2})\frac{@e_S\text{captured}=\texttt{false},\ P \xrightarrow{\alpha} P'}{P\|_{tc}\odot(@e_S \nearrow Q \searrow \odot) \xrightarrow{\alpha} P'\|_{tc}\odot(@e_S \nearrow Q \searrow \odot)}$$

$$(\text{INTER3})\frac{@e_S\text{captured}=\texttt{false},\ P \xrightarrow{\xi} P'}{P\|_{tc}\odot(@e_S \nearrow Q \searrow \odot) \xrightarrow{\xi} P'\|_{tc}\odot(@e_S \nearrow Q \searrow \odot)}$$

- *Sensory store* retains the sensory image for only a small part of a second, just long enough to develop a perception. This is stored in the *Sensory Buffer Memory* (SFM). Following [45], we also consider *Action Buffer Memory* (ABM) which is used as a buffer when recovering information.
- *Short Term Memory* (STM) lasts about 20 to 30 seconds when we do not consider rehearsal of the information. On the contrary, if rehearsal is used then short term memory will last as long as the rehearsal continues. Short term memory is also limited in terms of the number of items it can hold. Its capacity is about 7 items but can be increased by *chunking*, that is, by combining similar material into units.
  Let us remark that short term memory was originally perceived as a simple rehearsal buffer. However, it turns out to have a more complicated underlying process, being better modelled by using an analogy with a computer, which has the ability to store a limited amount of information in its cache RAM while performing other tasks. In other words, we can consider it as a kind of *working memory*.
- *Long Term Memory* (LTM) has been suggested to be *permanent*. However, even though no information is forgotten, we might lose the means of retrieving it.

Another interesting point regarding memory is to determine the mechanism to change the condition of a certain *memory*. In other words, how does short term memory *stuff* get into long term memory? We have to take into account the following:

- *Serial position effect*. Thus, *primacy* (i.e. first words get rehearsed more often and so that they move into long term memory) and *recency* (for instance, words at the

Table 8. Definition of function cond$(P, \Delta t)$

$$\mathrm{cond}(\textstyle\sum a_i \,;\, P_i, \Delta t) = \textstyle\sum a_i \,;\, \mathrm{cond}(P_i, \Delta t) \qquad \mathrm{cond}(\textstyle\sum \tau \,;\, P_i, \Delta t) = \textstyle\sum \tau \,;\, \mathrm{cond}(P_i, \Delta t)$$

$$\mathrm{cond}(\textstyle\sum \xi_i \,;\, P_i, \Delta t) = \textstyle\sum \xi'_i \,;\, P_i \qquad \mathrm{cond}(P \,;\, Q, \Delta t) = \mathrm{cond}(P, \Delta t) \,;\, Q$$

$$\mathrm{cond}(P >> Q, \Delta t) = \mathrm{cond}(P, \Delta t) >> Q$$

$$\mathrm{cond}(\,|\,(?\mathrm{expNUM} = i) \,;\, P_i, \Delta t) = \,|\,(?\mathrm{expNUM} = i) \,;\, \mathrm{cond}(P_i, \Delta t)$$

$$\mathrm{cond}((?\mathrm{expBL} = \mathtt{true}) \,;\, P \,|\, (?\mathrm{expBL} = \mathtt{false}) \,;\, Q, \Delta t) =$$
$$(?\mathrm{expBL} = \mathtt{true}) \,;\, \mathrm{cond}(P, \Delta t) \,|\, (?\mathrm{expBL} = \mathtt{false}) \,;\, \mathrm{cond}(Q, \Delta t)$$

$$\mathrm{cond}\left(\mathop{R}\limits_{i=1}^{\mathtt{n}} P, \Delta t\right) = \begin{cases} \mathrm{cond}(P, \Delta t) \,;\, \mathop{R}\limits_{i=1}^{\mathtt{n}-1} P & \text{if } n \geq 1 \\ \mathop{R}\limits_{i=1}^{\mathtt{n}} P & \text{otherwise} \end{cases}$$

$$\mathrm{cond}\left(\mathop{R}\limits_{\geq 1}^{\mathtt{expBL} \neq \mathtt{true}} P, \Delta t\right) = \mathrm{cond}(P, \Delta t) \,;\, \mathop{R}\limits_{\geq 0}^{\mathtt{expBL} \neq \mathtt{true}} P$$

$$\mathrm{cond}\left(\mathop{R}\limits_{\geq 0}^{\mathtt{expBL} \neq \mathtt{true}} P, \Delta t\right) = \begin{cases} \mathrm{cond}(P, \Delta t);\ \mathop{R}\limits_{\geq 0}^{\mathtt{expBL} \neq \mathtt{true}} P & \text{if } expBL = \mathtt{true} \\ \mathop{R}\limits_{\geq 0}^{\mathtt{expBL} \neq \mathtt{true}} P & \text{otherwise} \end{cases}$$

$$\mathrm{cond}(X := P, \Delta t) = \mathrm{cond}(P[X/X := P], \Delta t)$$
$$\mathrm{cond}(P \parallel_{tc} Q, \Delta t) = \mathrm{cond}(P, \Delta t) \parallel_{tc} \mathrm{cond}(Q, \Delta t)$$
$$\mathrm{cond}(P \oint Q, \Delta t) = \mathrm{cond}(P, \Delta t) \oint \mathrm{cond}(Q, \Delta t)$$
$$\mathrm{cond}(P \,|||\, Q, \Delta t) = \mathrm{cond}(P, \Delta t) \,|||\, \mathrm{cond}(Q, \Delta t)$$

$$\mathrm{cond}(@ti_{\mathtt{hh:mm:ss:ms}} \hookrightarrow P_i,\ i \in \{1, \ldots, n\}, \Delta t) = @ti_{\mathtt{hh:mm:ss:ms}} \hookrightarrow P_i,\ i \in \{1, \ldots, n\}$$
$$\mathrm{cond}(@eis \hookrightarrow P_i,\ i \in \{1, \ldots, n\}, \Delta t) = @eis \hookrightarrow \mathrm{cond}(P_i, \Delta t),\ i \in \{1, \ldots, n\}$$
$$\mathrm{cond}(P \parallel \odot(@es \nearrow Q \searrow \odot), \Delta t) = \mathrm{cond}(P, \Delta t) \parallel \odot(@es \nearrow \mathrm{cond}(Q, \Delta t) \searrow \odot)$$

end that are not rehearsed as often but that are still available in STM) affect long term memory.

- *Rehearsal* helps to move things into long term memory.
- According to the organizational structures of long term memory, we have also to consider:
  - Related items are usually remembered together.
  - Conceptual hierarchies are used as classification scheme to organize memories.
  - Semantic networks are less neatly organized bunches of conceptual hierarchies linked together by associations to other concepts.

- Schemas are clusters of knowledge about an event or object abstracted from prior experience with the object. Actually, we tend to recall objects that fit our conception of the situation better than ones that do not.
- A script is a schema which organizes our knowledge about common things or activities (if you know the script applicable to the event, you can better remember the elements of the event).

The process of storing new information in LTM is called *consolidation*.

4.3 Retrieval Process

Memory retrieval is not a random process. Once a request is generated the appropriate searching and finding processes take place. This process is triggered according to the organization structures of the LTM, while the requested information is provided via the Action Buffer Memory. The figure 1 captures this description.

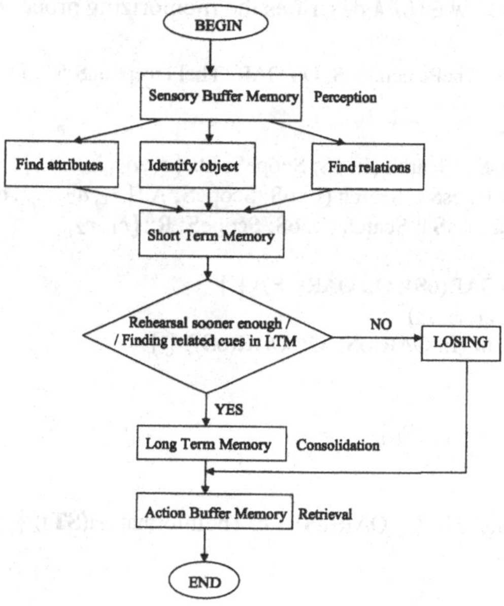

Fig. 1. The model of the memorizing process

5 Formal Description of the Memorizing Process

Taking as starting point the description of the memorization process given in the previous section, we present how RTPA describes in a rigorous way this cognitive process of the brain. It can also be found in [39] another description of the memorizing process, but the former is the one chosen in this comparative study since the original block diagram (Fig.1) has already been used for the specification in both process algebras syntax.

MemorizationProcess (I:: ThePerceptionS; O:: OAR (ThePerceptionS)**ST**, LmemorizationN) ≜
{
oS := ThePerceptionS
→ (ScopeS := ObjectsS ↳ Search (I:: oS; ScopeS; O::$\{o_1, o_2, \ldots, o_n\}$)
|| ScopeS := AttributesS ↳ Search (I:: oS; ScopeS; A::$\{a_1, a_2, \ldots, a_m\}$)
|| ScopeS := RelationsS ↳ Search (I:: oS; ScopeS; R::$\{r_1, r_2, \ldots, r_t\}$)
)
→ EncodingSTM (I:: OAR(oS); O:: OAR(oS)) { }
→ (? (REHEARSAL**RT** ≤ threshold**RT**) ∨ ($@_{Found-Related-Cues}$)
→ EncodingLTM (I:: OAR(oS); O:: OAR(oS)) { }
→ PL1S
| ? ~
→ LOSING (I:: OAR(oS); O::∅)
→ PL1S
)
→ PL1S ↳ DecodingABM (I:: OAR(oS); O:: TheInformation(**ST**)) { }
}

And now we present how STOPA describes the memorizing process.

MemorizationProcess (I:: ThePerceptionS; O:: OAR (ThePerceptionS)**ST**, LmemorizationN) ≜
{
(oS := ThePerceptionS
→ (ScopeS := ObjectsS ↳ Search (I:: oS; ScopeS; O::$\{o_1, o_2, \ldots, o_n\}$)
|| ScopeS := AttributesS ↳ Search (I:: oS; ScopeS; A::$\{a_1, a_2, \ldots, a_m\}$)
|| ScopeS := RelationsS ↳ Search (I:: oS; ScopeS; R::$\{r_1, r_2, \ldots, r_t\}$)
)
→ EncodingSTM (I:: OAR(oS); O:: OAR(oS)) { }
→ (? ($@_{Found-Related-Cues}$)
→ EncodingLTM (I:: OAR(oS); O:: OAR(oS)) { }
→ PL1?S
| ? ~
→ $\sum \xi_{REHEARSAL}$; P_{BL}
))
∮
(→ PL1!S ↳ DecodingABM (I:: OAR(oS); O:: TheInformation(**ST**)) { })
}

where P_{BL} can be:

- $P_1 = EncodingLTM(I :: OAR(o\mathbf{S}); O :: OAR(o\mathbf{S}))\{\}\ ; PL1?\mathbf{S}$
- $P_0 = LOSING(I :: OAR(o\mathbf{S}); O :: \emptyset)$

6 Conclusions

In this paper we have compared two timed process algebras, RTPA and STOPA which have been used for the specification and analysis of cognitive processes. The fundamental differences between them have been pointed out in sections 2 and 3.

As previously mentioned, STOPA is strongly based in RTPA with the only exception of time managing issues. To be more precise, in RTPA the notion of time follows a Uniform distribution, whereas in STOPA there is only one restriction when describing timing aspects, which is to know the probability distribution function of such a time. This is what we call "STOPA can represent *stochastic time*". A minor difference is the appearance of three choice operators in STOPA, one is needed for capturing the time difference stated, and there is one for *internal* and the other for *external choices*. Therefore the reason for choosing one against the other, can just fall on the amount and quality of time information of the events to be modelled/described, and in very rare cases on the possibility of founding pure non-deterministic behaviour (captured by *internal choices* of STOPA).

Section 5 shows that both languages are even closer when only seeing at the particular specification of a system, but the main differences appear when applying the semantics rules of the parallel and stochastic choice operators of STOPA, this fact becomes this formalism rather more complex than RTPA.

Our main line for future work in this theoretical field is to perform a more thorough study of the semantic framework. In particular, it would be very adequate to define both testing and bisimulations semantics over STOPA.

A comparison with some other formal model as timed Petri Nets could be an interesting field to explore.

Acknowledgments

We would like to thank Professor Yingxu Wang for his support for writing this paper. In particular, he suggested the idea of comparing RTPA and STOPA.

References

1. Baeten, J.C.M., Middelburg, C.A.: Process algebra with timing. EATCS Monograph. Springer, Heidelberg (2002)
2. Baeten, J.C.M., Weijland, W.P.: Process Algebra. Cambridge Tracts in Computer Science 18. Cambridge University Press, Cambridge (1990)
3. Bergstra, J.A., Ponse, A., Smolka, S.A. (eds.): Handbook of Process Algebra. North Holland, Amsterdam (2001)
4. Bernardo, M., Gorrieri, R.: A tutorial on EMPA: A theory of concurrent processes with non-determinism, priorities, probabilities and time. Theoretical Computer Science 202(1-2), 1–54 (1998)
5. Bravetti, M., Bernardo, M., Gorrieri, R.: Towards performance evaluation with general distributions in process algebras. In: Sangiorgi, D., de Simone, R. (eds.) CONCUR 1998. LNCS, vol. 1466, pp. 405–422. Springer, Heidelberg (1998)
6. Bravetti, M., Gorrieri, R.: The theory of interactive generalized semi-Markov processes. Theoretical Computer Science 282(1), 5–32 (2002)
7. Cazorla, D., Cuartero, F., Valero, V., Pelayo, F.L., Pardo, J.J.: Algebraic theory of probabilistic and non-deterministic processes. Journal of Logic and Algebraic Programming 55(1–2), 57–103 (2003)
8. Cleaveland, R., Dayar, Z., Smolka, S.A., Yuen, S.: Testing preorders for probabilistic processes. Information and Computation 154(2), 93–148 (1999)

9. D'Argenio, P.R., Katoen, J.-P., Brinksma, E.: An algebraic approach to the specification of stochastic systems. In: Programming Concepts and Methods, pp. 126–147. Chapman & Hall, Boca Raton (1998)
10. Davies, J., Schneider, S.: A brief history of timed CSP. Theoretical Computer Science 138, 243–271 (1995)
11. van Glabbeek, R., Smolka, S.A., Steffen, B.: Reactive, generative and stratified models of probabilistic processes. Information and Computation 121(1), 59–80 (1995)
12. Götz, N., Herzog, U., Rettelbach, M.: Multiprocessor and distributed system design: The integration of functional specification and performance analysis using stochastic process algebras. In: Donatiello, L., Nelson, R. (eds.) SIGMETRICS 1993 and Performance 1993. LNCS, vol. 729, pp. 121–146. Springer, Heidelberg (1993)
13. Harrison, P.G., Strulo, B.: SPADES – a process algebra for discrete event simulation. Journal of Logic Computation 10(1), 3–42 (2000)
14. Hillston, J.: A Compositional Approach to Performance Modelling. Cambridge University Press, Cambridge (1996)
15. Hoare, C.A.R.: Communicating Sequential Processes. Prentice Hall, Englewood Cliffs (1985)
16. López, N., Núñez, M.: A testing theory for generally distributed stochastic processes. In: Larsen, K.G., Nielsen, M. (eds.) CONCUR 2001. LNCS, vol. 2154, pp. 321–335. Springer, Heidelberg (2001)
17. López, N., Núñez, M., Pelayo, F.L.: STOPA: A STOchastic Process Algebra for the formal representation of cognitive systems. In: 3rd IEEE Int. Conf. on Cognitive Informatics, ICCI 2004, pp. 64–73. IEEE Computer Society Press, Los Alamitos (2004)
18. López, N., Núñez, M., Pelayo, F.L.: Specifying the memorization process with STOPA. The International Journal of Cognitive Informatics & Natural Intelligence 1(4), 47–60 (2007)
19. López, N., Núñez, M., Rodríguez, I., Rubio, F.: A formal framework for e-barter based on microeconomic theory and process algebras. In: Unger, H., Böhme, T., Mikler, A.R. (eds.) IICS 2002. LNCS, vol. 2346, pp. 217–228. Springer, Heidelberg (2002)
20. López, N., Núñez, M., Rubio, F.: An integrated framework for the analysis of asynchronous communicating stochastic processes. Formal Aspects of Computing 16(3), 238–262 (2004)
21. Milner, R.: Communication and Concurrency. Prentice Hall, Englewood Cliffs (1989)
22. Nicollin, X., Sifakis, J.: An overview and synthesis on timed process algebras. In: Larsen, K.G., Skou, A. (eds.) CAV 1991. LNCS, vol. 575, pp. 376–398. Springer, Heidelberg (1992)
23. Núñez, M.: Algebraic theory of probabilistic processes. Journal of Logic and Algebraic Programming 56(1–2), 117–177 (2003)
24. Núñez, M., de Frutos, D.: Testing semantics for probabilistic LOTOS. In: 8th IFIP WG6.1 Int. Conf. on Formal Description Techniques, FORTE 1995, pp. 365–380. Chapman & Hall, Boca Raton (1996)
25. Núñez, M., de Frutos, D., Llana, L.: Acceptance trees for probabilistic processes. In: Lee, I., Smolka, S.A. (eds.) CONCUR 1995. LNCS, vol. 962, pp. 249–263. Springer, Heidelberg (1995)
26. Núñez, M., Rodríguez, I.: PAMR: A process algebra for the management of resources in concurrent systems. In: 21st IFIP WG 6.1 Int. Conf. on Formal Techniques for Networked and Distributed Systems, FORTE 2001, pp. 169–185. Kluwer Academic Publishers, Dordrecht (2001)
27. Núñez, M., Rodríguez, I., Rubio, F.: Formal specification of multi-agent e-barter systems. Science of Computer Programming 57(2), 187–216 (2005)
28. Pelayo, F.L., Cuartero, F., Valero, V., Cazorla, D.: An example of performance evaluation by using the stochastic process algebra ROSA. In: 7th Int. Conf. on Real-Time Systems and Applications, pp. 271–278. IEEE Computer Society Press, Los Alamitos (2000)

29. Pelayo, F.L., Núñez, M., López, N.: Specifying the memorization process with STOPA. In: 4th IEEE Int. Conf. on Cognitive Informatics, ICCI 2005, pp. 238–247. IEEE Computer Society Press, Los Alamitos (2005)
30. Plotkin, G.D.: A structural approach to operational semantics. Technical Report DAIMI FN-19, Computer Science Department. Aarhus University (1981)
31. Reed, G.M., Roscoe, A.W.: A timed model for communicating sequential processes. Theoretical Computer Science 58, 249–261 (1988)
32. Solso, R.L. (ed.): Mind and brain science in the 21st century. MIT Press, Cambridge (1999)
33. Squire, L.R., Knowlton, B., Musen, G.: The structure and organization of memory. Annual Review of Psychology 44, 453–459 (1993)
34. Wang, Y.: On cognitive informatics. In: 1st IEEE Int. Conf. on Cognitive Informatics, ICCI 2002, pp. 34–42. IEEE Computer Society Press, Los Alamitos (2002)
35. Wang, Y.: The Real Time Process Algebra (RTPA). Annals of Software Engineering 14, 235–274 (2002)
36. Wang, Y.: Cognitive informatics: A new transdisciplinary research field. Brain and Mind 4, 115–127 (2003)
37. Wang, Y.: Using process algebra to describe human and software behaviors. Brain and Mind 4, 199–213 (2003)
38. Wang, Y.: On the mathematical laws of software. In: 18th Canadian Conf. on Electrical and Computer Engineering, CCECE 2005, pp. 1086–1089 (2005)
39. Wang, Y.: Formal description of the cognitive process of memorization. In: 6th IEEE Int. Conf. on Cognitive Informatics, ICCI 2007, pp. 284–293. IEEE Computer Society Press, Los Alamitos (2007)
40. Wang, Y.: A software science perspective, crc book series in software engineering. Sofware Engineering Foundations 2 (2007)
41. Wang, Y.: The theoretical framework of cognitive informatics. The International Journal of Cognitive Informatics & Natural Intelligence 1(1), 1–27 (2007)
42. Wang, Y.: Deductive semantics of rtpa. The International Journal of Cognitive Informatics and Natural Intelligence 2(2), 95–121 (2008)
43. Wang, Y.: A denotational mathematics for manipulating intelligent and computational behaviours. The International Journal of Cognitive Informatics and Natural Intelligence 2(2), 44–62 (2008)
44. Wang, Y., Dong, L., Ruhe, G.: Formal description of the cognitive process of decision making. In: 3rd IEEE Int. Conf. on Cognitive Informatics, ICCI 2004, pp. 124–130. IEEE Computer Society Press, Los Alamitos (2004)
45. Wang, Y., Wang, Y.: Cognitive models of the brain. In: 1st IEEE Int. Conf. on Cognitive Informatics, ICCI 2002, pp. 259–269. IEEE Computer Society Press, Los Alamitos (2002)
46. Wang, Y., Wang, Y.: Recent advances in cognitive informatics. IEEE Transactions on Systems, Man, and Cybernetics C 36(2), 121–123 (2006)
47. Yi, W.: CCS+ Time = an interleaving model for real time systems. In: Leach Albert, J., Monien, B., Rodríguez-Artalejo, M. (eds.) ICALP 1991. LNCS, vol. 510, pp. 217–228. Springer, Heidelberg (1991)

Author Index

Feng, Lin 133

Hirano, Shoji 161

Liu, Lan 84
Liu, Qing 84
Liu, Yong 133
Lopez, Natalia 224

Nakata, Michinori 180
Núñez, Manuel 224

Pelayo, Fernando L. 224
Polkowski, Lech 30

Sakai, Hiroshi 180

Terlecki, Pawel 118
Tsumoto, Shusaku 161

Walczak, Krzysztof 118
Wang, Guoyin 1, 133
Wang, Jue 100
Wang, Yingxu 1, 6, 46, 205

Yao, Yiyu 1, 100

Zhang, Du 145
Zhao, Yan 100

Lecture Notes in Computer Science

Sublibrary: SL 1 – Theoretical Computer Science and General Issues

For information about Vols. 1–4556
please contact your bookseller or Springer

Vol. 5205: A. Lastovetsky, T. Kechadi, J. Dongarra (Eds.), Recent Advances in Parallel Virtual Machine and Message Passing Interface. XVII, 342 pages. 2008.

Vol. 5202: P.J. Stuckey (Ed.), Principles and Practice of Constraint Programming. XVII, 648 pages. 2008.

Vol. 5153: A. Rausch, R. Reussner, R. Mirandola, F. Plášil (Eds.), The Common Component Modeling Example. VIII, 460 pages. 2008.

Vol. 5150: M.L. Gavrilova, C.J.K. Tan, Y. Wang, Y. Yao, G. Wang (Eds.), Transactions on Computational Science II. XI, 247 pages. 2008.

Vol. 5142: J. Vitek (Ed.), ECOOP 2008 – Object-Oriented Programming. XIII, 694 pages. 2008.

Vol. 5140: J. Meseguer, G. Roşu (Eds.), Algebraic Methodology and Software Technology. XIII, 432 pages. 2008.

Vol. 5136: T.C.N. Graham, P. Palanque (Eds.), Interactive Systems. IX, 311 pages. 2008.

Vol. 5135: R. de Lemos, F. Di Giandomenico, C. Gacek, H. Muccini, M. Vieira (Eds.), Architecting Dependable Systems V. XIV, 343 pages. 2008.

Vol. 5119: S. Kounev, I. Gorton, K. Sachs (Eds.), Performance Evaluation: Metrics, Models and Benchmarks. X, 323 pages. 2008.

Vol. 5111: Q. Chen, C. Zhang, S. Zhang, Secure Transaction Protocol Analysis. XI, 234 pages. 2008.

Vol. 5095: I. Schieferdecker, A. Hartman (Eds.), Model Driven Architecture – Foundations and Applications. XIII, 446 pages. 2008.

Vol. 5091: B.P. Woolf, E. Aïmeur, R. Nkambou, S. Lajoie (Eds.), Intelligent Tutoring Systems. XXI, 832 pages. 2008.

Vol. 5089: A. Jedlitschka, O. Salo (Eds.), Product-Focused Software Process Improvement. XIV, 448 pages. 2008.

Vol. 5082: B. Meyer, J.R. Nawrocki, B. Walter (Eds.), Balancing Agility and Formalism in Software Engineering. XI, 305 pages. 2008.

Vol. 5079: M. Alpuente, G. Vidal (Eds.), Static Analysis. X, 379 pages. 2008.

Vol. 5063: A. Vallecillo, J. Gray, A. Pierantonio (Eds.), Theory and Practice of Model Transformations. XII, 261 pages. 2008.

Vol. 5060: C. Rong, M.G. Jaatun, F.E. Sandnes, L.T. Yang, J. Ma (Eds.), Autonomic and Trusted Computing. XV, 666 pages. 2008.

Vol. 5055: K. Al-Begain, A. Heindl, M. Telek (Eds.), Analytical and Stochastic Modeling Techniques and Applications. XI, 323 pages. 2008.

Vol. 5052: D. Lea, G. Zavattaro (Eds.), Coordination Models and Languages. X, 347 pages. 2008.

Vol. 5051: G. Barthe, F.S. de Boer (Eds.), Formal Methods for Open Object-Based Distributed Systems. X, 259 pages. 2008.

Vol. 5048: K. Suzuki, T. Higashino, K. Yasumoto, K. El-Fakih (Eds.), Formal Techniques for Networked and Distributed Systems – FORTE 2008. XII, 341 pages. 2008.

Vol. 5047: K. Suzuki, T. Higashino, A. Ulrich, T. Hasegawa (Eds.), Testing of Software and Communicating Systems. XII, 303 pages. 2008.

Vol. 5030: H. Mei (Ed.), High Confidence Software Reuse in Large Systems. XII, 388 pages. 2008.

Vol. 5026: F. Kordon, T. Vardanega (Eds.), Reliable Software Technologies – Ada-Europe 2008. XIV, 283 pages. 2008.

Vol. 5025: B. Paech, C. Rolland (Eds.), Requirements Engineering: Foundation for Software Quality. X, 205 pages. 2008.

Vol. 5020: J. Barnes, Ada 2005 Rationale. IX, 267 pages. 2008.

Vol. 5016: M. Bernardo, P. Degano, G. Zavattaro (Eds.), Formal Methods for Computational Systems Biology. X, 538 pages. 2008.

Vol. 5014: J. Cuellar, T. Maibaum, K. Sere (Eds.), FM 2008: Formal Methods. XIII, 436 pages. 2008.

Vol. 5007: Q. Wang, D. Pfahl, D.M. Raffo (Eds.), Making Globally Distributed Software Development a Success Story. XIV, 422 pages. 2008.

Vol. 5002: H. Giese (Ed.), Models in Software Engineering. X, 322 pages. 2008.

Vol. 4989: J. Garrigue, M.V. Hermenegildo (Eds.), Functional and Logic Programming. XI, 337 pages. 2008.

Vol. 4970: M. Nagl, W. Marquardt (Eds.), Collaborative and Distributed Chemical Engineering. XII, 851 pages. 2008.

Vol. 4966: B. Beckert, R. Hähnle (Eds.), Tests and Proofs. X, 193 pages. 2008.

Vol. 4954: C. Pautasso, É. Tanter (Eds.), Software Composition. X, 263 pages. 2008.

Vol. 4951: M. Luck, L. Padgham (Eds.), Agent-Oriented Software Engineering VIII. XIV, 225 pages. 2008.

Vol. 4949: R.M. Hierons, J.P. Bowen, M. Harman (Eds.), Formal Methods and Testing. XIII, 367 pages. 2008.

Vol. 4937: M. Dumas, R. Heckel (Eds.), Web Services and Formal Methods. IX, 169 pages. 2008.

Vol. 4922: M. Broy, I.H. Krüger, M. Meisinger (Eds.), Model-Driven Development of Reliable Automotive Services. XVIII, 183 pages. 2008.

Vol. 4916: S. Leue, P. Merino (Eds.), Formal Methods for Industrial Critical Systems. X, 251 pages. 2008.

Vol. 4909: I. Eusgeld, F.C. Freiling, R. Reussner (Eds.), Dependability Metrics. XI, 305 pages. 2008.

Vol. 4906: M. Cebulla (Ed.), Object-Oriented Technology. VIII, 204 pages. 2008.

Vol. 4902: P. Hudak, D.S. Warren (Eds.), Practical Aspects of Declarative Languages. X, 333 pages. 2007.

Vol. 4899: K. Yorav (Ed.), Hardware and Software: Verification and Testing. XII, 267 pages. 2008.

Vol. 4895: J.J. Cuadrado-Gallego, R. Braungarten, R.R. Dumke, A. Abran (Eds.), Software Process and Product Measurement. X, 203 pages. 2008.

Vol. 4888: F. Kordon, O. Sokolsky (Eds.), Composition of Embedded Systems. XII, 221 pages. 2007.

Vol. 4880: S. Overhage, C.A. Szyperski, R. Reussner, J.A. Stafford (Eds.), Software Architectures, Components, and Applications. X, 249 pages. 2008.

Vol. 4849: M. Winckler, H. Johnson, P. Palanque (Eds.), Task Models and Diagrams for User Interface Design. XIII, 299 pages. 2007.

Vol. 4839: O. Sokolsky, S. Taşıran (Eds.), Runtime Verification. VI, 215 pages. 2007.

Vol. 4834: R. Cerqueira, R.H. Campbell (Eds.), Middleware 2007. XIII, 451 pages. 2007.

Vol. 4829: M. Lumpe, W. Vanderperren (Eds.), Software Composition. VIII, 281 pages. 2007.

Vol. 4824: A. Paschke, Y. Biletskiy (Eds.), Advances in Rule Interchange and Applications. XIII, 243 pages. 2007.

Vol. 4821: J. Bennedsen, M.E. Caspersen, M. Kölling (Eds.), Reflections on the Teaching of Programming. X, 261 pages. 2008.

Vol. 4807: Z. Shao (Ed.), Programming Languages and Systems. XI, 431 pages. 2007.

Vol. 4799: A. Holzinger (Ed.), HCI and Usability for Medicine and Health Care. XVI, 458 pages. 2007.

Vol. 4789: M. Butler, M.G. Hinchey, M.M. Larrondo-Petrie (Eds.), Formal Methods and Software Engineering. VIII, 387 pages. 2007.

Vol. 4767: F. Arbab, M. Sirjani (Eds.), International Symposium on Fundamentals of Software Engineering. XIII, 450 pages. 2007.

Vol. 4765: A. Moreira, J. Grundy (Eds.), Early Aspects: Current Challenges and Future Directions. X, 199 pages. 2007.

Vol. 4764: P. Abrahamsson, N. Baddoo, T. Margaria, R. Messnarz (Eds.), Software Process Improvement. XI, 225 pages. 2007.

Vol. 4762: K.S. Namjoshi, T. Yoneda, T. Higashino, Y. Okamura (Eds.), Automated Technology for Verification and Analysis. XIV, 566 pages. 2007.

Vol. 4758: F. Oquendo (Ed.), Software Architecture. XVI, 340 pages. 2007.

Vol. 4757: F. Cappello, T. Herault, J. Dongarra (Eds.), Recent Advances in Parallel Virtual Machine and Message Passing Interface. XVI, 396 pages. 2007.

Vol. 4753: E. Duval, R. Klamma, M. Wolpers (Eds.), Creating New Learning Experiences on a Global Scale. XII, 518 pages. 2007.

Vol. 4749: B.J. Krämer, K.-J. Lin, P. Narasimhan (Eds.), Service-Oriented Computing – ICSOC 2007. XIX, 629 pages. 2007.

Vol. 4748: K. Wolter (Ed.), Formal Methods and Stochastic Models for Performance Evaluation. X, 301 pages. 2007.

Vol. 4741: C. Bessière (Ed.), Principles and Practice of Constraint Programming – CP 2007. XV, 890 pages. 2007.

Vol. 4735: G. Engels, B. Opdyke, D.C. Schmidt, F. Weil (Eds.), Model Driven Engineering Languages and Systems. XV, 698 pages. 2007.

Vol. 4716: B. Meyer, M. Joseph (Eds.), Software Engineering Approaches for Offshore and Outsourced Development. X, 201 pages. 2007.

Vol. 4709: F.S. de Boer, M.M. Bonsangue, S. Graf, W.-P. de Roever (Eds.), Formal Methods for Components and Objects. VIII, 297 pages. 2007.

Vol. 4680: F. Saglietti, N. Oster (Eds.), Computer Safety, Reliability, and Security. XV, 548 pages. 2007.

Vol. 4670: V. Dahl, I. Niemelä (Eds.), Logic Programming. XII, 470 pages. 2007.

Vol. 4652: D. Georgakopoulos, N. Ritter, B. Benatallah, C. Zirpins, G. Feuerlicht, M. Schoenherr, H.R. Motahari-Nezhad (Eds.), Service-Oriented Computing ICSOC 2006. XVI, 201 pages. 2007.

Vol. 4640: A. Rashid, M. Aksit (Eds.), Transactions on Aspect-Oriented Software Development IV. IX, 191 pages. 2007.

Vol. 4634: H. Riis Nielson, G. Filé (Eds.), Static Analysis. XI, 469 pages. 2007.

Vol. 4620: A. Rashid, M. Aksit (Eds.), Transactions on Aspect-Oriented Software Development III. IX, 201 pages. 2007.

Vol. 4615: R. de Lemos, C. Gacek, A. Romanovsky (Eds.), Architecting Dependable Systems IV. XIV, 435 pages. 2007.

Vol. 4610: B. Xiao, L.T. Yang, J. Ma, C. Muller-Schloer, Y. Hua (Eds.), Autonomic and Trusted Computing. XVIII, 571 pages. 2007.

Vol. 4609: E. Ernst (Ed.), ECOOP 2007 – Object-Oriented Programming. XIII, 625 pages. 2007.

Vol. 4608: H.W. Schmidt, I. Crnković, G.T. Heineman, J.A. Stafford (Eds.), Component-Based Software Engineering. XII, 283 pages. 2007.

Vol. 4591: J. Davies, J. Gibbons (Eds.), Integrated Formal Methods. IX, 660 pages. 2007.

Vol. 4589: J. Münch, P. Abrahamsson (Eds.), Product-Focused Software Process Improvement. XII, 414 pages. 2007.

Vol. 4574: J. Derrick, J. Vain (Eds.), Formal Techniques for Networked and Distributed Systems – FORTE 2007. XI, 375 pages. 2007.